Regional and Long-range
Transport of Air Pollution

 ON CHEMICAL AND ENVIRONMENTAL SCIENCE

A series devoted to the publication of courses and educational seminars given at the Joint Research Centre, Ispra Establishment, as part of its education and training programme.

Published for the Commission of the European Communities, Directorate-General Information Market and Innovation.

Volumes already published

- APPLICATIONS OF MASS SPECTROMETRY TO TRACE ANALYSIS
  Edited by S. Facchetti

- ANALYTICAL TECHNIQUES FOR HEAVY METALS IN BIOLOGICAL FLUIDS
  Edited by S. Facchetti

- OPTICAL REMOTE SENSING OF AIR POLLUTION
  Edited by P. Camagni and S. Sandroni

- MASS SPECTROMETRY OF LARGE MOLECULES
  Edited by S. Facchetti

Regional and Long-range Transport of Air Pollution

Lectures of a course held at the Joint Research Centre, Ispra (Italy)
15–19 September 1986

Edited by

S. Sandroni

Joint Research Centre, Ispra, Italy

Published for the Commission of the European Communities
by
ELSEVIER Amsterdam – Oxford – New York – Tokyo   1987

Commission of the European Communities
Joint Research Centre, Ispra (Varese), Italy

Publication arranged by:
Directorate-General Information Market and Innovation
Luxembourg

Published under licence by:
Elsevier Science Publishers B.V.
Sara Burgerhartstraat 25
P.O. Box 211, 1000 AE Amsterdam
The Netherlands

Distributors for the United States and Canada:
Elsevier Science Publishing Company Inc.
52, Vanderbilt Avenue
New York, N.Y. 10017

© ECSC, EEC, EAEC, Brussels and Luxembourg, 1987

EUR 10832 EN

90   00646

Foreword

The emphasis of air pollution problems has recently moved from local and urban to regional, continental and global scales. Viruses, pollens, volcanic dust, industrial pollutants and radioactive materials released in nuclear plant disasters are but a few instances where atmospheric transport knowledge is important on a much larger scale. Hence the increase in residence time introduces chemical conversion processes which, in turn, increase the complexity of the monitoring requirements. These problems are sometimes associated with episodic releases but generally they occur over long periods, e.g., continuous emissions.

The increasing concern about long-range transport of airborne noxious material over Europe was made evident by the ratification of the Geneva Convention on Transboundary Air Pollutants Transport in March 1983. The nations of Watern and Eastern Europe were determined to control the problem of "acid rain" and special interest was focused on lake acidification in Scandinavia and on forest destruction in Central Europe. Recently, North American and European countries have performed experimental and model investigations on regional and continental scales in an attempt to describe trajectories and source-receptor relationships of air masses. Other studies are planned for the future. The ongoing analysis of dispersion and deposition data of radionuclides following Chernobyl accident is, for instance, a matter of debate in relation to the possibility of trans-Alpine exchange of atmospheric trace constituents.

In order to tackle these problems a combination of multidisciplinary subjects such as atmospheric physics, meteorology, atmospheric chemistry, ground-based remote sensing, airborne instrumentation and mathematical modelling is required. Further promotion and exploitation require a permanent scientific effort and multi-sided cooperation on an international basis.

The aim of the Course, in the frame of the Education and Training Programme of the Joint Research Centre, was to give an up-to-date overview of the present knowledge in the different fields involved in the atmospheric transport of trace constituents. The different aspects were dealt with in a series of 19 lectures by qualified experts from the various fields. Great effort was made to try to associate experimental techniques with modelling. Experiments on a wide area are a realistic but very expensive procedure. Hence measurements are usually limited in time and space distribution. On the other hand, modelling, which permits a systematic analysis of various parameters, must be physically and chemically realistic and fit the available experimental data. The integration of experiments with modelling is the most promising approach to the problem.

The Course lectures presented here are organised in four sections:

- atmospheric transport, conversion, deposition of atmospheric trace constituents and associated problems;
- conventional and sophisticated techniques for atmospheric sounding (e.g., Sodar, Lidar, Cospec, tetroons, instrument-carrying aircraft) and simulation techniques (non-reactive tracers);
- models available for various applications (long-range episodes, long-term averages, photochemical and deposition processes);
- a comparison of performances of different models and the linearity problem in the formation of acid deposition.

Finally I should like to express my gratitude to the lecturers for their excellent contributions and to all the participants for their active interest, which helped, I hope, to make the Course successful.

Ispra, March 1987   S. Sandroni

LIST OF CONTRIBUTORS

S. Beilke	Umweltbundesamt, Pilotstation Frankfurt, Frankfurterstr. 135, D-6050 Offenbach, F.R.G.
P. Bessemoulin	Centre National de Recherches Météoroliques, 42 Avenue G. Coriolis, 31057 Toulouse, Cedex, France.
R.N. Dietz	Department of Applied Science, Tracer Technology Center, Brookhaven National Laboratory, Bldg. 426, Upton, Long Island, NY 11973, U.S.A.
H. van Dop	KNMI, P.O. Box 201, 3730 AE De Bilt, The Netherlands.
F. Fiedler	Institut für Meteorologie und Klimaforschung der Universität, Kaiserstrasse 12, D-7500 Karlsruhe, F.R.G.
D. Fowler	Institute of Terrestrial Ecology, Bush Estate, Penicuik, Midlothian EH26 0QB, U.K.
J.L. Heffter	NOAA-Air Resources Laboratory, 8060 13th Street, Silver Spring, MD 20910, U.S.A.
Ø. Hov	NILU, Box 130, N-2001 Lillestrom, Norway.
W. Klug	Institut für Meteorologie der Technischen Hochschule, Hochschulstrasse 1, D-6100 Darmstadt, F.R.G.
A.R. Marsh	Central Electricity Research Laboratories, Kelvin Avenue, Leatherhead, Surrey KT22 7SE, U.K.
M.M. Millán	Ministerio de Industria y Energia, Centro de Investigaciones Energeticas, Medioambientales y Tecnologicas, JEN, Madrid, Spain
U. Schurath	Institut für Physikalische Chemie der Universität Bonn, Wegelerstrasse 12, D-5300 Bonn 1, F.R.G.
F.B. Smith	Meteorological Office (MET 014), London Road, Bracknell, Berkshire RG12 2SZ, U.K.
P. Thomas	Institut für Meteorologie und Klimaforschung, Kernforschungszentrum Karlsruhe, Universität Karlsruhe, Postfach 3640, D-7500 Karlsruhe, F.R.G.
R.H. Varey	Central Electricity Research Laboratories, Kelvin Avenue, Leatherhead, Surrey KT22 7SE, U.K.
T. Yamada	Los Alamos National Laboratory, Los Alamos, NM 87545, U.S.A.

CONTENTS

X

Regional and Long-range Transport of Air Pollution,
Lectures of a course held at the Joint Research Centre, Ispra, Italy,
15–19 September 1986, S. Sandroni (Ed.), pp. 1–42
© Elsevier Science Publishers B.V., Amsterdam — Printed in The Netherlands

PROBLEMS ASSOCIATED WITH LONG-RANGE TRANSPORT OF AIR POLLUTANTS

S.BEILKE

1. INTRODUCTION

The title of my lecture on problems associated with the long-range transport of air pollutants is a very general one comprising a lot of different meteorological,air chemical and environmental aspects.At first some essential restrictions have to be made.
In this report only some aspects of the long-range transport(LRT) of air pollutants can be discussed mainly from the point of view of air chemistry and environmental effects.Aspects related to pure meteorological modelling of long-range transport are not considered.There are a series of excellent summaries dealing with these aspects(for example refs.1-7).
In this lecture the main emphasis is placed on the problem of acid deposition on the regional scale of Europe including the current state of scientific understanding in North America.

In section 2 some more general definitions are given along with a brief historical look at problems of acid deposition in Europe which is useful to place the recent problem of LRT in the wider context of the physico-chemical behaviour of air pollutants in the lower atmosphere.

In section 3 an attempt has been made to give a survey of environmental effects in Europe and North America which are believed to be caused at least to some extent by air pollutants transported over long distances.

Section 4 provides a summary of some major processes involved in acid deposition under special consideration of their treatment in LRT models:emission,transformation and deposition.

In section 5 the central issue of concern is to discuss some aspects of a source-receptor relationship mainly from the point of view of atmospheric measurements.

My lecture on chemical and environmental aspects of LRT is based on the results of a series of conferences,reports and other activi-

ties both in Europe and in North America.The main emphasis is given
to the results of the important research projects COST 61a[1],
COST 61a bis[2],COST 611[3] and COST 612[4] of the Commission of
the European Communities.

The fact that air pollutants can be transported over thousands
of kilometers before being deposited at the earth's surface has
been recognized for a long time.
An impressive example for long-range transport of SO_2 and its reaction products via the stratosphere was given by Delmas and Graven-
horst(ref.8) who have shown that concentration peaks of sulphuric
acid and sulphates in deep Antarctic ice cores can clearly be
attributed to eruptions of some volcanoes like Mt.Krakatoa(1883) or
Mt.Agung(1963) located in the tropics(Indonesia).
Another example of long-range transport via the upper atmosphere
was the detection of the large scale SO_2 plume over Europe originating from the eruption of the Mount St.Helens volcano(USA) on 18
May 1980.On 26 May 1980 an 10-12 fold increase of SO_2 mixing ratios
was observed at the tropopause level over South Scandinavia(ref.9).
Normally transport of air pollutants via the upper atmosphere results in a strong dilution of concentrations and a rather homogeneous deposition over large surface areas.

In contrast to stratospheric transport,episodic long-range transport through the lower atmosphere normally leads to a rather inhomogeneous concentration field and deposition fluxes of material to
the earth's surface.
A recent example is the concentration and deposition field of radioactive material across Europe as a consequence of the Chernobyl reactor accident in the USSR in April 1986.Considering the meteorological situation it is reasonable to assume that most radioactive
material emitted by the Chernobyl reactor was transported within a
shallow layer of about 3 km thickness adjacent to the earth's sur-
face(refs.10-11).

	COST	: COoperation Scientifique et Technique(Scientific and Technical Cooperation)
(1)	61a	: Physico-chemical behaviour of SO_2 in the atmosphere (1972-1976)
(2)	61a bis	: Physico-chemical behaviour of atmospheric pollutants (1979-1983)
(3)	611	: Physico-chemical behaviour of atmospheric pollutants (1986-1990)
(4)	612	: Air pollution effects on ecosystems(1986-1990)

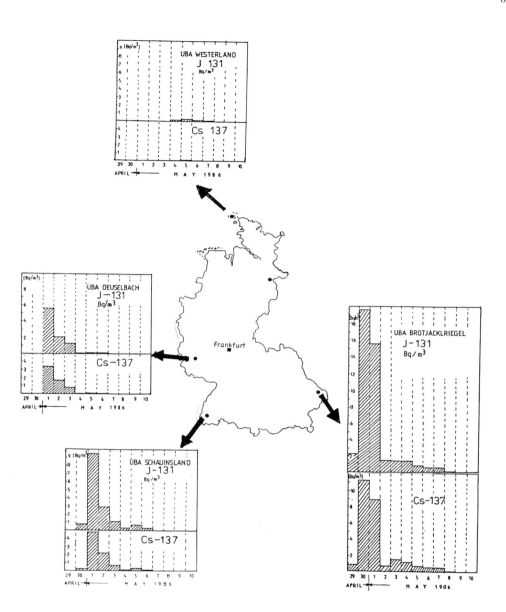

FIG. 1 :

Daily mean activity concentrations of J-131 and Cs-137
in aerosol particles in Bq/m^3 measured between 29 April
and 10 May 1986 at 4 stations of the UBA network:Westerland,
Deuselbach,Schauinsland and Brotjacklriegel.
Source: WEISS et al.,1986 (ref.12).

In figure 1 the distributions of aerosol bound activities of short-lived J-131 and long-lived Cs-137 are given as a function of time for four stations of the German background network of Umweltbundes-amt(ref.12).As seen in this figure the spatial distribution of daily mean activity concentrations is quite inhomogeneous and the time dependence is quite different for the stations.Maximum values were observed at the southeastern station Brotjacklriegel(19 Bq/m^3 for J-131 and 11 Bq/m^3 for Cs-137 on 30 April 1986) whereas the lowest values were detected at the northernmost site Westerland (North Sea;maximum values on 5 May 1986:0.26 Bq/m^3 for J-131 and 0.12 Bq/m^3 for Cs-137).

2. DEFINITIONS

A brief historical look at the problems of acid deposition is useful to show that long-range transport of acidic substances is a relatively recent problem in Europe.The fact that "acid rain" is a phenomenon that occurs in various European regions has been known for more than 120 years.Measurements of the chemical composition of rainwater began in the middle of the last century.
In 1872,R.Smith,the first British Chief Alkali Inspector,published a book entitled " Air and Rain-the Beginnings of a Chemical Clima-tology".This book contains a lot of results on chemical rainwater analyses measured at various European sites with different levels of air pollution(ref.13).
In a recent publication Schwela(ref.14)discussed the results of the rainwater analyses published by Smith in 1872(ref.13) and com-pared them with concentrations of rainwater constituents measured today.The comparision shows that the spatial distribution of rain-water components was much less uniform around 1870 than today.In industrial cities representing heavily polluted regions the rain-water pH s were found to fall between 3.5 and 4 i.e. the values were in the acidic range whereas in the then clean areas the pH was between 4.6-5.6 which rather resembles a more natural situation in Europe.
Similiar to pH,the spatial distributions of sulphates and nitrates seem to have been much more structurally differentiated a century ago,the actual values being of the same order of magnitude as those found in Europe at present(ref.14).Today there is a levelling off in the distribution in space of rain components mainly for two reasons: a.) a general geographical spreading of SO_2 and NO_x emission sources and a general increase of SO_2 and NO_x emission

rates since 1870 and b.) the release of a relatively larger fraction of SO_2 (and to some extent of NO_x) from tall stacks nowadays
increasing the potential of these gases for long-range transport
considerably compared to 1870 (refs.15-16). Therefore air pollutants
can be transported over hundreds of kilometers and may lead to harm-
ful effects on ecosystems in areas remote from the main emission
sources.

The spatial and time scales associated with pollutant transport
obviously depend on how long the pollutant resides in the atmosphere. In order to estimate the dimensions of the spatial and time
scales it is often helpful to use the concept of atmospheric residence times.
The residence time of a single molecule is the time between its
formation or introduction into the atmosphere and its removal either
by chemical conversion and/or by wet and dry deposition. The residence time of a single SO_2 molecule can be less than one second or
greater than one month depending on its emission and removal characteristics. In the atmosphere we are concerned with a frequency distribution of residence times of single molecules.
In the literature often a bulk quantity is used to describe a mean
residence time (frequantly just called residence time) which under
steady state conditions is identical with the average residence time
derived from the frequancy distribution of residence times of single
molecules (ref.17).
The mean residence time, τ, can be calculated as

$$\tau = \frac{M}{P} = \frac{M}{R} \tag{1}$$

M = total mass of a pollutant in a given reservoir
P = total production rate in a given reservoir
R = total removal rate from a given reservoir

The mean residence time, τ, resulting from the simultaneous occurrence of all the different removal mechanisms, is given by

$$\frac{1}{\tau} = \frac{1}{\tau_D} + \frac{1}{\tau_W} + \frac{1}{\tau_C} \tag{2}$$

τ = mean residence time
τ_D = M/D = residence time with respect to dry deposition D to the
          earth surface

τ_W = M/W = residence time with respect to wet deposition W to the
earth surface

τ_C = M/C = residence time with respect to removal by chemical conversion C.

In central Europe and in eastern North America, all of these removal
mechanisms have to be considered when calculating the mean residence times of substances relevant to the problem of acid deposition.
This makes modelling of acid deposition very difficult in these two
regions.

In table 1 mean residence times are given for some of the major
species relevant to the problem of acid deposition. The residence
times given there apply to the physico-chemical and meteorological
conditions pertaining to central Europe in the lower atmosphere and
include seasonal variations. The range of atmospheric residence times
may be different in other parts of the world. The residence times in
table 1 are estimated on the basis of the major removal processes
relevant to these substances.

TABLE 1

Species	Range of mean residence times
SO_2	1 - 2 days
NO_x	some hours - 2 days
HNO_3	some hours
HCl	some hours
NH_3	some hours - 1 day
O_3	ca. 1 week
NMHC[1]	1 hour - 1 month
sulphate	ca. 1 week
nitrate	ca. 1 week

Table 1 : Mean atmospheric residence times for some major species
relevant to the problem of acid deposition in central
Europe.
(1): NMHC = Non Methane HydroCarbons.

From a meteorological point of view it is sometimes useful to
classify transport behaviour by different scales on which the physi-
co -chemical processes take place: micro(or local), meso, synoptic, and
global scale(refs. 3,16).

The horizontal dimension of the micro or local scale is of the order of a kilometer,and the time scale on which the physico-chemical processes take place is of the order of tens of minutes.Examples for physico-chemical processes occurring on this scale are the oxidation of NO emitted by cars or by power plants under unstable atmospheric conditions(Pasquill classes B and C) at high ozone concentrations in the ambient air(ref.18).
Another example is the physico-chemical behaviour of HF(hydrofluoric acid) emitted by low stacks of brickworks,aluminium smelters and lead crystal factories.
The spatial dimension of the mesoscale extends to several hundred kilometers corresponding to a time scale of the order of a day.
An example for physico-chemical processes occurring on this scale is the NO-oxidation in power plant plumes under neutral or stable conditions at low ozone concentrations.
The synoptic scale is of the order of 1000 to 3000 km,the time scale on which physico-chemical processes are considered on this dimension is of the order of one day to 1 week.
Finally,the global and hemispheric scales reflect intercontinental to hemispheric transport with time scales larger than one week.

TABLE 2

Spatial scale	Time scale	Typical air pollutants pertaining to these scales
local scale 0 - 1 km	0-tens of minutes	NO emitted by cars,HF emitted by low stacks,large dust particles
mesoscale 1 - several hundred km	tens of minutes to 1 day	$NO,NO_2,HNO_3,HCl,SO_2,NH_3,$
synoptic scale	1 day-1 week	some $NMHC,O_3$,acidic aerosols in surface near air
hemispheric to global	some weeks to years	$COS,CO,CO_2,N_2O,CH_4,CCl_2F_2,$ CCl_3F etc.

Table 2: Classification of transport behaviour by different spatial and time scales on the basis of mean pollutant residence times(for comparision see table 1 ; refs.3,16)

As seen in table 1,mean residence times of substances which are relevant for acid deposition in central Europe are of the order of some hours to two days for NO_x and SO_2 and ca. 1 week for acidic aerosols.Thus,from an air chemical point of view the acid deposition problem is not a local or a global problem but rather a problem in the meso and synoptic(regional) scale of the order of some hundred to 3000 km.

As a very rough guide line Rodhe(ref.19) suggested a horizontal dimension which could characterize a size of a reservoir relevant for a certain species emitted at the centre of this reservoir as a function of the residence time:

$$D_{opt} \approx 4\, \tau\, \bar{v} \tag{3}$$

D_{opt} = diameter of a circular reservoir for an atmospheric constituent with its point source at the centre

$\bar{v}$ = average transport velocity in the reservoir

τ = mean residence time of the species concerned

As far as the vertical extention is concerned most air pollutants relevant to acid deposition are confined to the so-called planetary boundary layer which is of variable height ranging typically up to 3 km from the earth surface.For example,according to Jost and Georgii(ref.20) half of the atmospheric mass of SO_2 and NO_x is normally found in the lowest 1.5 km in central Europe.
Closest to the earth surface is the so-called mixing layer in which atmospheric turbulence very effectively mixes and dilutes air pollutants.The significant diurnal and seasonal variations of the height of the mixing layer(ref.21) has a direct impact on dry deposition pattern and chemical conversion processes which also show large diurnal and seasonal variations.
The appropriate treatment of these diurnal and seasonal variations seems to be a serious problem associated with episodic and long-term modelling of acid deposition.

3. ENVIRONMENTAL EFFECTS

In this section an attempt is made to give a survey of environmental effects which are believed to be caused at least to some extent by air pollutants transported over long distances.
It is recognized that damage to forest and freshwater ecosystems as well as to materials in various regions is attributed at least in part to air pollutants.

As far as materials are concerned it is generally accepted that increased air pollution levels lead to corrosion of buildings,and there is substantial evidence that sulphur compounds frequently in connection with high atmospheric humidity play an important role in this process(ref.22).

Damage to historic buildings and monuments has been recorded in many European cities.Extensive damage is for example reported for the Acropolis in Athens(ref.23),and for many historic buildings in Venice(ref.24).

As such damage to materials is largely caused by pollutant sources within certain urban areas extending to no more than 10-50 km,it is therefore a local problem rather than a regional one.This type of environmental effect is therefore not further considered here in the context of regional and long-range transport of air pollutants. Instead,this section summarizes the ecological effects on aquatic and terrestrial ecosystems and the extent of damage mainly in Europe that may be attributable to air pollutants such as acidic substances or photochemical oxidants transported over long distances.

As far as aquatic ecosystems are concerned,several review papers on the effects of acid deposition have been published in the last few years(see for example ref.25).

It is generally accepted that a series of lakes and streams in different European and North American regions became increasingly acidic during the last decades leading to a reduction and loss of fish populations,particularly salmon and trout,and to a damage of other aquatic flora and fauna(refs.22,26).

There is also general agreement among scientists that acidification of freshwater systems is mainly possible when the buffering capacity in and around lakes and streams is low making them particularly vulnerable to acidification(ref.26).

There is also a broad consensus,although not unanimity,that acidic inputs mainly in the form of acid rain generated during long-range transport have contributed to the considerable reduction of pH in certain lakes in Europe and North America over the last 3 decades (refs.22,26).

An important phenomenon associated with acid deposition in some ecologically sensitive areas(for example in southern Norway)is that a large fraction of the total annual amount of acidic inputs to these regions is due to a few episodes of high concentrations and deposition(ref.27).

As regards the geographical extent ,acidification of lakes and

rivers is mainly a problem in Scandinavia,southwestern Scotland,
northeastern USA,and eastern Canada.These regions have in common
mainly granitic bedrock and acidic rain with pH values below 4.6
(ref.22).For other European regions and countries an acidification
of lakes is less evident(ref.22).
The most affected areas of western Europe cover southern Norway
and part of south and central Sweden.Data from Norway indicate that
pH levels in rivers have decreased from 5.0-6.5 in 1940 to values
of 4.6-5.0 in 1976/78.Similiarly,pH levels in lakes also declined
from 5.5 to current levels of 4.7(ref.22).According to this reference,in Scotland some twenty lakes have experienced a significant
acidification as a result of acid deposition.

Fig.2 : Annual weighted mean pH in precipitation of 1980.
        The hatched areas provide regions with low buffering
        capacity.The black dots indicate important SO_2 emission
        areas i.e.SO_2 emissions greater than 100 kilotons per
        year(Source:Martin,1983;ref.28).

An acidification of surface waters was also observed in ca.100
lakes in eastern Canada located on the Canadian Shield(ref.28).
In the eastern United States,an acidification of some 40 lakes was
reported in the Adirondack Mountains over the period 1930-1975
(ref.22).

Figure 2 shows in a very simple form the regions in eastern North
America which are sensitive to acidification caused by atmospheric
inputs.The distribution of major SO_2 emission sources and the annual
volume weighted mean pH values in precipitation are also inserted
in this figure(ref.28).As a result of the mean meteorological flow
for major storm systems,which is towards the northeast,the isolines
of annual volume weighted mean pH in precipitation (ref.29)show
highest acidity over the northeast of the main SO_2 sources.
According to Martin(ref.28),acidification in eastern Canada has
been documented within the 4.5 isoline shown in figure 2.

As far as terrestrial ecosystems are concerned,the most important issue of public and scientific concern is the forest damage
observed recently in a series of countries in the northern hemisphere and particularly in some countries of central Europe.
It is evident that different species of trees(fir,spruce,pine,beech,
oak) at different locations show various symptoms of visible damage
the reasons of which are complex and traced back mainly to abiotic
stress factors.

In table 3 the damaged areas by regions and countries are summarized along with the main tree species concerned(ref.30).When comparing these figures one should keep in mind that the methods used to
estimate the extent of visible damage differ from year to year and
from country to country,so meaningful comparisions can hardly be
made.

Damage to forest systems is mainly observed in some countries of
central Europe like Poland,Czechoslovakia,Austria,Switzerland,the
German Democratic Republic and the Federal Republic of Germany
whereas in Scandinavia no such damage could be observed.

The recent forest damage has been particularly thoroughly followed
in the Federal Republic of Germany.The significance of the regional
decline of forest trees led to a country-wide assessment of forest
damages in 1982 identifying 8 % of West Germany's forested area as
being damaged(ref.31).Repeated and more systematic assessments
using unified and improved procedures during the following years
showed that 34 % of the total forested area was injured in 1983 and
ca. 50 % in 1984(refs.31,32).

Most of the damaged area lies in the central highlands of Europe
which are situated between 50 km and some hundred kilometers away
from major industrial emission areas.

The strongest forest damage in Europe is observed at the ridges of
the Ore Mountains(Erzgebirge) along the border between the German

TABLE 3

Country	Total forested area in % of land area	Area of forest damage in % of forested area	Tree species damaged
Northern Europe			
Denmark	12	< 3	Pine
Finland	76	< 3	Spruce,Pine
Norway	27	< 3	Pine
Sweden	64	<3	Spruce,Pine
Western Europe			
France	27	< 3	Pine,Spruce,Fir
Ireland	5	< 3	
Netherlands	9	35	Pine.Spruce,Beech
United Kingdom	9	< 3	
Southern Europe			
Spain	31	< 3	Pine
Central Europe			
Austria	45	9	Fir,Spruce
Czechoslovakia	36	10	Spruce,Fir,Pine
Fed.Rep.Germany	30	50	Fir,Spruce,Pine, Beech,Oak
Hungary	17	< 3	Black pine
Poland	29	7	Coniferous forest, especially pine
Switzerland	27	14	Fir,Spruce,Beech, Pine

Table 3 : Summary of damaged areas by regions and country in Europe
including the tree species concerned.The figures apply to
the situation in 1983 and 1984.
Source:National Acid Precipitation Assessment Program,
Annual Report 1984 to the President and Congress,prepared
by the Interagency Task Force on Acid Deposition,USA
(ref.30)

Democratic Republic and Czechoslovakia receiving air pollution mainly from the industrial regions of Saxony,the Bohemian basin and
Silesia.

Other central highlands showing more or less severe symptoms of
tree damage are the Black Forest,Frankenwald,Harz,Bavarian Forest,
the Vosges Mountains,the Iser Mountains and the Beskydy Mountains
(ref.31).

Table 4 shows damaged areas and percentages of different tree species affected in the Federal Republic of Germany in 1984(ref.32).
It has to be noted that damaged areas in table 4 include all categories of damage ranging from slight damage,to moderate,to severe
damage to dead.

TABLE 4

Tree species	Forest area covered by a certain tree species	Percent of area covered by tree showing damage
Spruce	2.89 mio of hectares	51
Pine	1.47	59
Fir	0.17	87
Beech	1.25	50
Oak	0.62	43
Others	0.97	31
Total	7.37	50

Table 4 : Damaged areas and percentages of different trees affected in the Federal Republic of Germany in 1984.
The damaged areas include all categories of damage
ranging from slight damage to dead(ref.32)

There is substantial agreement among scientists,although not unanimity,that the recent forest damages remote from the major industrial areas are to a large extent caused by anthropogenic air pollu-
tion.Among the different air pollutants which are thought to be
responsible for the severe damages of forest ecosystems acidic substances and photochemical oxidants are most emphasized in the
scientific literature.
As far as acid deposition is concerned,acidic inputs due to dry and
wet deposition affect the chemical status of the soil and therefore
change the nutriation of the trees(ref.33).
The deposition can reach the soil directly or the biomass above the
soil can interrupt and buffer the acidic input transferring the
acidic effect internally from the leaves to the roots(refs.33,34).
At present ozone as the most abundant photochemical oxidant is considered to be a decisive primary cause of forest decline in remote
areas,often combined with SO_2 and acid precipitation(refs.35,36).
    There is clear evidence that during summer months "anthropogenic"
ozone can be generated due to emissions of NO_x and reactive hydrocarbons in central Europe.As a result,ozone concentrations above
national air quality standards may occur over large areas and persist for several days.
An example for large scale anthropogenic generation and long-range
transport of ozone is given in figures 3a - 3d summarizing the results of an air sampling flight carried out on 4 July 1985 at low
altitude(50-150 m above the ground) along the border between the
Federal Republic of Germany and the German Democratic Republic for

Figure 3a:
Air sampling flight at low altitude
carried out on 4 July 1985 along the
border between the FRG and the GDR
for easterly incoming flow of air
masses which had been transported
some hundred km over polluted industrial areas.Flight route between
Heiligenhafen(Baltic Sea,checkpoint 1)
and Bayreuth(Bavaria,checkpoint 21)
representing a distance of 620 km.

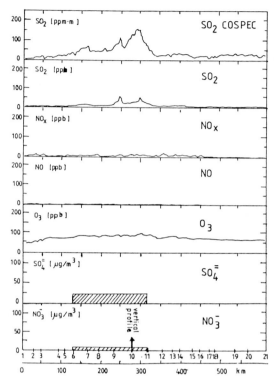

Figure 3c:
Horizontal concentration profiles along
the FRG-GDR border between checkpoints
1 to 21 (2.36p.m.-5.27p.m.).Transport
from east to west.

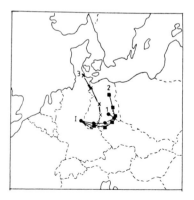

Figure 3b:
Backward trajectories arriving at Bad
Hersfeld(checkpt.14) on 4 July 1985.
1:1000 hPa; 2=925 hPa; 3=850 hPa.

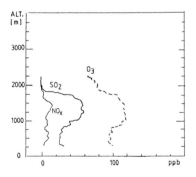

Figure 3d:
Vertical concentration profiles
measured at checkpoint 10 in the
center of the large scale plume

easterly incoming flow of air masses which had been transported
some hundred kilometers over polluted industrial areas(see trajectories in figure 3b).The horizontal concentration profiles in
figure 3c show elevated O_3-concentrations in the large scale SO_2
plume over a flight path of more than 250 km.Another indication for
ozone generation in the large scale SO_2 plume is shown in figure
3d indicating the parallel vertical profiles of SO_2 and O_3 in the
centre of the plume(ref.37).

4. ASPECTS OF EMISSIONS,TRANSFORMATION AND DEPOSITION

In this section some aspects are highlighted which are related
to the consideration of emissions,transformation,and deposition in
long-range transport models.
There are a large number of primary and secondary pollutants which
are involved in the problem of acid deposition and photochemical
oxidants.
As far as acid deposition is concerned the most important primary
acidifying and acid gases are:sulphur dioxide(SO_2),nitrogen oxide
(NO) and to a lesser extent nitrogen dioxide(NO_2),hydrochloric acid
(HCl),and hydrofluoric acid(HF).
The most important secondary pollutants formed from those acid generating precursor gases are:sulphuric acid(H_2SO_4) and nitric acid
(HNO_3) including their salts,sulphates and nitrates.
It has to be noted that ammonia is an acid when taken up by plants
and reacts according to $NH_3 \longrightarrow NH_2^- + H^+$.
The cation acid ammonium(NH_4^+) is also of importance for the acid
deposition problem.
As far as photochemical oxidants are concerned,they are mostly
secondary air pollutants formed when NO_x(NO_x = NO + NO_2)and hydrocarbons react in the sunlight photochemical cycle.The following
substances are addressed as photochemical oxidants in general:
ozone(O_3),peroxyacetyl nitrate(PAN = $CH_3COO_2NO_2$) including its ho-
mologes,aldehydes(for example formaldehyde = CH_2O),hydrogen per-
oxide(H_2O_2) etc.

4.1. EMISSIONS

Emission inventories are needed as an important input parameter
to models which try to establish source-receptor relationships in
acid deposition and photochemical oxidants.
Emission values on which regional and long-range transport models
are based are not known precisely.The lack of precision varies with

pollutant species and from region to region. In my opinion the uncertainties in national values in Europe are at best $\pm$ 20 % for SO_2 and much greater for other species.

On the following pages some aspects concerning emissions are discussed which are of importance with regard to modelling of long-range transport.

The principal precursors of acid deposition and photochemical oxidants (SO_2, NO_x, NH_3, NMHC) are emitted both by natural and man-made sources which are distributed very unevenly over the earth's surface.

As far as sulphur is concerned, over 90 % of the man-made S-emissions are produced in the northern hemisphere and there, too, the emissions are unevenly distributed (refs.38,39). Roughly 40 % of global anthropogenic sulphur emissions occur in Europe representing ca. 3 % of the earth's surface (refs.40,41).

On the regional scales of the industrial regions of Europe, man-made emissions of sulphur oxides are clearly dominant over natural ones. In Europe and eastern North America natural sources contribute less than 10 % to the total emissions (refs.16,41,42).

Anthropogenic emissions of SO_2 in Europe and eastern North America have risen substantially during industrial times (refs.43,44,45,46). In figure 4 long-term SO_2 emission curves are given for Europe without the territory of the USSR according to Fjeld (ref.46) and Bettleheim and Littler (ref.43). Both publications show an increase of SO_2 emissions by a factor of two between 1950 and 1970. However, there are considerable discrepancies about the absolute figures which are not acceptable with respect to long-range transport modelling.

A long-term SO_2 emission curve for the territory of the Federal Republic of Germany (ca.250 000 km^2, figure 5) is much more structurally differentiated than the long-term trends in figure 4 (ref.47). According to Häberle and Herrmann (ref.47), SO_2 emissions increased from ca. 0.2 mio tons per year in 1860 to ca. 3 mio tons per year in 1982.

Until the mid 1970s the SO_2 emission curve reflects directly the economic situation with high emissions before and during World War II and low emissions during inflation after World War I (1923) and after World War II (1945-1947).

In my opinion the relative long-term SO_2 emission trend in figure 5 is more realistic than the emission trends in figure 4 taking into account that most European countries were involved-either directly or indirectly-in World Wars I and II.

Since the 1970s the SO_2 emissions in a series of European countries including Belgium,Denmark,the Federal Republic of Germany,France, Ireland,the Netherlands,Norway,Sweden,Switzerland,United Kingdom have decreased(refs.5,47,48,49,50).

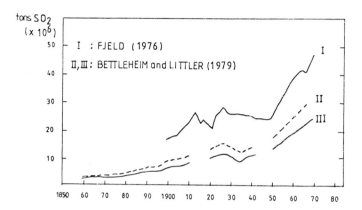

Fig.4 : Long-term SO2 emission curves in Europe between 1860 and 1970(Curve III:without the territory of the USSR) (refs.43,46)

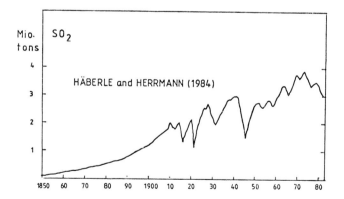

Fig.5 : Long-term emission curve for the territory of the Federal Republic of Germany(249 000 km2) (ref.47)

The development over time with decreasing emissions is illustrated in figure 6 showing man-made emissions for selected European countries presented by source over the time period 1965-1982(ref.50). When interpreting these long-term SO_2 trends,caution should be exercised because definitions of emission sources and procedures for

18

estimating emissions may vary from country to country(ref.50).
Most European countries have signed a protocol on reduction of sul-
phur(SO_2) emissions.Before 1993,the total national sulphur emissions shall be reduced by 30 % using the 1980 emission data as reference levels(ref.51).
In the years to come it will be an important task for modellers to
assess the influence of SO_2 emission reduction on acid deposition
in Europe.

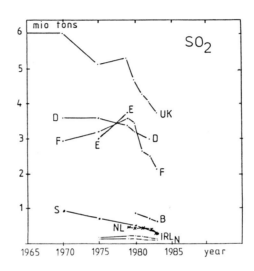

Fig.6 : Emissions of SO_2 in 103 tonnes/year,by source,for selected
        CEC countries(1965-1983)
        Source:OECD Environmental Data Compendium,1985(ref.50)

In figure 7 long-term SO_2 emission trends are shown for eastern
North America(refs.44,45).
Similiar to the long-term curves of Häberle and Herrmann(ref.47),
both trends reflect the economic situation showing a downward
trend when industrial activity decreased.
    As opposed to SO_2,the establshment of emission inventories for
NO_x is much more complicated and the range of uncertainties is
considerably greater.The main reason is that man-made emissions of
SO_2 originate mainly from the combustion of fossil fuels at stationary sources as a consequence of its sulphur content,whereas man-
made NO_x originates from the oxidation of atmospheric nitrogen
during combustion as well as from trace nitrogen compounds in
fossil fuels in stationary and mobile sources.
Similiar to SO_2,on the regional scale of the industrialized

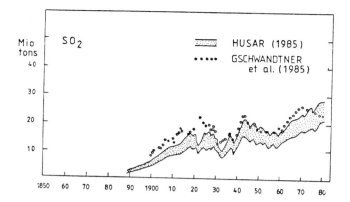

Fig.7 : Long-term SO2 emission trend estimates for the eastern
North America according to Husar(ref.44) and Gschwandtner
et al.(ref.45).The two lines indicate high and low
estimates of Husar.

regions of Europe and eastern North America man-made emissions of
NO_x are overwhelmingly dominant over natural ones(refs.2,16,41,52,
53,54).Unlike SO_2,NO_x emissions seem still to be on a rising trend
in some countries and are believed to have increased by 40-50%
over the last 10-15 years(ref.22).

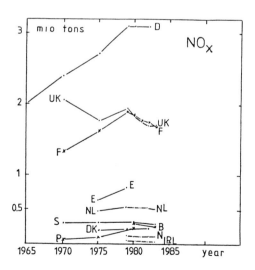

Fig.8 : Emissions of NO_x,by source,for selected CEC countries
(1965-1983).
Source:OECD Environmental Data Compendium,1985(ref.50)

As opposed to the situation in eastern North America,long-term
NO_x emission curves for Europe as a whole do not exist.

20

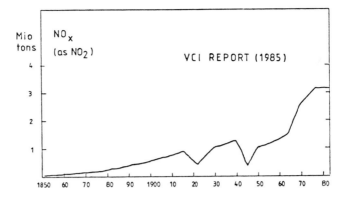

Fig. 9 : Long-term NO$_x$ emission curve for the territory of the
Federal Republic of Germany(249 000 km2).
(ref.47)

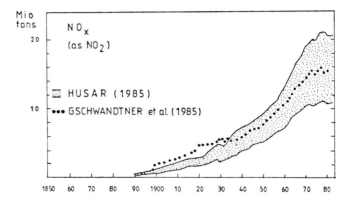

Fig.10 : Long-term NO$_x$ emission trend estimates for eastern North
America according to Husar(ref.44) and Gschwandtner et al.
(ref.45).The two lines indicate high and low estimates of
Husar(ref.44).

Instead,for some European countries NO$_x$ trends are reported over
the period 1965-1983(ref.50) which are shown in figure 8.
Similiar to the values in figure 6,these figures should also be
taken with caution because definitions of sources as well as the
measuring method may vary considerably from country to country.
In a recent publication(Häberle and Herrmann,ref.47) an increase
of man-made NO$_x$ emissions by a factor of 3 was assessed over the
time period of 30 years for the Federal Republic of Germany(from
1-1.2 mio tons in 1954 to ca. 3.1 mio tons in 1984,figure 9).
Extensive long-term NO$_x$ emission trends exist for eastern North

America(refs.44,45) which are shown in figure 10.The shaded area
representing the range of uncertainty is larger than that for SO_2
(see figure 7).Furthermore,the estimated long-term NO_x emission
trends in eastern North America indicate a strong and monotonic increase since the late 1800s.

As far as the emissions of other gases is concerned which are
relevant to the problems of acid deposition and photochemical oxidants,the situation is even more uncertain compared with NO_x.

With regard to hydrochloric acid(HCl) it is known that important
natural sources of gaseous HCl are the generation of this gas from
reactions of sea salt aerosols with strong acids and the direct
emission by volcanoes.The most important anthropogenic emissions of
HCl result from the combustion of coal and from waste incinerators.
A quantification of the HCl emission strenghts is,however,difficult.

Concerning hydrofluoric acid(HF),emissions by brickworks,alumi-
nium smelters and lead crystal factories can be important on a local scale and are therefore not important with respect to regional
and long-range transport.

Of considerable importance for the acid deposition problem is
ammonia(NH_3) which is produced by three processes:a.) volatilisation from natural soils and ocean water,b.) combustion processes
and c.) volatilisation from commercial fertilizers and animal ex-
crements.The main source of NH_3 in Europe is the decomposition of
organic material above all of domestic animal excrements(ref.55).
There is growing concern about increasing emissions of NH_3 from
agricultural activities in certain European regions(refs.56,57,58).
Examinations of domestic animal records imply that emissions of
ammonia are likely to have increased by ca.50 % between 1950 and
1980 over Europe as a whole.In some countries such as the Netherlands,Belgium,Poland,and GDR the emissions have about doubled,
whereas in other countries there is little change(ref.58).
Ammonia is very important for the acid deposition problem in Europe
because this trace gas can considerably increase the acidic inputs
to ecosystems.
A first long-range transport model for ammonia and ammonium was
recently presented for Europe within project EURASAP[1] (ref.59).

Hydrocarbons play an important role for both the generation of
acidic substances and the formation of photochemical oxidants.The
anthropogenic emissions of volatile hydrocarbons is mainly connected with the production and use of light petroleum distillates

---

(1): EURASAP = European Association of the Science of Air Pollution

and natural gas.Some man-made emission data are available for central Europe (refs.41,60,61),but these data are much more uncertain than the corresponding emissions of SO_2 and NO_x.Data on natural source strengths of hydrocarbons by vegetation,soils and water surfaces are insufficiently known.The large uncertainties of hydrocarbon emissions provide one of the major problems in any regional or long-range transport model which tries to simulate episodes of photochemical oxidant formation (for example the PHOXA project, ref.61).

Problems concerning the emissions of trace substances relevant to acid deposition are mainly:

- the different degree of certainty of SO_2 emission values for different European regions,
- the general large uncertainty of the emissions of NO_x and NH_3 as major precursors for acid deposition including seasonal and even diurnal variations of these gases and SO_2,
- the unknown amount of biogenic emissions of natural hydrocarbons and
- the influence of emission heights on long-range transport. For a slightly reactive gas like SO_2 the increasing potential for long-range transport with increasing emission height (refs.15,16) is mainly a meteorological problem and not a chemical one.As opposed to SO_2,the potential for long-range transport of NO_x is also determined by chemical processes. Nitrogen oxide(NO) emitted by cars is rather rapidly converted to NO_2 due to rapid mixing and fast oxidation of NO by O_3. On the other hand,NO emitted from tall stacks is normally oxidized rather slowly because the oxidation of NO by O_3 into NO_2 is in most cases diffusion controlled i.e.the oxidation rate is controlled by the rate at which the plume mixes with the ambient air rather than by the relevant chemical reaction of NO with O_3 within the plume(ref.18).The different NO oxidation rates with emission height have most likely to be taken into account in long-range transport models for NO_x.

4.2. CHEMICAL TRANSFORMATION

As chemical transformation processes are extensively treated in another lecture(Schurath,ref.62) I will confine myself to a discussion of some general aspects concerning mainly the transformation processes in atmospheric droplets.

Transformation processes are very important because deposition of

the primary pollutants and that of their reaction products are
governed by quite different mechanisms.

As far as the formation of atmospheric acidity is concerned,the
most important acids are H_2SO_4,HNO_3,and NH_4^+ which are generated to
a large extent by oxidation of SO_2 and NO_x and conversion of NH_3.
The formation of acids can proceed by homogeneous gas phase reactions,in aqueous droplets and on surfaces of aerosol particles.The
rates of these reactions depend on the specific environment
(meteorological conditions such as solar radiation,humidity,presen-
ce of other substances etc) being considered.

In central Europe all of these are involved in acid deposition
generation.

As far as the formation of photochemical oxidants is concerned,it
had previously been widely accepted that this was a process occurring almost entirely in the gas phase.However,there is recent evidence that photochemical oxidant generation is also an important
process in atmospheric droplets affecting acidity generation.

Compared with transformation in the liquid phase,homogeneous gas
phase reactions are relatively well known.Thermodynamic and kinetic
data of a series of key reactions in the generation of acid deposition and photochemical oxidants have been determined and satisfy
the requirements for long-range transport model inputs.

Regarding the chemical transformation of SO_2 and NO_x there is considerable knowledge about the formation of H_2SO_4 and HNO_3(ref.63).
It is now accepted that the oxidation by OH radicals due to reactions 4 and 5 is the dominant mechanism for generation of sulphuric
acid in general and for nitric acid during daytime:

$$SO_2 \;+\; OH \xrightarrow{\;M\;} HSO_3 \xrightarrow{\;O_2\;} SO_3 + HO_2 \qquad (4)$$

$$NO_2 \;+\; OH \xrightarrow{\;M\;} HNO_3 \qquad\qquad (5)$$

An important problem for modelling is the uncertainty in the concentrations of OH radicals under different meteorological condi-
tions.During nighttime the most important NO_x removal process is
reaction 6 in combination with reaction 7 which ultimately leads
to the generation of nitrates at quantities which are most likely
higher than those produced by reaction 5 for most of the year especially in winter,spring and autumn(ref.64):

$$NO_2 \;+\; O_3 \longrightarrow NO_3 + O_2 \qquad\qquad (6)$$

24

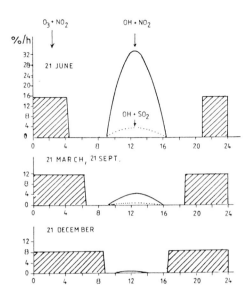

Fig.11a: Daily variations of NO_x removal rates in %/h for different
seasons according to the rate-determing. reactions OH + NO_2
(reaction 5) and O_3 + NO_2 (reaction 6).The rates were calculated for the following conditions:OH maximum concentration is 8 x $10^6/cm^3$;O_3 concentration is 30 ppb for all
cases.Mean temperatures:20° C,10° C and 0° C for summer,
spring/autumn and winter.
Source:Platt,1986(ref.64)

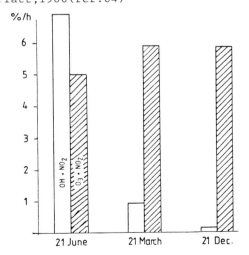

Fig.11b: Mean NO_x removal rates in %/h averaged over 24 hours calculated for the conditions specified under fig.11a.
Source:Platt,1986(ref.64)

$$NO_3 + NO_2 \rightleftharpoons N_2O_5 \xrightarrow{H2O} 2\ HNO_3 \qquad (7)$$

In figures 11a and 11b typical daily variations of NO_x removal rates (in %/h) are given calculated by Platt(ref.64) for NO_x removal according to the rate-determining reactions 5 and 6(NO_2 + OH and NO_2 + O_3).The corresponding SO_2 removal rates according to the SO_2-OH-reaction 4 is inserted in figure 11a.

Another problem which was raised recently is the so-called linearity issue.As the linearity problem will be treated in another lecture(Smith,ref.7) I will confine myself to a few comments on a possible interaction of reactions 4 and 5 which could lead to a generation of H_2SO_4 being non-linear with respect to the SO_2 gas phase concentrations.Non-linear effects were predicted by Rodhe et al.(ref.65) in their transformation-transport model simulating the formation of sulphuric and nitric acid during long-range transport through the atmosphere.The authors concluded that,mainly because of the common dependence of the gas phase oxidation of SO_2 and NO_x on the concentration of the OH radical concentration,the concentration of NO_x tends to reduce the levels of OH close to the source of SO_2 and NO_x retarding the oxidation of SO_2 to H_2SO_4 resulting in a non-linear dependence of concentrations on SO_2 emissions and concentrations,respectively.
However,theoretical calculations(ref.66) taking account of the latest results of realistic laboratory studies have greatly reduced the non-linear effects predicted by Rodhe et al.(ref.65).
Computer simulations of the chemistry of the polluted atmosphere have shown that only ca.10% of the OH-HO_2-radical termination occurs through the NO_2-OH-reaction 5 for air masses typical of an urban,polluted area.As a result of a more realistic treatment of the homogeneous tropospheric chemistry,the non-linear effects in the model of Rodhe et al.(ref.65) were greatly reduced i.e.a quasi-linear relationship between changes in ambient concentrations of SO_2 and changes in gas phase generation of sulphuric acid was obtained(refs.16,66).

As opposed to homogeneous gas phase reactions,chemical conversion processes in cloud droplets are much more difficult to treat in long-range transport models.
Sulphuric acid formation due to SO_2 oxidation by strongly oxidizing agents(H_2O_2,O_3) according to reactions 8 and 9 seems to be the dominant mechanisms in cloud droplets under most atmospheric

conditions (see for example ref.67).

$$S(IV) + H_2O_2 \longrightarrow S(VI) + O_2 \qquad (8)$$

$$S(IV) + O_3 \longrightarrow S(VI) + H_2O \qquad (9)$$

According to some authors(refs.16,68) the possibility cannot be
ruled out that,in contrast to SO_2 gas phase conversion,a significant nonlinear conversion of SO_2 to sulphuric acid can result from
the liquid phase oxidation of SO_2(HSO_3^- and/or $SO_3^=$) in cloud water
by H_2O_2.The theory of hydrogen peroxide generation predicts that
the concentrations of H_2O_2 may be significantly less than those of
S(IV) in cloud water especially in winter.As a result,only a fraction of sulphur IV is oxidized and reaction 8 becomes oxidant
limited.
It is also possible that even with sufficient oxidant supply(H_2O_2)
in cloud water,other substances such as formaldehyde(CH_2O) may in-
hibit the S(IV)-H_2O_2-reaction(refs.69,70).
Whether or not such an oxidant-limited or formaldehyde-inhibited
S(IV)-H_2O_2-reaction can lead to an appreciable non-linear conversion of SO_2 to sulphuric acid cannot be tested yet on the basis of
laboratory and field measurement results.
Figure 12 shows a simplified reaction scheme for SO_2 absorption
and generation of sulphurous and sulphuric acid in a cloud droplet
along with the major sources for a non-linear conversion of SO_2 to
sulphate.As seen in this figure,for a given SO_2 gas phase concentration,the amount of HSO_3^- and $SO_3^=$ formed depends strongly on droplet pH with HSO_3^- being proportional to $[H^+]^{-1}$ and $SO_3^=$ to $[H^+]^{-2}$.
Droplet pH values are determined by all the different substances
dissolved(aerosol particles,ammonia,etc,refs.71,72).
Figure 13 shows sulphuric acid formation rates in cloud water as a
function of droplet pH calculated for the following conditions:
constant gas phase concentrations for SO_2 = 1 ppb; O_3 = 5 and 50
ppb.Equilibrium between gas and liquid phase was assumed.For H_2O_2
measured concentrations in cloud water were used(1 to 100 μ molar).
As seen in this figure,the S(IV)-H_2O_2 reaction which is only
slightly dependent on droplet pH seems to be the dominant reaction
in the pH range of cloud droplets in central Europe(pH: 3 to 6).
    As far as the formation of nitric acid and nitrate in the
aqueous phase is concerned there is little hard information about
the origin of these species.Indications are strong that the rather

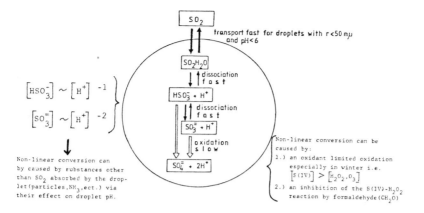

Fig.12:Simplified reaction scheme for SO2 absorption and generation
of sulphurous and sulfuric acid in cloud droplets along with
the major sources for a nonlinear SO2 conversion

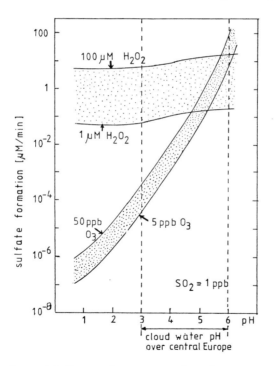

Fig.13:Sulphuric acid formation rates in cloud droplets due to
S(IV) oxidation by H2O2 and O3 as a function of droplet pH.
The calculated rates apply to the following conditions:
Constant SO2 gas phase concentration: 1 ppb; O3 = 5 and 50
ppb.H2O2 concentrations in droplets: 1 to 100 μ mole/litre
The cloud water pH range over central Europe(pH between 3
and 6) is also inserted in this figure.

high nitrate concentrations found in cloud and rainwater have their
origin in processes such as absorption and chemical reactions of
gaseous HNO_3, NO_3/N_2O_5 and in an incorporation of nitrate containing
aerosols during condensation rather than in a direct absorption
and chemical reaction of NO_x.
Figure 14 shows a highly simplified reaction scheme for the generation of nitrous acid(HNO_2) and nitric acid(HNO_3) due to absorption of NO, NO_2, HNO_2, N_2O_5, and HNO_3 by a cloud droplet.

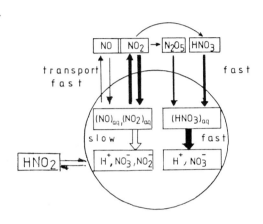

Fig.14:Simplified reaction scheme for the generation of nitrous
acid(HNO_2) and nitric acid(HNO_3) due to absorption of NO_x,
HNO_3, HNO_2, and N_2O_5 in cloud droplets.

Of the different processes by which nitrogen oxides are converted,
only the generation of acids by dissolved NO_x can be quantified:

$$2\ NO_2(aq.) + H_2O \longrightarrow 2\ H^+ + NO_3^- + NO_2^- \tag{10}$$

$$NO(aq.) + NO_2(aq) + H_2O \longrightarrow 2\ H^+ + 2\ NO_2^- \tag{11}$$

Using the thermodynamic and kinetic data published by Schwartz and
White(ref.73) NO_x removal rates of the order of 10^{-4} - 10^{-5} %/h
were calculated for a cloud liquid water content of 1 g/m^3.
However, the possibility cannot be ruled out that NO_x oxidation to
HNO_3 in the interstitial air may be considerably increased under
conditions of high solar radiation(Winkler, P., personal communication, 1986).
These processes are, however, difficult to quantify.

Some of the major problems concerning the chemical transformation
of substances in clouds being relevant to acid deposition are
   - the general limited knowledge of acidity generation in clouds
     compared with our knowledge about the corresponding acid for-
     mation processes in the gas phase,
   - the origin of oxidizing agents in cloud water which involve
     both gas and liquid phase generation processes,
   - the efficiency and seasonal variations of transformation pro-
     cesses which are no doubt different for gas phase and aqueous
     phase processes,
   - the supply of substances into the cloud droplets on different
     diffusion scales.Dynamic and micro-physical parameters are very
     important for acidity generation in clouds(air flow in the
     vicinity of clouds,entrainment of air,residence times of cloud
     droplets,cycles of droplet formation and evaporation etc.).

4.3. DEPOSITION

   As removal by dry and wet deposition is extensively treated in
another lecture(Fowler,ref.74) I will confine myself to a few aspects of this subject.
There is considerable uncertainty about the relative proportions
of acid contributed by dry and wet deposition.This uncertainty is
mainly due to a lack of knowledge about dry deposition processes
which make estimates of this removal mechanism much less accurate
than for wet deposition.The integrated wet deposition is compara-
tively simple to measure.As a consequence there exists a substantial data base on wet deposition from a series of European net-
works.On the other hand,dry deposition is extremely difficult to
measure and the corresponding data base is small.
In principle,the chemical reactive nature of SO_2,NO_2,HNO_3,HCl,and
HF makes dry deposition an important removal process.On the basis
of literature release data,the following deposition velocity range
for gas transfer to natural surfaces can be estimated(table 5).
   A serious problem associated with long-range transport of acidic substances is that often mean deposition velocities are used
and weighted by concentrations in order to get dry deposition
fluxes.The major uncertainty arises from the wide variations of
deposition velocities and concentrations with season,time of day,
weather and surface properties.These variations are not sufficiently well known .As a result, the estimates of the mean fluxes for

SO_2,NO_x,HNO_3,and other relevant gases are subject to considerable
uncertainties probably of the order of 50% or more(ref.77).
Instead of mean deposition velocities different classes of deposition velocities should be defined for certain atmospheric stability classes and different surface properties.A first attempt to use
such classes in models was made recently(for example ref.6).

TABLE 5

Acidic gas	Range of deposition velocities to natural surfaces			Remarks
SO_2	0.1	-	2 cm/s	maximum values for complete absorber(ocean water) and small surface resistance
NO_2	0.01	-	1 cm/s	maximum values over vegetation, mid-day ,stomata open
NO	0.001-	0.1	cm/s	soils are often sources rather than sinks
HNO_3	1	-	20 cm/s	There is only little surface re-sistance i.e.v_d depends almost entirely on the aerodynamic re-sistance.The very large values
HCl	1	-	20 cm/s	apply to conditions where atmos-pheric turbulence is enhanced e.g. over mature wheat or forests (refs.75,76)
HF	1	-	20 cm/s	No measurements available but deposition behaviour should be similiar as for HNO_3 and HCl

Table 5: Range of deposition velocities for gas transfer to natural
surfaces according to literature release data.

Another important problem in the modelling of long-range transport
arises from the uncertainty of particle dry deposition fluxes.As
can be seen in figure 15,considerable controversy remains concerning the values of the deposition velocity for particles in the
size range between 0.1 to 1 μm where most sulphate,nitrate and acidity is found.Some authors(for example ref.77) advocate the continued use of 0.1 cm/s or less,whereas other autors have reported
values of the order of 1 cm/s in this size range(for example refs.
16,78).The reason for the considerable disagreement of particle deposition could be the influence of the roughness height and wind
velocity.There is good evidence that when surface roughness increases,deposition velocities increase as well.If the higher values in

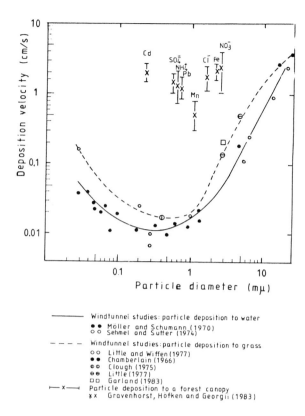

Fig15: Measured deposition velocities of aerosol particles from
wind tunnel measurements over relatively smooth surfaces
(water and short grass).
The deposition velocities for particle deposition on to
beech and spruce forests(Gravenhorst et al.,1983;ref.78)
were determined for mass mean diameter(MMD) of an aerosol
component distributed over the large particle size spectrum.

figure 15 were representative for European regions with rough surfaces like areas covered with forests(see for example table 3),dry
deposition of particles would be an important removal process for
atmospheric acidity.

5. SOURCE-RECEPTOR RELATIONSHIP

Two basic approaches are generally used to establish a source-
receptor relationship: a.) the application of theoretical models
accounting for material balance,and b.) the interpretation of observations of wet deposition patterns in the light of emissions.

On the following pages some aspects regarding approach b.) are
discussed.

As discussed in section 4.3.,there is a substantial data base on
wet deposition from a series of European and North American net-
works.On the other hand,a substantial data base exists also in both
continents on the emission of gases,especially of SO_2.
There are some problems if we want to relate specific emission
sources with observed wet deposition fluxes.
This begins with a clear definition of the term "background" concentration of rain constituents.Often this term is used to indicate
background concentrations at the border of the area considered for
example Europe,a specific country or city.This concentration has
clearly to be distinguished from the term "natural background"
which is the concentration level in the case when all anthropogenic
influences can be neglected(refs.8,79).For certain European regions
the background concentrations at the edges may be determined
whereas the "natural background" values can most likely not be determined correctly.
It must be emphasized that the natural composition of rainwater and
hence the natural acidity and pH is variable depending on the relative concentrations of the various trace constituents incorporated
in water drops.An important point in this context is that a pH of
5.6 is not the natural reference value of rain which is not influenced by anthropogenic activities.A pH of 5.6 is simply the value
of distilled water in equilibrium with the ca. 330-340 ppm CO_2,
whereas the pH of natural rain is also determined by naturally
emitted particles(soil,dust,sea spray) and naturally occurring
gases(refs.8,71,72).
Based on extensive measurements of rain constituents in Europe between 1850-1870(ref.13) it can be concluded that rain pH values in
central Europe outside industrial urban areas were between 4.6 and
5.6(ref.14) which is most likely not very far from the natural
background pH in rainwater over Europe.A similiar pH range was also
reported by Holdgate(ref.80).
These considerations show that it might be very difficult to assess
man's contribution to the acidity in some areas which lie outside
the centres of high anthropogenic rain acidity.It is therefore hardly possible to assess by measurements the influence of emission reductions on air and precipitation quality in such regions.
The conclusions which are drawn on the basis of trends of rainwater
composition and acidity over large areas for long periods of time
are also limited because often those trends are derived from

inconsistent measurements.

In Europe these problems still exist today because there are a
series of different national networks working to some extent with
different methods.For the years to come,an important task of EMEP
should be to harmonize and standardize the instrumentation for
sampling,storage,and chemical analyses of precipitation on a European scale.

There are only a few publications investigating an emission-wet deposition relationship in acid deposition on the basis of measure-
ments(refs.16,49).

In eastern North America analyses of the spatial distribution of
the molar ratio of sulphate to nitrate in precipitation do not vary
substantially over a large region.For annual averages,this ratio is
similiar to the average molar ratio of SO_2 and NO_x in emissions.
The similiarity of the ion ratios in precipitation suggests that,
regardless of the specific physico-chemical processes involved,
there is no significant loss of one component relative to the other
(ref.16).In this report it was furthermore concluded on the basis
of other observations that within the uncertainties and for annual
averages and summed over large areas of eastern North America,there
is no evidence that the relation between emissions and deposition
is substantially non-linear i.e. the emission-deposition relationship can be taken for practical purposes as essentially linear.

The investigation of the linearity issue in acid deposition on
the basis of emission-wet deposition measurements seems to be more
complicated for Europe than for eastern North America despite the
fact that considerably more data on precipitation chemistry to evaluate trends are available in Europe,where monitoring programmes
habe been in existence for longer times than in North America.The
reasons are most likely the greater uncertainties of the emission
and wet deposition data in Europe compared to North America as discussed in the previous sections.Another reason could be a somewhat
different physico-chemical behaviour of SO_2 and NO_x in Europe due
to different climatological conditions on both continents.

In some recent publications(refs.49,81) long-term changes of
sulphate concentrations in precipitation at a number of Scandinavian stations from the EACN network have been compared with the
corresponding changes in anthropogenic SO_2 emissions in Europe
(EACN = European Air Chemistry Network).

Most stations show an increase in sulphate concentrations between
the late 1950s and the early 1970s by ca. 50%,followed by a decline

since then.A detailed analysis of data from 12 Swedish stations
between 1972 and 1985 shows an average decrease during the 14-year
period by ca. 30 percent.According to these authors,the increase
and the following decline of sulphate concentrations in precipitation in Scandinavia is by and large consistent with the long-term
trends of SO_2 emissions in a series of European countries.
In contrast to the Scandinavian sulphate precipitation data,the
corresponding data from the UK and the European Continent seem to
show less systematic variations.
It must be noted that,on the basis of this work,it is hardly possible to conclude that for the period 1950-1970 SO_2 emissions and
sulphate concentrations in rainwater changed relatively by the same
amount.The serious limitations associated with the quality of both
SO_2 emission and sulphate concentration data(see for example figures 4 and 5)prevent such interpretations.
After 1970 the quality of both the emission and wet deposition data
has improved.The average decrease of sulphate concentrations observed at most Scandinavian stations between the early 1970s and
the early 1980s is by and large consistent with the decrease in SO_2
emissions in most countries of western Europe by a similiar percen-
tage during the same period of time.Taking into account that these
countries contribute considerably to sulphate wet deposition in
Scandinavia there is at least some qualitative evidence that the
relation between wet sulphate deposition and SO_2 emission is not
substantially non-linear in this area when averaged over this
period of time.

6. CONCLUSIONS AND RECOMMENDATIONS

From the point of view of environmental effects which may be
attributed to acidic substances and/or photochemical oxidants transported and/or formed over long distances,the problem of long-range
transport is a relatively recent issue.Besides long-term inputs of
air pollutants into ecosystems episodic inputs with short term high
stress situations may also damage aquatic and terrestrial ecosystems.
From a chemical point of view the following more general conclusions and recommendations are given regarding the input parameters
of source strengths,chemical transformation,and deposition rates
into long-range transport models.

Emission inventories which are needed as an important input parameter into both long-term and episodic transport models need

improvement with respect to their accuracy in Europe.

As far as long-term(annual) deposition of sulphate is concerned, even the simlest models seem to give reasonably good estimates in the term of magnitude and geographical distribution enabling estimates of the contribution one country or region makes to the deposition in another(refs.27,82).

Such long-range transport models describe the transport and deposition with an accuracy of a factor of two or better(ref.2).The accuracy of these models is obviously dependent on the quality of the SO_2 emission data which are for western Europe probably within $\pm$ 50 % of the true contributions for each country(ref.27).

On the other hand it is recognized that large SO_2 emissions occur in some central European countries east of the FRG(refs.37,83)which have to be included in any model which tries to establish source-receptor relationships in Europe.In order to reduce the uncertainties in emission data,it is strongly recommended to measure fluxes of SO_2,NO_x,NMHC,and other pollutants originating from certain main area sources and to determine transboundary fluxes using a combination of airborne and ground based measurements.

With respect to emissions in Europe there is also a need to determine the fluxes of the background concentrations across the borders of the European budget areas.

In some European regions or countries outside the industrial areas of central Europe the so-called "undecided background" contributions originating from both natural and anthropogenic emissions may contribute considerably to acid deposition.

Apart from the need to establish emission inventories for compounds other than SO_2 and to improve the quality of the latter,there is a need to establish inventories for different European regions(similiar to the estimates of Husar,ref.47,for the eastern North America) taking into account the different trend development.

Another point is the investigation of seasonal emission trends.

As far as chemical transformation as an input to long-range transport models is concerned its treatment in models should be based on our present stage of knowledge which is completely different for chemical processes in the gas and liquid phase.

For gas phase reactions of SO_2 and NO_x the key reactions regarding acidity generation are known and should be included in models along with the thermodynamic and kinetic quantities.A source of uncertainty is probably the computation of the true atmospheric OH radical concentration.A comparision of measured OH radical concentrations

with calculated ones shows a tendency for OH concentrations to be
systematically overestimated when calculated by chemical kinetic
models(ref.84).

Regarding the conversion of SO_2 and NO_x in atmospheric droplets,the
situation is completely different and much more complicated than in
the case of homogeneous gas phase reactions.

Laboratory experiments have shown that SO_2 oxidation by H_2O_2 proceeds very fast but the importance of this reaction is not really
known under atmospheric conditions.

The same is true and even more uncertain for other chemical reactions converting the precursor gases SO_2 and NO_x into H_2SO_4 and
HNO_3 in cloud droplets.

Taking into account the present stage of knowledge about acidity
generation in cloud droplets,I think it is not yet possible and
justified to treat this type of transformation in long-range transport modelling in such a manner that detailed chemical equations
with their kinetic and thermodynamic data can be used.

At present the use of such an approach in episodic long-range transport models pretends an accuracy which is not achieved at all.

On the other hand,modelling of episodic long-range transport with
a rather simple parametrisation regarding conversion processes in
clouds is also dangerous since,for example,SO_2 conversion alone can
vary by at least two orders of magnitude depending on the special
physico-chemical conditions encountered(see for example figure 13).

In order to solve this problem and to provide a more realistic
chemical input to episodic LRT models I suggest the measurement of
SO_2 and NO_x conversion in clouds under different meteorological and
chemical conditions as a function of distance and time from well
defined emission sources.First aircraft measurements to study the
physico-chemical conversion in clouds were carried out during the
last few years(refs.85,86,87 .

Similiar to the treatment of dry deposition processes in LRT models
different oxidation classes for SO_2 and NO_x conversion should be
defined on the basis of aircraft measurements and used in episodic
long-range transport models as input data.

   As far as physico-chemical processes of dry and wet deposition
are concerned it was suggested by the CEC Task Force on Acid Deposition in 1986 that deposition velocities of species other than SO_2
which are relevant to the acid deposition problem should be determined with priority(for example HNO_3,HCl,NO_x,NH_3) and their dependence on vegetation and orographic characteristics be considered.

For gases which undergo a rapid chemical conversion like NO, NO_2 and O_3 caution should be exercised when using the gradient method in order to determine deposition velocities because the observed concentration gradients may be influenced by chemical reactions (ref.88).

The CEC Acid Deposition Task Force(1986) expressed the general point of view that co-ordinated research programmes where a scientific question is addressed both in terms of instrument development laboratory work, field studies and model calculations, seems more promising scientifically than carrying out research according to a shopping-list of single items, where the interrelationships are not well thought out(Hov,1986,personal communication).

Finally, it is recommended that modellers and atmospheric chemists should co-operate more closely than was the case in the past. The European projects EURASAP and COST 611 and 612 could provide a forum to practice such a cooperation.

Acknowledgement

The author would like to express appreciation to Prof.G.Gravenhorst (University of Göttingen) for his valuable suggestions and useful discussions.

In addition, the author thanks Prof.W.Weiss(Bundesamt für Zivilschutz, Freiburg), Prof.U.Platt(KFA Jülich) and Dr.L.Werner (Umweltbundesamt, Berlin) for their support.

7. REFERENCES

1 A.Eliassen , The OECD study of long-range transport of air pollutants:long-range Transport modelling,Atm.Env.12,479,1978

2 H.Van Dop, The residence and transport of pollutants in the atmosphere-a meteorological problem.Proceedings:CEC Symposium on Acid Deposition-A Challenge for Europe,Karlsruhe,19-21 Sept. 1983,pp.48-57.

3 J.Pankrath, Großräumiger Transport von Luftverunreinigungen in Europa:Anforderungen an Modelle zur Simulation der Ausbreitung und Deposition von säurebildenden Luftverunreinigungen,1983 Proceedings:Acid Precipitation,Lindau,VDI-Berichte 500,43-50.

4 W.Klug,P.J.H.Builtjes,H.Van Dop,N.D.Van Egmond,H.Glaab,D.Gömer, B.J.De Haan,K.D.Van d.Hout,N.Kesseboom,R.Röckle,C.Veldt, A comparision between different interregional air pollution models.NATO report No.155,Committee on the Challenges of Modern Society, 1984

5 B.C.A.Fisher,The long-range transport of air pollutants-some thoughts on the state of modelling.Atm.Env.18,pp.553-562,1984

6 R.Stern,The Transport-Deposition Model of Acidifying Pollutants (TADAP).Proceedings:Symposium on Air Pollution,Berlin,Reichstag,22-24 Januar 1986,pp.661-692.

7   F.B.Smith,The atmospheric turbulent transport and linearity
    problems in the formation of acid deposition.Lectures given
    during the Ispra course on Regional and Long-Range Transport
    of Air Pollution,15-19 September 1986.

8   R.J.Delmas and G.Gravenhorst,Background precipitation acidity,
    Proceedings.CEC Workshop on Acid Deposition,9 Sept.1982,
    Reichstag Berlin,Reidel Publ.Comp.,pp.82-108.

9   F.Meixner,H.W.Georgii and R.Ockelmann,Eruption des Vulkans Mt.
    St.Helens:Nachweis der stratosphärischen Rauchfahne über Euro-
    pa.Annalen der Meteorologie,Neue Folge 16,pp.249-251,1980

10 Finnish Report on Radioactive Fallout,Finnish Centre for
    Radiation and Nuclear Safety,STUK-B-VALO 44:Interim Report on
    Fallout Situation in Finland from April 26 to May 4,1986.
    33 pages.Helsinki,Finnland,ISN 078-2868.

11 BMU-Report,Bericht über den Reaktorunfall in Tschernobyl,Der
    Bundesminister für Umwelt,Naturschutz und Reaktorsicherheit.
    RS I 1-518042-SOW/2,Bonn,18 June 1986.

12 W.Weiss,H.Sartorius,H.Stockburger,A.Sittkus and K.H.Rath,
    Die Radioaktivität der bodennahen Luft der Bundesrepublik
    Deutschland im Zeitraum 29 April bis 10 Mai 1986(4.bis 15.
    Tag nach Tschernobyl).Meßdaten und erste Bewertung,For-
    schungsbericht im Auftag des Bundesministers des Innern.

13 R.A.Smith,Air and rain-the beginnings of a chemical climatology,
    London,Longmans,Green,and Co.,1872(600pages).

14 D.Schwela,Wet deposition of air pollutants in about 1870 com-
    pared with today's impact,Staub-Reinhaltung der Luft 43,Nr.4,
    April 1983.

15 W.M.Koerber,Trends in SO$_2$ emissions and associated release
    height for Ohio River valley power plants.Proc.:75th Annual
    Meeting of the Air Poll.Control Association,New Orleans,1982

16 NAS Report,Acid Deposition-Atmospheric Processes in eastern
    North America,US National Academy Sci.,Nat.Academic Press,
    Washington,D.C.,1983.

17 B.Bolin and H.Rodhe,A note on the concepts of age distribution
    and transit time in natural reservoirs,Tellus 25,pp.58-62,
    1973.

18 A.J.Elshout and S.Beilke,Oxidation of NO to NO$_2$ in flue gas
    plumes of power stations,Proceedings:Third European Symposium
    COST 61a bis,Varese,Italy,10-12 April 1984,pp.535-543.

19 H.Rodhe,Budgets and turn-over times of atmospheric sulphur com-
    pounds,Atm.Env. 12,No.1-3,1978,pp.671-680.

20 D.Jost and H.W.Georgii,Untersuchungen über die Verteilung von
    Spurengasen in der freien Atmosphäre.Pure and Appl.Geoph.59,
    p.217,1964

21 F.Fiedler,Erfassung der Ausbreitung und der Konzentration von
    Luftschadstoffen,Waldschäden,Theorie und Praxis auf der Suche
    nach Antworten,R.Oldenbourg Verlag,pp.307-334 1985

22 CEC Report on Acid Rain,A review of the phenomena in the CEC
    and Europe.Published by Graham and Trotman Limited,London,
    1983,For the Commission of the European Communities,EUR 8684

23 T.N.Skoulikidis,Effects of primary and secondary air pollutants
    and acid deposition on buildings and monuments.Proc.:CEC
    Symp.on Acid Rain,Karlsruhe,23-25 Sept.1983,pp.193-228.

24 G.G.Amoroso and V.Fassina,Stone decay and conservations.Materials Science Monographs 11,Elsevier Science Publishers B.V., 1983,453 pages.

25 H.Hultberg,Effects of acid deposition on aquatic ecosystems, Proceedings:CEC Symposium on Acid Deposition-A Challenge for Europe,Karlsruhe,19-21 September 1983,pp.167-186.

26 Sandefjord Conference:Ecological impact of acid precipitation, Proceedings of an international conference,Sandefjord,Norway, March 11-14,1980,pp,383 pages.

27 F.B.Smith,Conditions pertaining to high acid concentrations in rain,Proceedings:VDI Colloquium Acid Deposition,VDI-Berichte 500,Lindau(W-Germany),7-9 June 1983,pp.67-76.

28 H.C.Martin,Acidification of the Environment:A Canadian perspective,Proceedings:CEC Symposium on Acid Deposition-A Challenge for Europe,Karlsruhe,19-21 September 1983,pp.186-192.

29 L.A.Barrie and J.M.Hales,The spatial distribution of precipitation acidity and major ion wet deposition in North America during 1980,Tellus 36b,p.333,1984.

30 NADAP-Report,National Acid Precipitation Assessments Program, Annual Report 1984 to the President and Congress,Prepared by the Interagency Task Force on Acid Deposition,USA.

31 F.Scholz,Report on Effects of Acidifying and Other Air Pollutants on Forests.Mitteilungen der Bundesforschungsanstalt für Forst-und Holzwirtschaft,Hamburg,Nr.143,April 1984.

32 Forest Damage Survey for 1983 and 1984 published by the Federal Ministry of Food,Agriculture and Forestry,Bonn.

33 B.Ulrich,effects of accumulation of air pollutants in forest ecosystems,Proceedings:CEC Symposium on Acid Deposition-A Challenge for Europe,Karlsruhe,19-21 September 1983,pp.48-57.

34 K.E.Rehfuess,Über die Wirkungen der sauren Niederschläge in Waldökosystemen,Forstwissenschaftliches Zentralblatt,100, Jahrgang h 6,p.363,1981.

35 R.Guderian,Impact of photochemical oxidants on vegetation in the Federal Republic of Germany,Proceedings:Workshop on Ozone, Göteborg,Sweden,29 Feb.- 2 March 1984,pp.76-91.

36 M.Wagner,Effects of photochemical oxidants on the environment, Proceedings:International Workshop on the Evaluation and Assessment of the Effects of Photochemical Oxidants on Human Health,Agricultural Crops,Forestry,Materials and Visibility, Göteborg,Sweden,29 Feb.-2 March 1984,pp.31-35.

37 S.Beilke,R.Berg,W.Grosch,A.H.Blommers,F.W.Jansen and G.Lelieveld, Air sampling flights at low altitudes along the border between the Federal Republic of Germany and its neighbors,Paper presented at the 4th European Symposium on Physico-Chemical Behaviour of Atm.Pollutants,Stresa,Italy,23-25 September 1986, Project COST 611.

38 C.F.Cullis and M.M.Hirschler,Emissions of sulphur into the atmosphere,Proceedings:International Symp.on Sulphur Emissions and the Environment,London,8-10 May 1979,The Chemical Society, London,pp.1-24.

39 G.Varhelyi,Continental and global sulphur budgets-I:Antropogenic SO_2 emissions,Atmospheric Environment 19,1985,pp.1029-1040

40

40 H.W.Georgii,The atmospheric sulphur budget.In:Chemistry of the
    unpolluted and polluted atmosphere,1983,Reidel Publ.Comp.,
    Editors:Georgii and Jaeschke,pp.295-324.

41 B.Ottar,Air pollution emissions and ambient concentrations.
    Proceedings:CEC Symposium on Acid Deposition-A Challenge for
    Europe,Karlsruhe,19-21 September 1983,pp.33-43.

42 D.F.Adams,S.O.Farwell,M.R.Pack and E.Robinson,Biogenic sulphur
    gas emissions from soils in eastern and southeastern United
    States.J.of Air Poll.Contr.Ass.31(10),1981,pp.1983.

43 J.Bettelheim and A.Littler,Historic trends in sulphur oxide
    emissions in Europe since 1865.GEGB Report PL-GS/E/1/79,1979

44 R.B.Husar,Man made SO_x and NO_x emission trends for eastern
    North America.Chapter prepared for NAS Acid Deposition Trend
    Committee,USA,1985,47 pages.

45 G.Gschwandtner,K.C.Gschwandtner and K.Eldridge,Historic emis-
    sions of sulphur and nitrogen oxides in the USA from 1900-
    1980,Volume I,US EPA-600/7-85-009a,1985,Research Triangle
    Park,NC,USA,1985.

46 B.Fjeld,Forbruk av fossilt brensel i Europa og utslipp av SO_2
    i perioden 1900-1972.Norwegian Institute for Air Research,
    Teknisk Notat No. 1/7,1976.

47 M.Häberle and K.Herrmann,Entwicklung von Emissionen und Immis-
    sionen wichtiger Luftschadstoffe,WBL Zeitschrift für Umwelt-
    technik,August 1984.

48 CITEPA Report 1984,Report of the French National Institute of
    Science.

49 H.Rodhe and L.Granat,An evaluation of sulphate in European
    precipitation 1955-1982,Atm.Env.12,No.1-3,1984,pp.2627-2639.

50 OECD Environmental Data Compendium 1985,Paris 1985,297 pages.

51 H.Dovland,Concluding remarks,Proceedings:Workshop on experien-
    ces with the application of advanced air pollution assess-
    ment methods and monitoring techniques,NATO/CCMS Pilot Study
    on Air Pollution Control Strategies and Impact Modelling,
    Nr,153,December 1985,pp.1-3.

52 K.Bonis,E.Meszaros and M.Putsay,On the atmospheric budget of
    nitrogen compounds over Europe,Idojaras,Vol.84,Nr.2,March-
    April 1980,Budapest,pp.57-68.

53 OTA report,Acid rain and transported air pollutions:implica-
    tions for public policy,Washington,D.C.,US Congress,1984,
    Office of Techn.Assessment,OTA-O-204,June 1984.

54 G.Gravenhorst,Natural NO_x emissions in the Federal Republic of
    Germany,Proceedings:Meeting of Working Groups 4 and 5 of
    project COST 611,23-25 Sept.1985,Bilthoven,The Netherlands.

55 U.Lenhard and G.Gravenhorst,Evaluation of ammonia fluxes into
    the free atmosphere over Western Germany,Tellus 32,1980,p.48.

56 H.Ott,The acid deposition problem:scientific background,re-
    search priorities and abatement technologies.Paper presented
    at Symposium on Acid precipitation in Italy,Venice,24-25
    October 1984.

57 E.Buijsman,J.F.M.Maas and W.A.H.Asman,Some remarks on the
    ammonia emission in Europe,Proceedings:Meeting of Working
    Groups 4 and 5,project COST 611,23-25 Sept.1985,Bilthoven.

58 H.M.ApSimon,N.Bell and M.Kruse,Emissions of ammonia and their
    role in acid deposition.Paper presented at the Symposium on
    Interregional Air Pollutant Transport,Budapest,22-24 April
    1986.Inaugural Meeting of EURASAP.

59 W.A.H.Asman and A.J.Janssen,A long-range transport model for
    ammonia and ammonium for Europe.Paper presented at the Sym-
    posium on Interregional Air Pollution Transport,Budapest,
    April 22-24,1986.Inaugural Meeting of EURASAP.

60 P.Bruckmann,Cycles of organic gases in the atmosphere.Second
    European Symposium on Physico-Chemical Behaviour of Atm.
    Pollutants,Varese,Italy,29 Sept.-1 Oct.1981,pp.336-348,
    project COST 61a bis.

61 P.J.H.Builtjes,The PHOXA project,photochemical oxidants and
    acid deposition model application.Proceedings:Workshop of
    Working Groups 4 and 5 of project COST 611,Bilthoven,The
    Netherlands,23-25 September 1985,pp.147-150.

62 U.Schurath,Chemical processes in the atmosphere,Lecture given
    during this Ispra course.

63 R.A.Cox and S.A.Penkett,Formation of atmospheric acidity.
    Proceedings:CEC Workshop on Acid Deposition,9 September 1982,
    Berlin,Reichstag,Reidel Publ.Comp.,pp.56-81.

64 U.Platt,Oxidierte Stickstoffverbindungen in der Atmosphäre,
    Proceedings:Seminar on Atmospheric Processes,Air Chemistry
    and Deposition,Berlin,Reichstag,22-24 Jan.1986,pp.193-210.
    Editor:Umweltbundesamt,1000 Berlin 33.

65 H.Rodhe,P.Crutzen and A.Vanderpol,Formation of sulphuric and
    nitric acid in the atmosphere during long-range transport,
    Tellus 33,p.132,1981.

66 J.G.Calvert and W.G.Stockwell,Acid generation,Env.Sci.Techn.,
    Vol.17,No.9,pp.428A-443A,1983.

67 S.A.Penkett,B.M.R.Jones,K.A.Brice and A.E.J.Eggleton,The im-
    portance of atmospheric ozone and hydrogen peroxide in oxi-
    dizing sulphur dioxide in cloud and rainwater,Atm.Env.13,
    1979,pp.123-137.

68 A.T.Cocks,Chemical aspects of the SO_2 proportionality issue.
    Paper presented at the Symposium on Interregional Air Pollu-
    tant Transport,Budapest,22-24 April 1986,Inaugural Meeting
    of EURASAP.

69 L.W.Richards,J.A.Anderson,D.L.Blumenthal,J.A.Mc Donald,G.L.Kok
    and A.L.Lazrus,Hydrogen peroxide and suphur IV in Los Ange-
    les cloud water,Paper presented at Precipitation Scaveng.
    Symposium,Santa Monica,USA,29 Nov- 3 Dec.1982.

70 J.W.Munger,D.J.Jacob and M.R.Hoffmann,The occurrence of bisul-
    phite-aldehyde addition products in fog and cloud water,
    J.Atm.Chemistry 1 ,1984,pp.335-351.

71 L.A.Barrie,S.Beilke and H.W.Georgii,SO_2 removal by cloud and
    rain drops as affected by ammonia and heavy metals.Proc.:
    Precipitation Scavenging,Champaign,Illinois,USA,14-18 Oct.
    1974,pp.151-166.

72 R.J.Charlson and H.Rodhe,Factors controlling the acidity of
    natural rainwater,Nature,Vol.295,No.5851,pp.683-685,1982

73 S.E.Schwartz and W.H.White,Kinetics of reactive dissolution of
    nitrogen oxides into aqueous solution.Advances in Env.Sci.
    and Technology,Vol.12,pp.1-116,1983,John Wiley and Sons,Inc.
    New York,Editor:S.Schwartz

74 D.Fowler,Removal of sulphur and nitrogen compounds from the
    atmosphere in rain and by dry deposition.Proc.Int.Conf.on
    Ecol.Impact of Acid Precipitation,Sandefjord,Norway,March
    11-14,1980,pp.22-32.

75 G.J.Dollard,J.J.Davies and J.P.C.Lindstrom,Measurements of the
    dry deposition rates of some trace gas species,Paper presen-
    ted at 4th European Symposium on Physico-Chem.Behaviour of
    Atm.Pollutants,Stresa,Italy,23-25 September 1986.COST 611.

76 F.X.Meixner,Paper to be published in the Proceedings of 2nd
    meeting on Atmosphere Biosphere Interactions,Mainz,FRG,1986.

77 J.A.Garland,Principles of dry deposition:Application to acidic
    species and ozone.VDI-Berichte 500,Acid Precipitation,Lindau,
    FRG,7-9 June 1983,pp.83-97.

78 G.Gravenhorst,K.D.Hoefken and H.W.Georgii,Acidic input to a
    beech and spruce forest.Proc.:CEC Workshop on Acid Deposition,
    Berlin,9 Sept.1982,pp.155-172,Reidel Publ.Company.

79 P.J.H.Builtjes,Eurasap Newsletters Nr.1,May 1986.

80 M.W.Holdgate,The ecological effects of deposited sulphur and
    nitrogen compounds,Proceedings:CEC Symposium on Acid Deposi-
    tion-A Challenge for Europe,Karlsruhe,19-21 Sept.1983,p.338.

81 H.Rodhe,Trends of sulphate in Scandinavian precipitation agree
    with trends in European SO_2 emissions.Paper presented at the
    Symposium on Interregional Air Pollution Transport,Budapest,
    April 22-24,1986.Inaugural Meeting of EURASAP.

82 A.Eliassen and J.Saltbones,Modelling of long-range transport
    of sulphur over Europe:a two year model run and some model
    experiments.Atm.Env.,Vol.17,No.8,p.1457,1983.

83 M.Melzer,C.Schwartau,M.Lodahl and W.Steinbeck,Pilotstudie zur
    Erstellung eines Emissionskatasters für SO_2 für die Fläche
    der DDR(mit einem Exkurs über die CSSR).Gutachten im Auftrag
    des KFZ Karlsruhe,DIW Berlin,1983(unveröffentlicht).

84 U.Platt,Messung der OH-Radikal Konzentration in der Atmosphäre,
    Proceedings:Seminar on Atmospheric Processes,Air Chemistry
    and Deposition,Berlin,Reichstag,22-24 Jan.1986,pp.225-236.
    Editor:Umweltbundesamt,1000 Berlin 33.

85 A.R.W.Marsh,Studies of the acidity and chemical composition of
    clouds,CEC Workshop on Acid Deposition,Berlin,Reichstag,
    9 Sept.1982,Proceedings,pp.185-194,COST 61a bis.

86 F.G.Roemer,J.W.Viljeer,L.Van den Beld,H.J.Slangewal,A.A.Veld-
    kamp,H.F.R.Reijnders,Preliminary measurements from an air-
    craft into the chemical composition of clouds,Proc.:CEC
    Workshop on Acid Deposition,Berlin,Reichstag,9 Sept.1982,
    pp.195-203.COST 61a bis.Reidel Publ.Comp.

87 H.F.R.Reijnders,J.Van Zijl,W.A.J.Van Pul,F.G.Römer and
    H.J.Slangewal,Characterisation of cloud and rainwater chemi-
    cal composition,Paper presented at 4th European Symposium
    of project COST 611,Stresa,Italy,23-25 Sept.1986.

88 G.Kramm,Numerische Modellrechnungen zur Best.der trock.Dep. atm.
    Spurenst.,Proc.:Trock.Dep.,Neuherberg,9-10 Sept.1985,BPT Be-
    richt 6/85.Editors:Hoefken and Bauer,pp.79-94.

Regional and Long-range Transport of Air Pollution,
Lectures of a course held at the Joint Research Centre, Ispra, Italy,
15–19 September 1986, S. Sandroni (Ed.), pp. 43–69
© Elsevier Science Publishers B.V., Amsterdam — Printed in The Netherlands

ATMOSPHERIC TURBULENT TRANSPORT

F.B. SMITH

1. INTRODUCTION

The release of potentially hazardous material as a gas, an aerosol, or as fine particulates, into the atmosphere, whether deliberately or accidentally, poses the question as to how will it affect people and the environment. Will it do any significant damage? Do special measures have to be taken and if so where and how quickly? The recent accidental release of radionuclides into the atmosphere from the Chernobyl Nuclear Reactor in late April, 1986, is a disturbing example. The first relevant question is "where is the material going?". It is almost always going somewhere since the atmosphere is rarely at complete rest. At times the determination of the wind, its speed and direction, as experienced by the plume of the emitted material, is far from easy. The wind may vary in a quite complex way with height, with horizontal position (especially in complex terrain) and with time. These situations are usually the light wind situations and the plume will not travel very far in a few hours. Concentrations will tend to be relatively high, partly because of the low wind speed and partly because vertical mixing is often very limited in these situations. This is particularly true at night. The dangers to the local environment are therefore at their greatest. More normally the wind speed is high enough for a reasonable degree of synoptic control; that is, the movement of the plume is controlled by winds determined implicitly by the larger scale weather patterns and their associated pressure gradients. That is not to say that the underlying surface ceases to be important; on the contrary it remains of considerable importance in the lower atmosphere, but on a much broader scale so that every tree, house and hill, whilst having its own localised wake, only has a relatively small influence in the integrated effect of the surface on the winds carrying the plume.

This lecture discusses how the wind affecting the plume can be assessed from theoretical considerations and from measurements. It includes a discussion of the possible accuracy of these assessments and the importance of scale (spatial and temporal). The final sections discuss the importance and nature of horizontal mixing and of synoptic swinging of the plume that becomes so important in longer releases.

2. ADVECTION

2.1 The geostrophic wind and boundary layer deviations

The lower parts of the atmosphere are directly affected, in a variety of ways, by the underlying surface. For example, the drag of this surface slows down the wind, the temperature of the surface affects the stability of the air and often creates motions driven or controlled by buoyancy forces. This part of the atmosphere is called the atmospheric boundary layer.

Above the boundary layer the influence of the ground in only indirect. Consequently the equations of motion, which determine the wind vector in space and time, are simpler and do not contain terms which represent the surface's direct influence. In the simplest situation these equations reduce to a simple balance between the local pressure gradient and the so-called Coriolis force. The Coriolis force has its origins in the Earth's rotation. Air moving over the rotating Earth acts as if it were subject to a force proportional to its speed but acting at right angles to it. Let us see how this arises. Every point on the Earth's surface is undergoing rotation about the local vertical axis, with a magnitude equal to the resolved part of the angular velocity Ω of the Earth about its North Pole to South Pole axis, i.e. the local angular velocity $\Omega' = \Omega \sin \varphi$, where φ is the latitude. Consider the curious game of football displayed in Fig. 1. The game is being played on a circular field which is made to rotate about the goal with an angular velocity Ω'. The team has a new player B who has never played this variety before. When B is a distance r from the goal he kicks the ball straight towards the open goal with velocity V thinking to score. Poor B! He has forgotten the ball already has an absolute velocity $\Omega'r$ in the tangential direction. To his amazement, after a small time Δt, he sees the ball curling away to the right whilst an observer off the rotating pitch would see the ball travel in a simple straight line as expected. The combined effect of $\Omega'r$ and V has carried the ball to C. Meanwhile B has moved, unknown to him, to B' since he too has the tangential velocity $\Omega'r$. To him the ball has moved a distance CD away from the line B'G that he tought he had kicked the ball along

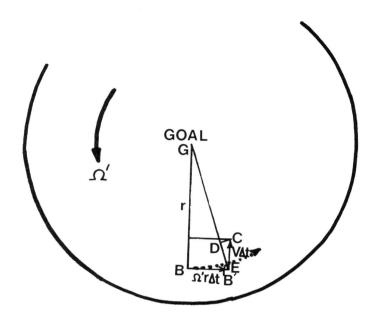

Fig. 1. The rotating football field.

towards the goal. Now a little simple algebra. The triangle B'CD is similar to triangle BEG.

Therefore

$CD = V \Delta t \sin(\widehat{CB'D}) = V \Delta t \sin(\widehat{BGE}) = V \Delta t \sin(\Omega' \Delta t).$

The acceleration of the ball at right angles to V is therefore:

$$f = \frac{CD}{1/2 \, \Delta t^2}$$ remembering the familiar Newtonian equation distance $s = ut + 1/2 at^2$

Thus

$$f = \frac{2V}{\Delta t} \sin(\Omega' \Delta t),$$ which for small Δt becomes:

$$f = 2 \, \Omega'V = 2 \, \Omega \sin \varphi \cdot V$$

This is the Coriolis acceleration, and as we see it acts to the right of V in the northern hemisphere where Ω' is positive (anticlockwise) and to the left of V in the sothern hemisphere where Ω' is clockwise.

Above the boundary layer then, in simple meteorological situations where the pressure field changes only slowly in time and space, we have air in equilibrium when the tendency for the air to accelerate down the pressure gradient $P = 1/\rho \; dp/dy$ (where p = pressure, ρ = density, y = horizontal distance) is exactly balanced by the tendency to accelerate due to the Coriolis effect:

i.e. $P = \dfrac{1}{\rho} \dfrac{dp}{dy} = 2 \; \Omega \; V \sin \varphi$

Thus $V = \dfrac{P}{2 \; \Omega \; \sin \varphi}$. We call this velocity the geostrophic velocity G. Note that G must have a direction along the isobars (or should this now be the iso-Pascals?!) - the lines of equal pressure.

Within the boundary layer, the wind deviates from its geostrophic value because a third acceleration is present. This results from the drag of the surface which sets up a profile of shearing stress with height. In physical terms it can be explained as follows. Air descending as a result of turbulence (the boundary layer tends to be full of turbulent motions) moves faster than its new environment because the wind speed increases with height from zero at the solid ground to the geostrophic value at the top of the layer. This descending air then tends to accelerate the environment. Ascending air conversely tends to decelerate the environment. The relative magnitudes of these opposing effects must create a nett positive or negative acceleration which now must balance the other two main accelerations arising from pressure and the Coriolis effect. No simple relation emerges for the wind speed since it is fairly obvious the shearing stress term must depend in a rather complex way on the roughness of the surface and on the profile and magnitude of the turbulent motions (which in turn must depend on the stability of the air).

When buoyancy effects are small, we say the boundary layer is neutrally stratified and vertical motions are neither accelerated or slowed down by forces of thermal origin. This state is not unusual in maritime climates where winds are frequently brisk and stratus-cloud is common. For example Western Europe knows these conditions only too well. In contrast, in more continental climates, neutral stability is quite unusual and when it occurs it is often only transient in nature, while the atmosphere moves from being stable to being unstable (or vice versa). Daytime sunny conditions warm the ground relative to the air and the resulting upward flux of heat makes the atmosphere unstable. At night, the ground cools by radiation and the heat flux is then downwards. Then vertical movements of air are inhibited because the density of the surrounding air is such as to generate a restoring force to the vertically moving air. The atmosphere is then said to be stably stratified.

In neutral conditions over rather uniform terrain, the wind speed at 10 metres above ground is typically between about 0.4G and 0.6G over land and about 0.7G over the sea. Its direction is rotated anticlockwise by about 15°-20° over land and by about 5°-10° over the sea. In the lowest 150 metres or so, the speed varies logarithmically with height:

$$u(z) = \frac{u_*}{k} \ln \frac{z}{z_0}$$

where:

u_* is the so-called surface friction velocity ($u_* \cong 0.04\ G$ over land) and is simply related to the surface shearing stress.

k is von Karman's constant $\cong 0.4$

z_0 is a length-scale of the surface roughness elements called the roughness length (see Table 1). z_0 is strictly defined by the logarithmic wind profile.

TABLE 1

Approximate values of the roughness length z_0 over different types of terrain.

Terrain	z_0 (metres)
sea	10^{-5}
desert	0.001
long grass	0.03
mixed agricultural country	0.25
suburbia	0.8
woods	1.0
cities	2.0

In unstable conditions, the wind within the boundary layer is a somewhat larger fraction of the geostrophic wind at the same normalised height z/h (where h is the boundary layer height) because the enhanced vertical mixing more readily transports momentum downwards from aloft to balance the sink at the ground due to surface drag. However h is greater in unstable conditions than in neutral conditions (all other things being equal) so that at some real heights the wind may actually be less. In stable conditions, on the other hand, the magnitude of h is often much smaller and the wind speed falls off more quickly with decreasing z/h than in neutral conditions. If the stabilising effect is too great, the wind near the surface becomes very small. However the air is very rarely at rest. Even on a clear cold winter's night, bonfire smoke will usually reveal some gradual drift which carries the smoke off in a broad shallow plume across neighbouring gardens and fields. Those of us who have experimented with small home-made hot-air balloons to interest the children (not to mention our-selves) will have observed, I expect, that on an apparently calm night the as-cending balloon will often pass through one or more shallow layers of turbulent motion aloft, sometimes bringing unexpected ruin to the balloon as it tilts rapidly first one way than another and allows the flames to catch the fabric of the balloon. Should it survive this drama it ultimately passes into a much more quiescent deeper layer with positive coherent motion.

Returning to neutral conditions, a very simple model of the wind profile throughout the boundary layer comes from assuming

$$u(z) = \frac{u_*}{k} \ln \frac{z}{z_0} + az \qquad \text{where a is some constant.}$$

If we require $du/dz = 0$ at the top of the boundary layer $z = h$, and $u = G$ there then writing $y = u_*/G$ and $x = G/fz_0$ (f being the Coriolis parameter $2 \Omega \sin \varphi$) we obtain with a little algebra.

$$\ln \frac{x}{4} = 1 + \frac{k}{y} - \ln y$$

This yields the results in Table 2, which are roughly correct and may be used as guides to real values in the atmosphere.

2.2 The depth h of the boundary layer

The depth of the boundary layer may be determined by one or more of the following:

(i)   inversions (that is, marked jumps or gradients of temperature which inhib-it vertical motions) generated by subsidence of the air on a synoptic scale, for example, in anticyclones.

(ii)   inversions generated by the air flowing from warmer to colder terrain.

(iii) the magnitude of the surface fluxes of heat and momentum.

In near neutral conditions over land:

TABLE 2.

Boundary layer relationships in neutral conditions according to a very simple boundary layer model.

u_*/G	$\log_{10} \dfrac{G}{fz_o}$	$\bar{u}(10m)/G$	u/G	α_o degrees
0.02	11.4	0.85	0.975	12.9
0.025	9.6	0.81	0.969	14.5
0.03	8.3	0.74	0.962	15.9
0.035	7.5	0.70	0.956	17.2
0.04	6.8	0.64	0.950	18.4
0.045	6.2	0.57	0.943	19.6
0.05	5.8	0.52	0.937	20.7
0.055	5.5	0.48.	0.931	21.8
0.06	5.2	0.41	0.925	22.8

u(10m) is the wind speed at 10 metres, $\bar{u}$ is the average wind speed in the boundary layer, α_o is the backing (clockwise turning) of the wind in degrees at the surface. The average backing of the wind throughout the boundary layer will be roughly half α_o.

$$h \cong \frac{1}{4} \frac{u_*}{f} \qquad \text{or} \qquad \cong 100 \ G$$

if (iii) is the main determiner.

In stable conditions, considerable uncertainty must prevail due to the common occurrence of effects due to upwind heterogeneous terrain. However on very uniform terrain, the simple relationship

$$h = 21500 \frac{u^2}{\sqrt{H}}$$

frequently seems to hold (see Fig. 2). u_* is the friction velocity and H the downward surface heat flux.

In unstable conditions during the day, several theories have been developed to represent the evolutionary process in which the turbulent upward flux of heat is re-distributed into a deepening layer with a quasi dry adiabatic lapse rate (i.e. a temperature profile corresponding to a well mixed layer in which the temperature falls at almost 1°C per 100 metres). Carson (1973) has given the following formula:

$$\frac{dh^2}{dt} = 2 (1 + 2A)\frac{H}{\rho c_p \gamma}$$

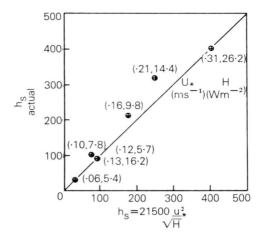

Fig. 2. Data from the Minnesota experiment confirming the formula $h = 21500 \frac{u_*^2}{H^{1/2}}$ for the depth of the stable mixed layer.

where A is typically in the range 0.2-0.5 and represents the effective fraction of total heat entering the layer at the top by entrainment, ρ is the density of the air, c_p is the specific heat at constant pressure ($\rho c_p = 1305 - 4.3$ T, where T = the temperature in °C) and γ is the gradient of potential temperature in the over-riding stable layer.

If A = 0.25, $\rho c_p = 1250$, γ takes a typical values of 0.006 °K m^{-1} and if $H = H_m \sin \Omega t$, then the boundary layer depth h takes the following values in light wind conditions:

$h(t) = 105 \, H_m^{\frac{1}{2}} \sin \frac{\pi}{24}(t-6)$ where t is time in hours.

TABLE 3

Values of boundary layer heights at different times given a sinusoidal heat input at the surface with amplitude H_m. A = 0.25, $\rho C_p = 1250$ and $\gamma = 0.006$.

H_m (Wm^{-2})	h (06Z)	h (09Z)	h (12Z)	h (15Z)	h (18Z) metres
50	0	280	520	680	740
100	0	400	740	970	1050
150	0	490	910	1190	1285
200	0	570	1050	1370	1480
250	0	630	1170	1530	1660
300	0	700	1290	1685	1820

In stronger winds, the value of h is increased as a result of the additional effects of "mechanical" turbulence (i.e. arising from the shearing profile of the wind and its instability to perturbations). Fig. 3 gives a nomogram for deriving an approximate value for h under these conditions, and an example illustrating how the nomogram is to be used.

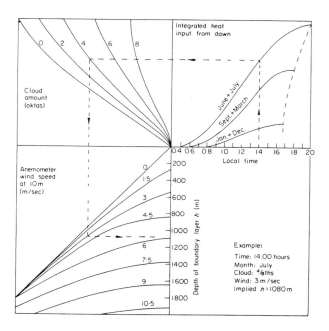

Fig. 3. A nomogram for estimating the depth of the boundary layer in the absence of marked advective effects or basic changes in weather conditions. The marked example shows how the diagram is to be used.

2.3 Winds near complex topography

(a) The speed-up of the mean wind. Experience tells us that windspeed $(U + \Delta U)$ encountered on a hill can be much higher than that obtaining nearby on level low ground (U). The speed-up factor $s = \Delta U/U$ depends on the size and shape of the hill, the position of measurement on the hill, the height above the surface and the atmospheric surface stability. When air flows over an isolated hill or ridge of height h in a neutrally stratified boundary layer, an inner flow region of strong shear is produced, whose thickness d depends mainly on the lengthscale of the hill but also on the surface roughness z_o. Theory and experiment show that d can be found from:

$$d = \frac{k^2 L}{\ln(d/z_o)}$$

where L is the half width of the hill at half its height. Thus we have the numbers in the following Table 4:

TABLE 4

Values of the thickness of the near-surface shear layer on an isolated hill.

L/z_o	d/z_o	d/L
10^2	7.8	0.078
10^3	42.6	0.043
10^4	283	0.028
10^5	2090	0.021

As a rough guide $d = L/20$ is typical of many hills and is therefore several tens of metres thick. Within this sub-layer, S_{max} occurs on the summit and is approximately $2h/L$. This can exceed 1 (resulting in a more than doubling of the windspeed) for hills with slopes of 25% gradient or more. A speed-up factor of 1.3 was found on a small isolated hill called Brent Knoll in S.W. England. The hill has a height of 130 metres above the surrounding plain and has an approximate value of 220 metres for L, giving a predicted factor of 1.2 and a maximum summit wind of 22 m s^{-1} in neutral conditions with an up-stream wind of only 10 m s^{-1}.

(b) Stable conditions. In stable conditions over the same hill windspeeds at 2 metres were found to be 6 or 7 times greater than at the same height above the surface over level ground upstream.

Air can only rise over a hill or ridge in stable conditions if it has enough kinetic energy of motion to overcome the work needed to rise. This defines a certain height below the crest, above which sufficient energy is available and the air can travel over the hill, but below which the air has insufficient energy and must find some way round the hill or stagnate. This goes some way to explaining the high speed-up factor recorded on Brent Knoll, since the two measurements were not in the same air, the air on the summit probably originated at a much higher level relative to the surface upstream of the hill.

This critical height is $h_c = h (1 + Fr)$
where Fr is the Froude number of the hill defined as:

$$Fr = \frac{U}{h\left(\frac{g}{T}\left(\frac{dT}{dz} + 0.01\right)\right)^{1/2}}$$

TABLE 5

Values of the critical depth ($h - h_c$) below the summit below which the air has insufficeint kinetic energy to rise over the summit, as a function of the vertical temperature gradient and the wind speed U.

dT/dz	U = 1	3	5	10 m/s
0	53	160	267	534
0.01	38	113	189	378
0.02	31	93	154	308
0.04	24	72	119	239

(c) <u>Effect on wind direction</u>. Topography can have a profound effect on the wind. Wind roses differ considerably from position to position, reflecting the channelling effects of hills and valleys. Figs. 4 and 5 illustrate this in rather simple regimes. In more complex terrain, predicting the likely flow pattern can be very difficult and speculative.

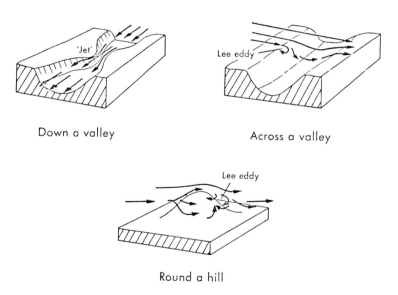

Down a valley

Across a valley

Round a hill

Fig. 4. Schematic flows in valleys and around hills.

3. DETERMINATION OF THE WIND

As we have seen the wind can vary in speed and direction and with height in a sometimes quite complex way. Generally if vertical mixing is good then wind shears are rather small when viewed over a large area, but if the mixing becomes

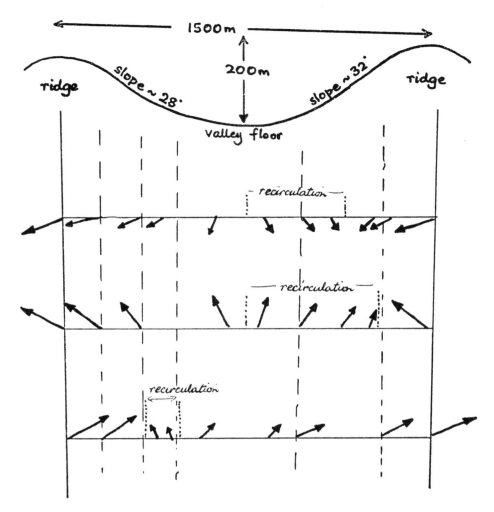

Fig. 5. Plan views of the flow in the Sirhowy valley in south Wales for different wind directions on the ridge. Note the rather critical effect the slope angle has on the flow.

small as in stable conditions then considerable shears can arise. Even jets of super-geostrophic winds can appear. When tracking hazardous material it becomes imperative to have as good an assessment of the advecting wind as possible. Boundary layer winds can be estimated in a variety of ways:

(i)   from radiosondes or pilot balloons,

(ii)   by tracking tetroons,

(iii)   from instruments on towers or captive balloons,

(iv)   by instrumented aircraft,

(v)   by doppler radar,

(vi)   by acoustic sounders,

(vii)   by satellites,

(viii) from Rossby similarity theory, or other boundary layer models, using winds at just one level,

(ix)   from weather prediction models.

Let us consider these in somewhat more detail.

3.1 Radiosondes

As a radiosonde ascends, the pressure and temperature of its changing environment is measured and relayed back by radio to the ground station, and from this information its height is inferred. At the same time a metallized nylon mesh reflector reflects radio-waves transmitted and later received by a radar at base, yielding the position on the sonde. From these the horizontal displacement, and hence the winds, can be evaluated. Radiosondes are not particularly good at determining low-level winds, and because of the rather rapid rate of climb, sharp gradients in wind are smeared out over larger height ranges. Furthermore the very short time-averaging involved means that implied boundary layer winds are subject to the uncertainty associated with any spot reading within a turbulent medium.

Finally radiosonde stations are often widely spaced; in Europe the spacing is of the order of 200 km, whilst elsewhere it may greatly exceed this. Interpolation to the site of the release of pollution may therefore be a very uncertain procedure, especially in complex terrain (see later).

Pilot balloons being cheaper and requiring only a theodolite to track them may be used where they are directly required, provided the sky is sufficiently clear of low cloud and the balloon etc. and trained personnel are rapidly available. Here an assumption has to be made concerning the rate of climb, remembering that such an assumption is fraught with uncertainty in a highly turbulent unstable boundary layer.

3.2 Tetroons

Tetroons are balloons with a tetrahedron shape made of Du Pont Mylar material. They are usually about 1 m^3 in volume and are inflated with hydrogen or helium so that they float at, or near, a constant density level which can be predicted. The floating level can be adjusted by adding or subtracting weights to the system. In practise the tetroon exhibits quite large vertical oscillations. Some of the major causes for these seems to be dew formation on the balloon surface, precipitation and updraughts and downdraughts in the air.

The most successful way of tracking a tetroon is to fit it with a 403-MHz transponder which is triggered into action by a radar signal transmitted from a ground station and the 403 MHz signal received back by the radar station. The signal can then be interpreted in terms of range, azimuth and elevation out to well over 100 kilometres range.

3.3 Towers and Captive balloons

Towers provide an excellent platform for mounting anemometers and vanes provided care is taken to minimise tower-interference on the measurements. Obvious drawbacks include the limited height of the tower compared to daytime boundary layer depths, its fixed position in space and its high capital cost.

Captive balloons avoid some of these problems at the expense of creating fresh problems. The U.K. Meteorological Office has used captive balloon systems for many years and the following points may be briefly mentioned:

(i) The system is reasonably mobile, but requires hard standing for the balloon winch and a few skilled personnel to operate the launch and retrieval. Balloons may not be flown without the agreement of air traffic control because of the obvious risk to aircraft of all types.

(ii) Balloons can reach up to more than 3 km according to size and load, and hence can easily encompass even the deepest boundary layers with instruments making continuous measurements at a number of heights.

(iii) Balloon systems do tend to exhibit marked swinging (like an inverted pendulum) and special care has to be taken to eliminate these effects from the measurements of wind and turbulence, as far as possible.

(iv) Special care has to be taken to avoid the risk of lightning and thus any measure of deep convective cloud has to be avoided. The balloon system cannot fly in strong winds either since stresses within the system become too great, and the authorities do not take kindly to owners of breakaway balloons (with trailing steel cables) endangering life and property as they cross roads and housing.

A schematic diagram of the current U.K. system is shown in Fig. 6.

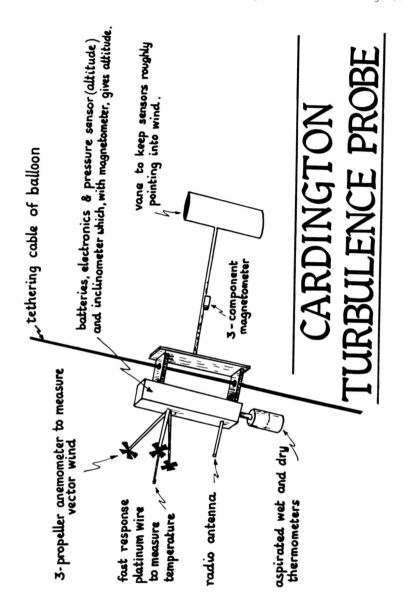

tethering cable of balloon

batteries, electronics & pressure sensor (altitude) and inclinometer which, with magnetometer, gives altitude.

vane to keep sensors roughly pointing into wind.

3 - component magnetometer

3-propeller anemometer to measure vector wind

fast response platinum wire to measure temperature

radio antenna

aspirated wet and dry thermometers

CARDINGTON TURBULENCE PROBE

Fig. 6.

3.4 Instrumented Aircraft

Average winds can be inferred from the vector difference of the true airspeed

and direction, as measured by pitot tubes on the aircraft, and the deduced track
and speed relative to the ground obtained either from knowing the precise position of the plane in time from one other of the various navigation aid systems
or from doppler radar reflections from the ground.

The use of aircraft gives good horizontal coverage but is very expensive,
yields winds at only one level at a time, and has to be within the aircraft's
range of a suitable airfield.

3.5 Doppler Radar

This technique has rather limited applicability, requiring as it does reflecting "particles" in the volume where the wind speed is needed. Usually the
"particles" are raindrops, hail or snowflakes falling with a velocity which has
to be unravelled from the true wind speed if the precipitation is occurring at
an angle significantly greater than zero to the horizon. This is not easy unless
two radars are used in conjunction.

3.6 Acoustic Sounders

Although acoustic sounders are mainly used for remote sensing the temperature
structure of the lower atmosphere, they can be used in a Doppler mode to measure
wind speeds. The speaker produces a narrow beam sound wave in a series of short
pulses, and these pulses are scattered by turbulence in the atmosphere. Whilst
the amount of energy scattered directly back to the sounder's receiver depends
mainly on temperature fluctuations, the amount scattered in other directions
depends also on wind fluctuations. The scattered waves undergo a Doppler shift
Δt of the transmitted frequency f if there is a windspeed U:

$$U = 1/2 \; \frac{V \Delta f}{f} \; \sin \frac{\theta}{2}$$

where V is the propagation velocity and θ is the angle between the transmitter
and the receiver as viewed by the reflecting eddies.

3.7 Satellites

Satellites have the capability of measuring the speed of visible low cloud by
inferring movements between successive scans whenever distinctive parts of the
cloud can be identified, and also of inferring low-level winds over the sea by
association with the state of the sea surface and the amount of reflected solar

energy. This latter technique is not yet available but will be in a few years time when wind scatterometers will be carried by new satellites. It is anticipated that inferred winds will be usefully accurate with typical errors in the 10 m wind being about 2 m s^{-1} in speed and 15° in direction.

3.8 Winds from boundary layer models

Measured winds are most common at very low level using a variety of techniques: cup anemometers, drag plates, hot-wire and hot-film anemometers, sonic anemometers which depend on the variation of the speed of sound with temperature and the difference of travel times in sound pulses travelling in opposite directions, dynamic pressure anemometers, propeller anemometers and vortex sensing anemometers. In general the achievable accuracy is quite high, although some are better than others in light winds.

Conversely geostrophic winds can be inferred directly from the spacing of the isobars on synoptic charts and corrections applied for isobaric curvature and for issallobaric effects (changing pressure gradients experienced by the air in time).

The problem with the former is that low-level winds are very much influenced by immediate upwind terrain and are not necessarily a good measure of an areally-averaged wind. The use of both low-level and geostrophic winds to infer mean boundary layer winds implies a simple structure within the boundary layer. For example Rossby Similarity Theory implies a rather simple relationship (see Pasquill and Smith, 1983) between the surface friction velocity u_*, the surface heat flux H and the geostrophic wind G which can be invoked in the estimation of mean boundary layer winds. However it is well known that Rossby Similarity Theory requires a high degree of surface homogeneity over a wide area, an ideal rarely achieved in most parts of the world, especially in the most populated parts. Other simple boundary layer model approaches are subject to the same difficulty.

3.9 Weather Prediction Models

In principle, future states of the atmosphere can be predicted from earlier states out to a few days ahead by solving the basic dynamic and thermodynamic equations if sufficient input data of high quality is available. The wind fields so obtained are at least consistent with the physics of these equations and can

be made reasonably consistent with actual observed winds at earlier times. However the solution of the equations within the boundary requires a degree of em-
piricism (for example in relating the wind speed at the lowest level to the surface drag) that means any implied mean wind is only as good as the physical
validity of these empirical assumptions, and the adequacy of the height resolution given by the number of grid-levels within the boundary layer itself.

4. CURRENT OPERATIONAL TECHNIQUES FOR OBTAINING ADVECTIVE WINDS

This section will provide brief notes on the techniques used in various programmes to determine winds within the boundary layer suitable for plume advection.

4.1 NOAA's Atmospheric Transport and Dispersion (ATAD) model

The ATAD method (Heffter, 1980) is a well-tried and reasonably successful
method which derives layer-average radiosonde winds and then interpolates these
to the point-in-space where winds are required using a $1/R^2$ weighting (R = distance from the radiosonde station to the point-in-space).

$$u = \frac{\sum(u_i/R_i^2)}{\sum(1/R_i^2)}$$

The method has been used by Clarke et al. (1983) to estimate the likely trajectories of tetroons released in a series of major field experiments in the U.S.A.
Some 32 actual tetroon trajectories were available to the authors which travelled for more than 250 km or for more than 12 hours. After a distance of travel D
the "error" e between D and the nearest point on the ATAD trajectory was evaluated. They found that the average error was e = 0.08 D to the left of the
tetroon with a standard deviation σ_e = 0.25 D. Compared to other trajectory
methods tested, Clarke et al. found the ATAD method gave the smallest mean error.

4.2 The NCAR Method

The National Center for Atmospheric Research (NCAR) in the U.S.A. has a method which involves constructing trajectories on isentropic surfaces (surfaces of
constant potential temperature which small air parcels are supposed to remain on
if no heat is added or subtracted from the parcel). Trajectory winds are ob-

tained by using linear interpolation between observations at radiosonde stations in time and in the vertical and by fitting quadratic surfaces to neighbouring winds to interpolate in the horizontal. In the Clarke et al. (1983) test the NCAR method did almost as well as the ATAD method having the same σ_e/D but had a somewhat larger mean error $\bar{e} = - 0.2\ D$, with the NCAR trajectories lying now to the right of the tetroon trajectories. How the two methods would rate compared to a vertically diffusing plume is not clear.

4.3 The CAPITA Method

The CAPITA method exchanges observations of wind throughout the depth of the boundary layer for surface winds at a much greater spatial density. The method uses surface wind speed and direction observations, with a resolution of 1 knot ($= 0.5\ m\ s^{-1}$) and $10°$, at 200 synoptic stations in the U.S.A. The method then interpolates these on to a grid of points 127 km apart using a $1/R^2$ weighting as in the ATAD method. North and east wind components are interpolated independently. Winds in time are then inferred using linear interpolation between successive midday measurements, and finally boundary layer winds are obtained by veering the "surface winds" by $20°$ and by multiplying the speeds by 2.5.

To simulate plume behaviour, a constant horizontal diffusivity K is chosen which implies that after every time step Δt a single particle, representing a part of the plume, is advected by the vector sum of the interpolated boundary layer wind plus a random vector whose radius is $(2K \Delta t)^{1/2}$. Many such particles are released to represent the whole plume, each undergoing this advection plus random walk.

4.4 Hoecker's Method

Hoecker (1977) describes the adjustments he found necessary to various winds to minimise the errors between derived trajectories and tetroon paths. He analysed 13 tetroon trajectories which travelled more than 400 km from Oklahoma City, and recommended the following adjustments:

(i)   sea-level geostrophic winds: in S or W flows: back winds by $20°$

in N flows: back winds by $50°$

(ii)   boundary-layer averaged winds: in S flows: back by $10°$

in W flows: back by $20°$

in N flows: no adjustment

(iii) surface wind data: in S flows: multiply speeds by 2

in W flows: 30° veering, multiply by 3

in N flows: 10° veering, multiply by 2.

No advice is given for easterly flows.

4.5 Cressman's Method

Cressman's (1959) method involves using a first guess field, from a previous forecast or analysis, and modifying this by observations using a weighting function, ranging from 0 to 1, which produces a circular area of influence around each observation point:

the weighting $w_i = \dfrac{R2 - d_i^2}{R^2 + d_i^2}$ where $d_i < R$, a pre-defined radius of influence.

$$= 0 \quad \text{if } d_i > R.$$

4.6 Elliptic-weighting Methods

Various authors (see Benjamin and Seaman (1985)) have modified Cressman's method to give more weighting to observations along the line of flow than to those across the line, since analysis shows a larger correlation between winds along the line.

One way of doing this is to replace d_i by d_m where

$$d_m^2 = ax_i^2 + y_i^2$$

where:

$a = (1 + \beta |V|)$, $x_i = d_i \cos\theta$, $y_i = d_i \sin\theta$

β = constant in the range $(0.02 - 0.2 \text{ s m}^{-1})$

V = the observed wind speed

θ = the angle between the separation vector and the wind vector.

4.7 "Banana-shaped" Weighting Methods

Benjamin and Seaman (1985) carry the idea of elliptic-weighting one stage further and make allowance for the curvature of the isobars, so that maximum weighting is given along the curved isobars and minimum weighting in the orthogonal direction.

4.8 Optimum (statistical) Interpolation

These methods (see for example Schlatter (1975)) employ empirically observed correlations between winds in different orientations to optimise the weighting functions to be employed in interpolation procedures. This method is now fairly frequently used but is computationally rather costly compared to some of the simpler schemes.

4.9 Polynomial Interpolation

Sykes and Hatton (1976) describe a technique which involves fitting ortho-gonal polynomials to the whole of an area, like Europe, which contains a distri-bution of observed pressures (or winds). They advocate first fitting a poly-nomial in space for each 3 hourly time-level, and then the time-dependence is treated separately. The method involves generating its own appropriate set of polynomials which minimise the errors at each observation station in a lest-squares sense. A polynomial of order 7 is found to fit a typical pressure field sufficiently accurately, although essential difficulties are found near sharp frontal surfaces. The time interpolation is carried out by fitting a fourth-order polynomial to the nine pressures evaluated over a 24-hour period at each point of interest.

4.10 Modern numerical-weather-prediction techniques

Within the U.K. two similar methods are employed at the Meteorological Office and the European Centre for Medium range Weather Forecasting (ECMWF) for deriv-ing wind fields. The Met. Office scheme uses a 6-hour forecast field as a back-ground and then takes new observations at their respective geographical points and compares these with the background values. The difference (or anomaly) at each is then spread to the 4 surrounding gridpoints using a simple spatial in-terpolation scheme. However since some observations are known to have lower re-liability than others, they are given less weighting than good quality observa-tions. Weighting is also higher along the wind than acrosswind. The forecast is then repeated over the previous 6 hours, in which the derived grid-point anoma-lies are introduced gradually in ever-increasing amounts at each small time-step. This process generates small gravity waves and associated local diver-gences which are damped out exponentially by a special process in the scheme, but allows the adjustment to take place in a smooth manner and for its effect to spread out over a number of neighbouring squares. However the final winds are not necessarily equal to the observed wind at the station and differences of 2.5 m s^{-1} are typical at, for example, 850 mb. Such differences are comparable to the quality of the observed winds.

At ECMWF, new observations are introduced all at once at the time of observation. The field then is subject to a process of initialisation which creates approximate geostrophic balance throughout the field. This is a complex process and again the final analysis winds are not identical to the observed winds.

Finally, it should be said that both schemes lead to generally good forecasts!

4.11 Mass consistent non-divergent fields

Whilst flows are not exactly non-divergent, even on isentropic surfaces, the divergence is usually small on the scale of synoptic features. Simply-interpolated wind fields do not always have this property and can lead to plumes losing mass in a significant way. One way to avoid this is to separate the field into a rotational non-divergent component and a non-rotational potential flow. The latter can then artificially be put to zero, or reduced in size, and a new wind field reconstructed.

Alternatively the observations can be fitted by a subset of orthogonal non-divergent functions:

$$u_o(x) = a_1 u_1(x) + a_2 u_2(x) + \ldots\ldots + a_i u_i(x)$$

in which the coefficients $a_1 \ldots\ldots a_i$ are found by fitting $u_o(x)$ to the observed $u(x)$ by a least-squares technique. Sometimes a further requirement of minimum total kinetic energy is applied.

4.12 Meso-scale model wind fields

Meso-scale models are now becoming fairly common in which the gridscale is much smaller (20 km or less) than that of conventional weather prediction models, including so-called "fine-mesh" models (gridlengths typically 80-100 km). Such meso-scale models are often non-hydrostatic and non-anelastic. They use the equations for momentum, heat and continuity and provision is made for such physical processes as radiation, convection and condensation. Fluxes within the boundary layer are often derived using simplified 2nd order closure schemes and are consistent with Monin-Obukhov similarity concepts, although sometimes applied at a hieght where the concepts strictly no longer apply.

Such detailed models inevitably involve a marked degree of parametrization and are limited in their performance by the rather broad spacing and limited quality of the input data. Whilst the model are useful for understanding physical processes and for predicting the implications of the broad features of complex terrain on the flow (but only within bounds set by the limited physical understanding embodied within the various parametrizations), it has yet to be shown that meso-scale models generally give better wind fields (especially in convective conditions) than can be obtained with more conventional fine-mesh models, using not only a larger gridscale but also simplified physics.

5. LATERAL DISPERSION

At short range, vertical dispersion is very important in determining ground-level concentrations. Detailed descriptions of how this vertical dispersion can be assessed is given in many books and papers, for example in Pasquill and Smith (1983) and in Randerson (1984). At longer range the boundary layer becomes almost uniformally filled with the emitted pollutant, and the rate of vertical mixing becomes rather unimportant. The same is not true for lateral spread. Because of the broad spectrum of horizontal motions, stretching from the micro-scale up to global scales, lateral spread continues rather rapidly over a very large range of travel distance. Fig. 7 shows an example of a spectrum due to Hess and Clarke (1973) for the eastward component of wind derived from hourly observations of wind deduced from measurements of pilot balloons at various heights made during the Wangara Experiment in Australia. The peak is at about 7 days, corresponding roughly to the typical passage time of large synoptic features.

The effect of this great range of motions on the dispersion of plumes out to 1000 km from the source has been studied over the last few years. In particular, measurements of the width of the Mt. Isa smelter plume in Australia (Brigg et al. 1978 and Gifford 1982) have been very revealing. Fig.8 shows these data together with measurements by Crabtree (1982) taken by an aircraft flying across a power-station plume crossing the North Sea, together with other measurements from tracer studies and from small free balloon releases.

In general the width appears to grow at a rather consistent rate

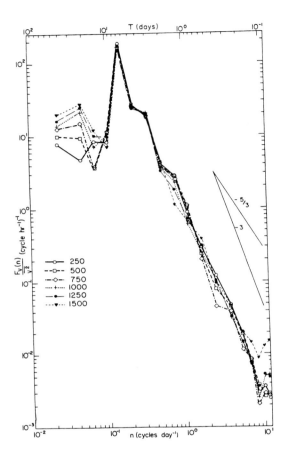

Fig. 7. Normalized u-spectra at various heights averaged over bandwidths.

$$W = 0.62 \ x^{0.875}$$

where W and x are in kilometres. However it is interesting to note that in some circumstances, for example over the sea in Crabtree's measurements, where meso-scale turbulence is relatively rather weak, the rate of growth seems to be much slower than the average and approaches the $x^{1/2}$ behaviour expected when the plume has grown larger the largest energy-containing eddies.

Conversely it is likely that in more complex terrain and in more variable meteorological situations, deliberately avoided in these measurements the rate of growth may be much more rapid than shown in the figure, and may even be exponential with time for a while.

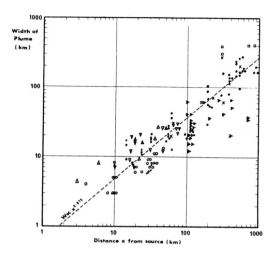

Fig. 8. Data on the width of plumes as a function of distance from the source. x Richardson and Proctor (1925); ∇ Porton data (Pasquill, 1974); + Gifford (see Slade, 1968); □ Classified project (see Slade, 1968); ∆ Braham et al; (1952); o Smith and Heffernan (1956); ● Mt. Isa data (Bigg et al., 1978; Carras and Williams, 1981); ∋ Crabtree (1982); data collected over the sea.

6. EFFECT OF RELEASE ON SAMPLING DURATION

The instantaneous width of a plume of hazardous material becomes relatively unimportant if the plume continues for a long time, provided it is the <u>dosage</u> rather than the instantaneous concentration that poses the real threat. Plumes are subject to significant meandering due to large-scale eddies and synoptically-induced changes in wind direction that swamp the influence of the instantaneous width. Generally the stronger the wind the smaller is the meandering. Fig. 9 illustrates this with data collected from the Belmont tower in England. The ordinate A is shown to be a simple function of sampling time t, where the average wind direction swing is given by $\theta = A/\bar{u}$.

This dependence on t and u is reflected in an expression given by D.J. Moore (1976), in which the width of a plume is given by:

$$\sigma_y = 0.08 \ (7t/\bar{u})^{1/2} \ x$$

where t lies between 1 and 24 hours, x is in km as is σ_y, and $\bar{u}$ is in m s^{-1}. Smith (1980) has given a similar set of expressions for the synoptic swinging of a plume out to a 1000 km range and over sampling durations of up to 100 hours (4 days). The expressions come from an analysis of trajectories derived to follow sulphur pollution across Europe. His results are expressed in terms of probabil-

68

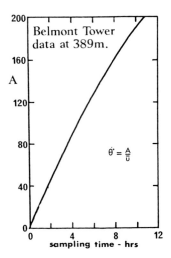

Fig. 9. Hourly-average wind direction measurements from the vane at the top of the Belmont Tower in Lincolnshire show that the magnitude of the average directional swing is inversely proportional to wind speed and increases with sampling time.

ities. He states that there is a 10% probability the angular spread of the plume due to synoptic swinging will exceed Θ_{10} (in degrees), where:

$$\Theta^{\circ}_{10} = 27\ t^{0.64}\ x^{-0.125}$$

a 50% probability it will exceed:

$$\Theta^{\circ}_{50} = 4.7\ t^{0.85}\ x^{-0.125}$$

and a 90% probability it will exceed:

$$\Theta^{\circ}_{90} = 0.53\ t^{1.16}\ x^{-0.125}$$

Thus for x = 1000 km and t = 24 hours Θ°_{10} = 87°, Θ°_{50} = 30° and Θ°_{90} = 9°.

REFERENCES

1   S.G. Benjamin and N.L. Seaman, A simple scheme for objective analysis in curved flow, Mon. Wea. Rev., 113, (1985) p. 1184.
2   E.K. Bigg, G.P. Ayers and D.E. Turvey, Measurements of the dispersion of a smoke plume at large distances from the source, Atmos. Environ., 12 (1978) p. 1815.
3   D.J. Carson, The development of a dry inversion-capped convectively unstable boundary layer, Quart. J. R. Met. Soc., 99, (1973) p. 450.
4   J.F. Clarke, T.L. Clark, J.K.S. Ching, P.L.Haagenson, R.B. Husar and D.E. Patterson, Assessment of model simulation of long-distance transport, Atmos. Environ., 17, (1983) p. 2449.

5   J. Crabtree, Studies of plume transport and dispersion over distances of travel up to several hundred kilometres, Proc. 13th NATO/CCMS Conf. on Air Pollution Modelling and its Applications, Plenum Press, New York, 1982.

6   G.P. Cressman, An operational objective analysis system, Mon. Wea. Rev., 87 (1959) p. 367.

7   F.A. Gifford, Long-range plume dispersion: comparisons of the Mt. Isa data with theoretical and empirical formulas, Atmos. Environ., 16, (1982) p. 883.

8   J.L. Heffter, Air Resources Laboratories atmospheric transport and dispersion model, NOAA Tech. Memo., ERL ARL-81, 1980, pp. 24.

9   G.D. Hess and R.H. Clarke, Time spectra and cross-spectra of kinetic energy in the planetary boundary layer, Quart. J. R. Met. Soc., 99 , (1973) p. 130.

10 W.H. Hoecker, Accuracy of various techniques for estimating boundary layer trajectories, J. Appl. Met., 14, (1977) p. 374.

11 D.J. Moore, Calculation of ground-level concentrations for different sampling periods and source locations, Atmospheric Pollution, Elsevier, Amsterdam, 1976.

12 F. Pasquill and F.B. Smith, Atmospheric Diffusion, 3rd Ed., Ellis Horwood, Chichester, 1983.

13 D. Randerson, Atmospheric Science and Power Production, DOE/TIC-27601 Tech. Info. Center, Off. of Sci. & Tech. Info., U.S. Dept. of Energy, 1984.

14 T.W. Schlatter, Some experiments with a multivariate statistical objective analysis scheme, Mon. Wea. Rev., 103, (1975) p. 246.

15 F.B. Smith, The influence of meteorological factors on radioactive dosages and depositions following an accidental release, Proc. of CEC Seminar on Radioactive Releases, Riso, Denmark, 1980 p. 22.

16 R.I. Sykes and L. Hatton, Computation of horizontal trajectories based on the surface geostrophic wind, Atmos. Environ., 10, (1976) p. 925.

Regional and Long-range Transport of Air Pollution,
Lectures of a course held at the Joint Research Centre, Ispra, Italy,
15–19 September 1986, S. Sandroni (Ed.), pp. 71–93
© Elsevier Science Publishers B.V., Amsterdam — Printed in The Netherlands

CHEMICAL PROCESSES IN THE ATMOSPHERE

U. SCHURATH

1 INTRODUCTION

Stable constituents of (dry) tropospheric air (lifetimes in excess of 5 years) are N_2 (78.08 %), O_2 (20.95 %), the noble gases (0.97 %), CO_2 (0.034 %), H_2 (0.5 ppm), and N_2O (0.33 ppm). Their mixing ratios are fairly constant throughout the troposphere. Numerous other much shorter living trace gases of natural and/or anthropogenic origin have also been detected, but at much lower mixing ratios, typically in the ppb- and sub-ppb range. Their mixing ratios are extremely variable in space and time. The cause of this variability is the delicate interplay between emission, transport, chemical transformation and destruction in the gas phase, and loss by dry and wet deposition.

Different types of computer models of the polluted atmosphere have been developed in recent years, with the ultimate goal of predicting the temporal and spatial distribution of trace constituents, e.g. of ozone and other photooxidants (ref. 1), or of acidic products (ref. 2), as function of the input of primary pollutants into the atmosphere.

A rigorous equation of mass conservation for each chemical species forms the basis of all conventional diffusion models. It describes the rate of change of the concentrations of each species i at a point x,y,z in space, as function of the wind at this point with components u,v,w, of the molecular diffusivity D_i, of the chemical reactions forming and destroying species i with a net rate R_i, and of its local sources and sinks S_i, if present at this point:

$$\frac{\partial c_i}{\partial t} =$$

$$- u\frac{\partial c_i}{\partial x} - v\frac{\partial c_i}{\partial y} - w\frac{\partial c_i}{\partial z} + D_i\left(\frac{\partial^2 c_i}{\partial x^2} + \frac{\partial^2 c_i}{\partial y^2} + \frac{\partial^2 c_i}{\partial z^2}\right) + R_i + S_i \qquad (1.1)$$

It is, however, necessary to replace the rigorous equation of mass conservation, which cannot be solved in closed form, by an approximation more suitable for numerical integration, which refers to volume elements and time steps of finite size:

$$\frac{\partial \bar{c}_i}{\partial t} = - u\frac{\partial \bar{c}_i}{\partial x} - v\frac{\partial \bar{c}_i}{\partial y} - w\frac{\partial \bar{c}_i}{\partial z}$$

$$+ \frac{\partial}{\partial x}\left(K_h\frac{\partial \bar{c}_i}{\partial x}\right) + \frac{\partial}{\partial y}\left(K_h\frac{\partial \bar{c}_i}{\partial y}\right) + \frac{\partial}{\partial z}\left(K_z\frac{\partial \bar{c}_i}{\partial z}\right) + \bar{R}_i + \bar{S}_i \tag{1.2}$$

In this equation molecular diffusion is neglected altogether, and time- and space averages are introduced. u,v,w are now the components of the mean wind, while turbulent transport is parameterized by the horizontal and vertical eddy diffusivities K_h and K_z, which are assumed to be functions of space and time only.

A detailed discussion of the mathematical background of, and the approximations involved in equation 1.2, or of other practically useful versions of the mass conservation equation can be found in the literature (refs. 3,4). Only the chemical rates R_i appearing in these equations will be considered in this chapter. It is, however, important to note that the averages $\bar{R}_i$ of the chemical rates figuring in equation 1.2 cannot be correctly calculated in any of the existing models, and must be approximated by the chemical rates calculated from the concentration averages:

$$\bar{R}_i(c_1,c_2,\ldots c_N) \cong R_i(\bar{c}_1,\bar{c}_2,\ldots\bar{c}_N) \tag{1.3}$$

This introduces errors owing to the square dependence of the reaction rates R_i on reactant concentrations, cf. equations 1.4 and 2.5 below.

In this chapter such difficulties are circumvented by making the simplifying assumption that all terms on the right hand side of equations 1.1 or 1.2 are zero, with the only exception of the chemical terms R_i. This is equivalent with assuming that the concentration gradients of the reactants are zero in all space. Clearly this situation does not exist in the real world. However, one tries to approximate the situation in smog chambers. This is not entirely possible, since the source-sink terms S_i are non-zero at the chamber walls for some of the more reactive species (refs. 5,6).

Anyway, the reaction rates R_i, actually a set of i coupled differential equations governing chemical change of i species,

$$\frac{dc_i}{dt} = \underbrace{\sum_{l,m} k_{lm} c_l c_m}_{\substack{\text{chemical} \\ \text{formation}}} - \underbrace{\sum_{j} k_{ij} c_i c_j}_{\substack{\text{chemical} \\ \text{destruction}}} + \underbrace{Ph_i}_{\substack{\text{photochemical} \\ \text{format. and destruct.}}} \tag{1.4}$$

or the underlying set of chemical reactions, is often called the
"chemical model". Powerful numerical methods are available for integrating fairly large sets of coupled differential equations 1.4
(ref. 7). When the initial conditions and other parameters (rate
constants, temperature, pressure, light intensity) are specified,
the concentrations c_i of the reactants and products can be obtained
with the required accuarcy as function of time, provided that the
chemical reaction mechanism is correct and complete (cf. section
4.3 below).

In reality atmospheric chemical models are never complete and
rigorously correct, because the number and accuracy of available
rate constants (ref. 8) is limited. It is our intention to mediate
the basic principles of chemical kinetics in the gas phase, which
are essential for an understanding of atmospheric chemistry, and
provide the tools for assembling useful chemical "models" from laboratory data on elementary reactions, and for judging their degree
of completeness and reliability under atmospheric conditions.

2 BASIC CHEMICAL KINETICS

2.1 The "reaction vessel"

The vertical dimension of the chemical reaction vessel to be considered in regional and long range transport of pollutants is that
of the planetary boundary layer, ca. 1 - 2 km, or if necessary and
desirable, of the entire troposphere, ca. 12 km (ref. 9). The vertical pressure drop amounts to - 11.2 %/km, while the mean vertical
temperature gradient is approximately constant, - 6.5 $^{\circ}$C/km. The
partial pressure of water, which is an important source of OH radicals (cf. section 3.2), decreases rapidly from around 20 hPa at
ground level to less than 0.1 hPa above 6 km. The reaction vessel is
periodically illuminated by the sun, depending on time of day and
season, and on weather conditions. The chemically most important
section of the solar spectrum at the surface of the earth is shown
in figure 1. It is highly structured, and is cut off slightly below
300 nm by the ozone layer. These are in essence the physical conditions under which chemical transformations occur in the troposphere.

A microscopic "snapshot" of a small volume element of tropresphe-
ric air at 295 K, 1 atm, is shown in figure 2. It contains 25 randomly distributed N_2 and O_2 molecules in 1000 nm^3. Clearly the average distance between the molecules is much larger than their molecular diameter, and although the molecules propagate at an average
speed of 460 m/s, the mean distance travelled between collisions

amounts to more than 300times the molecular diameter of ca. 0.3 nm. It is therefore not surprising that simple gas kinetic laws can be used to calculate collision numbers and reaction rates in air of 1 atm pressure.

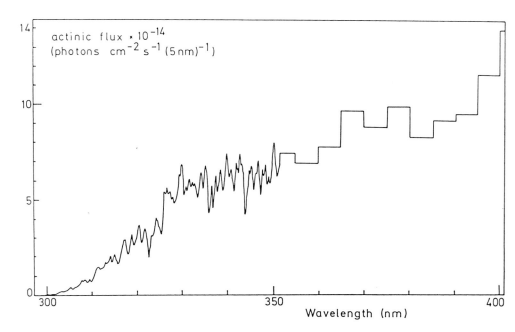

Figure 1. Spectrum of actinic solar flux at ground level, solar zenith angle z approximately 50°. Stepfunction spectrum: after data from ref. 17.

2.2 Bimolecular reactions

What is the microscopic mechanism leading to a chemical reaction between atmospheric constituents? In order to answer this question, imagine an air mass containing trace amounts of NO and O_3, which are known to react by transfer of an oxygen atom:

$$NO + O_3 \longrightarrow NO_2 + O_2 \tag{2.1}$$

If the mixing ratio of ozone (of NO) is 40 ppb (10 ppb) on the average, only four (only one) out of 4×10^6 randomly selected volume elements of the size shown in figure 2 will contain one mole-cule of ozone, or one molecule of NO. The number of NO_2 molecules formed per cm^3 per s by reaction 2.1 is the reaction rate R_1 by de-finition (cf. equation 1.4). It equals the number of collisions Z between NO and O_3 per cm^3 per s, times the probability $p \leq 1$ of a

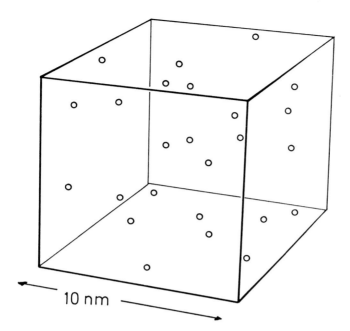

Figure 2. "Snapshot" of nitrogen and oxygen molecules in a small cubic volume of air.

collision being a reactive one:

$$R_1 = + d(NO_2)/dt = - d(NO)/dt = - d(O_3)/dt$$
$$= z_1 \cdot p_1 . \tag{2.2}$$

Gas kinetic collision theory yields a simple expression for the collision rate Z:

$$Z_1 = \text{coll. cross section} \times \text{mean relative velocity} \times (O_2)(NO)$$
$$= z_1(O_3)(NO) = \pi \left(\frac{\sigma_A + \sigma_B}{2}\right)^2 \cdot \left(\frac{8 R T}{\pi \mu_{AB}}\right)^{1/2} \cdot (O_3)(NO) \tag{2.3}$$

The collision rate Z is proportional to the absolute number densities (NO) and (O_3) of the reactants, times a gas kinetic constant z which is only weakly dependent on the molecular collisional diameter σ_{AB}, the reduced mass μ_{AB} of the colliders, and on temperature. For atmospheric conditions z is typically in the range

$$z = 1.5 - 3 \times 10^{-10} \text{ cm}^3 \text{ molecule}^{-1} \text{ s}^{-1} . \tag{2.4}$$

The probability factor p, on the other hand, is extremely variable, from close to unity for reactions between certain radicals, to practically zero for reactions involving the primary pollutants and other stable atmospheric constituents. Unfortunately there is no simple way of predicting the probability factor p of a bimolecular reaction, which can be viewed as an index of reactivity. Table 1 lists a few examples of bimolecular reactions, which exhibit no correspondence between the thermodynamic free enthalpy change ΔG of the reactions, and the reaction probabilities p, at room temperature.

Comparison with equation 1.4 shows that the collision constant z times the reaction probability p equals the reaction rate constant k:

$$R_1 \quad = \quad z_1 \cdot p_1 \cdot (O_3)(NO) \quad = \quad k_1(O_3)(NO) \tag{2.5}$$

Although more sophisticated theories of bimolecular rate constants are available (ref. 10), it is sufficient for all practical purposes to memorize that the temperature dependence of available rate constants is extremely well reproduced by a two-parameter equation, better known as the Arrhenius equation:

$$k_{(T)} \quad = \quad A \exp(-E/RT) \quad = \quad A \exp(-E^*/T) \tag{2.6}$$

This equation presents a convenient way of tabulating experimentally determined rate constants: A is termed the pre-exponential factor, while E in kJoule/mol (or E^* in K^{-1}) is the activation energy (or the reduced activation energy) of the reaction.

TABLE 1

Free enthalpy changes ΔG and probability factors p of hypothetical and observed bimolecular reactions

reaction	ΔG (kJ/mol)	probability factor p
$CH_4 + O_2 \rightarrow CH_2O + H_2O$	- 281	0, no direct reaction
$SO_2 + O_3 \rightarrow SO_3 + O_2$	- 234	0, no direct reaction
$CO + O_3 \rightarrow CO_2 + O_2$	- 421	0, no direct reaction
$NO_2 + O_3 \rightarrow NO_3 + O_2$	- 98.3	1.8×10^{-7}
$NO + O_3 \rightarrow NO_2 + O_2$	- 199	1.0×10^{-4}
$NO + HO_2 \rightarrow NO_2 + OH$	- 24.3	0.05
$O + NO_2 \rightarrow O_2 + NO$	- 197	0.06
$O(^1D) + H_2O \rightarrow 2 OH$	- 124	ca. 1

A plausible interpretation of the activation energy of reaction 2.1, which can be generalized for other bimolecular reactions as well (refs. 11,12), is given in figure 3. Reaction 2.1 is known to be exothermic by $\Delta H \cong -200$ kJ/mol. This amount of energy is released when an oxygen atom is transferred between ozone and nitric oxide in a reactive collision. It represents the difference between the $O-O_2$ bond energy in ozone and the $O-NO$ bond energy in nitrogen dioxide, which is newly formed in the reaction. It happens that some of the energy needed to pull one oxygen atom away from the O_3 molecule must be invested <u>before</u> an equivalent amount of energy is gained back by forming the new $O-NO$ bond. This produces a barrier in the reaction path, which has the height E of the experimental activation energy, equation 2.6. The barrier between reactants and products can be crossed, (a) if the colliders have a suitable orientation , e.g. the orientation shown on top of the barrier in figure

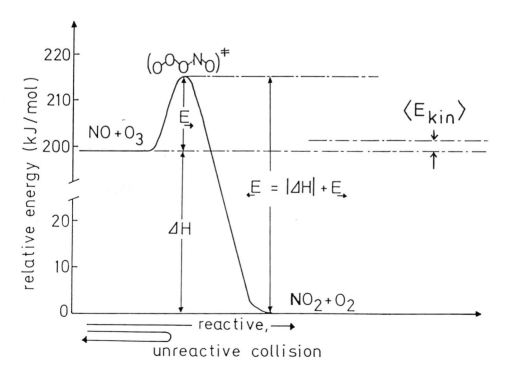

Figure 3. Energy along the reaction coordinate of reaction 2.1 and its reverse 2.7, showing energy barriers $E_\rightarrow$ and $E_\leftarrow$ as well as reaction enthalpy ΔH. The mean relative kinetic energy $\langle E \rangle$ available in a bimolecular collision at 300 K is indicated on the right.

3, and (b) if the relative kinetic energy of the colliders exceeds
the barrier height. Otherwise the colliders rebounce from the
barrier and separate again without having reacted. As already mentioned, the probability for crossing the barrier of reaction 2.1 is
1 in 10,000 collisions at room temperature!

Now consider the reverse of reaction 2.1:

$$NO_2 + O_2 \longrightarrow NO + O_3 \tag{2.7}$$

This reaction (which, by the way, is the stoichiometric equivalent
of the well known photochemical source of ozone in the troposphere,
cf. reactions 3.14 and 3.15 below) does not occur. The reason for
this is shown in figure 3: The barrier height $\overleftarrow{E}$ of this endothermic
reaction is the sum of the enthalpy change ΔH and the barrier height
$\overrightarrow{E}$ of reaction 2.1. Activation barriers of endothermic reactions are
always at least as high, and often higher than the corresponding
enthalpy change ΔH. Reactions with activation barriers in excess of
30 kJ/mol are too slow under atmospheric conditions to be important
in tropospheric models.

2.3 Elementary versus complex reactions

The reaction rate expression 2.5 can be readily generalized to
apply to other elementary bimolecular reactions. A reaction is
elementary if the species figuring in the stoichiometric equation,
and the reactants and products of the reactive encounter on the
molecular level, are one and the same. This is true in reaction 2.1.
Only elementary reactions may be safely included in chemical models
of the atmosphere!

To illustrate this point, consider the complex oxidation of
methane, which follows the stoichiometry

$$CH_4 + O_2 \ ---\rightarrow \ CH_2O + H_2O \tag{2.8}$$

Although methane and oxygen do not react when mixed at room temperature, methane is known to undergo slow oxidation in the troposphere,
following the stoichiometry of equation 2.8. However, the oxidation
reaction is by no means an elementary event, but consists of several consecutive elementary reactions:

$$CH_4 + {}^{\bullet}OH \longrightarrow {}^{\bullet}CH_3 + H_2O \tag{2.9}$$

$${}^{\bullet}CH_3 + O_2 \longrightarrow CH_3O_2^{\bullet} \tag{2.10}$$

$$CH_3O_2^{\bullet} + {}^{\bullet}NO \longrightarrow CH_3O^{\bullet} + NO_2^{\bullet} \tag{2.11}$$

$$CH_3O^{\bullet} + O_2 \longrightarrow CH_2O + HO_2^{\bullet} \tag{2.12}$$

$$HO_2^{\bullet} + {}^{\bullet}NO \longrightarrow {}^{\bullet}OH + NO_2^{\bullet} \tag{2.13}$$

The overall effect of these elementary reactions can be summarized in a single stoichiometric equation:

$$CH_4 + 2 \; NO + 2 \; O_2 \quad - - - \rightarrow \quad CH_2O + H_2O + 2 \; NO_2 \qquad (2.14)$$

Note that the oxidation of methane in the troposphere is linked with the oxidation of NO. Note furthermore that in each of the elementary steps at least one of the reactants is a free radical. (In reactions 2.9 - 2.13, and only in that example, dots are used to indicate the free radical character of a species. It must be emphasized that, unlike other pollutant gases, NO and NO_2 are odd electron species like other free radicals, which is the main reason for their exceptional reactivity and complex chemistry in the atmosphere). Only OH radicals are sufficiently reactive to start the chain oxidation of methane, and of many other volatile organics, by abstracting a hydrogen atom. Reaction 2.13 has been added to the mechanism to indicate that the OH radical invested in 2.9 is recovered in the presence of NO, and acts as a catalyst in the oxidation chain.

The following useful generalizations can be made: 1. Elementary gas phase reactions in the troposphere are only fast enough to contribute to chemical change if at least one of the reactants is a free radical; 2. in the large majority of homogeneous chain oxidations involving gaseous pollutants (volatile organics, SO_2, but not NO), an OH radical is required for chain initiation.

2.4 Monomolecular dissociations and association reactions

A few elementary reactions do not follow bimolecular kinetics. E.g. peroxyacetyl nitrate (PAN), which is a well-studied photooxidant (ref. 13), undergoes "spontaneous" dissociation into radical fragments:

$$CH_3C\!\!\begin{array}{c} {\nearrow}^{O} \\ {\searrow}_{OONO_2} \end{array} \quad \longrightarrow \quad CH_3C\!\!\begin{array}{c} {\nearrow}^{O} \\ {\searrow}_{OO} \end{array} + NO_2 \qquad (2.15)$$

Since a single molecule is involved in the elementary dissociat step, the reaction is said to be monomolecular. The rate equati appropriate for the monomolecular decay is

$$\frac{d(PAN)}{dt} = - k_{15}(PAN) \; , \qquad (2.16)$$

which can be rearranged and integrated to yield

$$\frac{d \; \ln(PAN)}{dt} = - k_{15}$$

$$(PAN)_t = (PAN)_o \exp(-k_{15}t) . \qquad (2.17)$$

However, unless the peroxy radicals are rapidly removed from the system (e.g. by reaction with NO which is fast, comparable with HO_2 + NO), the monomolecular decay 2.15 is counteracted by reassociation:

$$CH_3C\diagup_{OO}^{O} + NO_2 \longrightarrow CH_3C\diagup_{OONO_2}^{O} \qquad (2.-15)$$

The association/dissociation rate constants are linked together in a thermodynamic equilibrium constant:

$$K = \frac{(PAN)}{(CH_3CO_3)(NO_2)} = \frac{k_{-15}}{k_{15}} \qquad (2.18)$$

A similar pair of association/dissociation reactions of nitrogen oxides is probably very important in the nighttime and winter atmosphere:

$$N_2O_5 \rightleftharpoons NO_2 + NO_3 \qquad (2.19,-19)$$

Heterogeneous loss of N_2O_5, which is the anhydride of nitric acid, constitutes a potential loss mechanism for the nitrogen oxides, cf. section 4.2 and figure 6.

Reaction 2.19 and some other important association/dissociation reactions of medium-sized molecules exhibit a complex pressure dependence (ref. 8 and references given there), which must not be neglected in models which exceed the planetary boundary layer.

An important property of dissociation/association reactions is the strong temperature dependence of the equilibrium constant K, and of the dissociation rate constant (association reactions show little or no temperature dependence), which is well represented by the Arrhenius equation 2.6. The pre-exponential factor of monomolecular reactions has the dimension of a frequency, and is typically in the range 10^{12} - 10^{14} s^{-1} (ref. 10).

A few irreversible association reactions of atoms and small molecules are also important in the atmosphere:

$$O + O_2 (+ M) \longrightarrow O_3 (+ M) \qquad (2.20)$$

$$H + O_2 (+ M) \longrightarrow HO_2 (+ M) \qquad (2.21)$$

The rate constants of these reactions exhibit a linear pressure dependence, because a third collider is required to stabilize the associating fragments while they are close together in a collision. This is shown by symbolically adding the third collider M, which is N_2 or O_2 in the atmosphere, on both sides of the equation.

2.5 Units and conversion factors

To this point mixing ratios (mol-%, molar mixing ratios expressed in ppm or ppb) have been used to characterize pollution levels, while true concentrations, which appear in rate equations, have been expressed in "molecular units". The latter are readily converted to other absolute concentration units:

$$1 \text{ molecule cm}^{-3} \; \widehat{=} \; (1/N_A) \text{ mol cm}^{-3} \; \widehat{=} \; (1000/N_A) \text{ M} \qquad (2.22)$$

$N_A = 6.0223 \times 10^{23}$ is Avogadro's number. It follows from the discussion of bimolecular reactions and reaction rates that the corresponding rate constants may be expressed in the following units (we use a typical rate constant of a fast radical reaction, $k = 10^{-11} \text{ cm}^3$ molecule^{-1} s^{-1}, as an example):

$$10^{-11} \text{ cm}^3 \text{ molecule}^{-1} \text{ s}^{-1} \; \widehat{=} \; 6.022 \times 10^{12} \text{ cm}^3 \text{ mol}^{-1} \text{ s}^{-1}$$

$$\widehat{=} \; 6.022 \times 10^9 \text{ M}. \qquad (2.23)$$

Unfortunately bimolecular rate constants are often expressed in units of ppm^{-1} min^{-1} (at 298 K, 1 atm) by atmospheric modelers (refs. 1,2). We use the symbol k^* to distinguish rate constants expressed in these units from rate constants k expressed in "molecular" units. The relation between k and k^* is found as follows: Consider a bimolecular elementary reaction,

$$A + B \; \longrightarrow \; C + D$$

with the appropriate rate equation (k in molecular units)

$$\frac{d(A)}{dt} = -k(A)(B).$$

Using the relationship $(A) = 10^{-6} \gamma_A (M)$, where γ_A denotes the mixing ratio of A in ppm, and (M) is the number density of air in (molecules /cm^3), we find

$$\frac{d\gamma_A}{dt} = -k \gamma_A \gamma_B \cdot 10^{-6} (M) \qquad \text{(t in seconds)}$$

$$= -k \gamma_A \gamma_B \cdot 6 \times 10^{-5} (M) \qquad \text{(t in minutes)}$$

$$= -k^* \gamma_A \gamma_B \qquad \text{(t in minutes)}$$

Comparison of the last equations yields the conversion factor:

$$k^* \text{ (ppm}^{-1} \text{ min}^{-1}) = 6 \times 10^{-5} (M) k$$

$$= 4.40 \times 10^{17} (p/T) k \qquad \text{(p in atm, k in "molecular" units)}$$

This and other useful conversion factors for atmospheric chemists are discussed in ref. 14.

3 ATMOSPHERIC PHOTOCHEMISTRY

3.1 Basic considerations

We have pointed out in section 2.3 that chemical change in the atmosphere is mediated by radicals, particularly by OH radicals, which are extremely reactive species and act as catalysts in chain oxidation processes. However, radicals are rather imperfect catalysts, because their effective lifetime is limited by irreversible losses. E.g., the OH radical which is reproduced in the chain oxidation of methane, reactions 2.9 - 2.13, can be lost before being recycled:

$$OH + NO_2 \; (+ M) \longrightarrow HNO_3 \; (+ M) \tag{3.1}$$

This reaction is irreversible in the lower troposphere, because HNO_3 is rapidly removed from the gas phase by wet and dry deposition.

Still there exists ample evidence (ref. 15), and some direct proof (ref. 16), that in the order of 10^5 up to several 10^6 OH/cm^3 are permanently present in the atmosphere during the day. Clearly, the permanent radical losses must be compensated by some source or soruces of radicals.

Radicals can be created from stable molecules by bond breaking (cf. reactions 2.15 and 2.19). However, typical bond energies of stable molecules amount to several 100 kJ/mol, which must be compared with an average collision energy of RT = 2.5 kJ/mol at 300 K. Clearly, thermal dissociation of stable molecular bonds cannot be an important source of free radicals in the troposphere. But the energy of photons in the near ultraviolet solar spectrum at ground level, figure 1, is commensurate with some weaker molecular bonds:

$h\nu(\lambda = 400$ nm$) \; \triangleq \; 300$ kJ/mol,

$h\nu(\lambda = 300$ nm$) \; \triangleq \; 400$ kJ/mol.

An important requirement is that photons can be absorbed, i.e. the absorption spectrum of the molecule to be photodissociated must overlap with the solar spectrum:

$$AB + h\nu \longrightarrow A + B \tag{3.2}$$

The rate of photodissociation is given by the following expression, which must be incorporated in chemical models for all photochemically active trace gases:

$$\frac{d(A)}{dt} = j(AB) = (AB) \int_\lambda I_\lambda \sigma_\lambda \emptyset_\lambda^A \, d\lambda \qquad (3.3)$$

j is the equivalent of a first order rate constant (cf. equation 2.16), and is called the photolysis frequency, or simply the photolysis rate. σ_λ is the absorption cross section of the absorber AB in (cm^2 molecule^{-1}), and $\emptyset_\lambda^A$ is the quantum yield of the photodissociation 3.2 yielding A as product. $\emptyset_\lambda^A$ can be less than unity if collisional quenching of the excited molecule AB* is competitive with photodissociation, and becomes zero below the energy threshold of the dissociation process. I_λ is the spectrally resolved actinic light intensity in units of (photons cm^{-2} s^{-1} nm^{-1}), which is a function of the solar zenith angle z. I_λ is available in tabulated form (ref. 17), and absorption cross sections as well as quantum yield functions of important tropospheric absorbers have been reviewed (ref. 8).

3.2 Important photochemical reactions in the atmosphere

Only few spectra of atmospheric trace gases overlap weakly with the near uv spectrum of the sun at ground level (figure 1). Important examples, which will be discussed in this section, are ozone, formaldehyde, and NO_2. We begin with ozone:

$$O_3 + h\nu(\lambda \leqslant 315 \text{ nm}) \longrightarrow O_2 + O(^1D) \qquad (3.4)$$

$$O_3 + h\nu(\lambda \leqslant 800 \text{ nm}) \longrightarrow O_2 + O(^3P) \qquad (3.5)$$

Reaction 3.5 does not lead to permanent chemical change in the troposphere, because ozone is instantly regenerated via reaction 2.20. However, below about 315 nm the quatum yield $\emptyset_\lambda$ of electronically excited oxygen atoms, $O(^1D)$, increases rapidly to near unity. Figure 4a shows the overlap between the actinic spectrum I_λ and the function $\sigma_\lambda \emptyset_\lambda$ of reaction 3.4. This yields the integrand $I_\lambda \sigma_\lambda \emptyset_\lambda$ of equation 3.3, which is called the photoaction spectrum of reaction 3.4, figure 4b. Graphical integration yields a photolysis frequency j = 1.4 x 10^{-5} s^{-1} for this particular example of $O(^1D)$ production by ozone photolysis.

The metastable $O(^1D)$ atom can either react with water vapor (last reaction in TABLE 1), or is physically deactivated by collisions with air:

$$O(^1D) + H_2O \longrightarrow OH + OH \qquad (3.6)$$

$$O(^1D) + M \longrightarrow O(^3P) + M \, . \qquad (3.7)$$

Reaction 3.6 is globally the most important source of OH radicals in the troposphere.

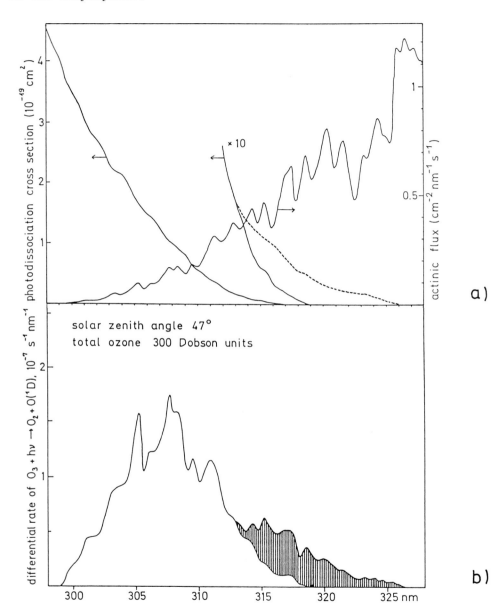

Figure 4. a) superposition of actinic flux $I\lambda$ at ground level and photodissociation cross section $\sigma_\lambda \emptyset_\lambda$ of the reaction

$$O_3 + h\nu(\lambda \leqslant 315 \text{ nm}) \longrightarrow O_2 + O(^1D). \tag{3.4}$$

b) resulting photoaction spectrum of reaction 3.4. - Solid line and white area: $\emptyset_\lambda$ after JPL recommendation (ref. 24); dashed line and hatched area $\emptyset_\lambda$ modified according to ref. 25.

The production rate of OH radicals in the troposphere due to the photolysis of ozone in the presence of water vapor is easily shown to be

$$\frac{d(OH)}{dt} = j(O_3) \frac{2k_6 (H_2O)}{k_7(M) + k_6 (H_2O)}$$

$$\cong 3.7 \times 10^8 \, j \, \gamma_{O_3} \, \gamma_{H_2O} \, p \qquad \text{(radicals cm}^{-3} \text{ s}^{-1}\text{)} \qquad (3.8)$$

The pressure p is in atm, while the γ_i are mixing ratios of the species i in ppm. Equation 3.8 illustrates that this OH source is most productive at low altitudes, where the pressure p and the mixing ratio of water vapor are both high. A typical noon-time production rate of OH radicals via 3.4 in ambient air (40 ppb ozone) is in the order of 2×10^6 radicals cm^{-3} s^{-1}.

In polluted air containing NO in large excess, ozone is often completely destroyed, and OH production via 3.4 is very ineffective. However, polluted air usually contains other traces gases which photodissociate, e.g. formaldehyde:

$$H_2CO + h\nu \, (\lambda < 340 \text{ nm}) \longrightarrow H + HCO \qquad (3.9)$$

$$H_2CO + h\nu \, (\lambda < 355 \text{ nm}) \longrightarrow H_2 + CO \qquad (3.10)$$

$$H + O_2 \, (+ M) \longrightarrow HO_2 \, (+ M) \qquad (3.11)$$

$$HCO + O_2 \longrightarrow HO_2 + CO \qquad (3.12)$$

$$HO_2 + NO \longrightarrow OH + NO_2 \qquad (3.13)$$

Reaction 3.10, which has been included for completeness, does not contribute to radical formation, but is an important loss reaction of formaldehyde. We have added important consecutive reactions of 3.9 to illustrate that formaldehyde is a source of HO_2 radicals, which are in turn rapidly converted to OH by NO. Clearly, OH and HO_2 radical sources are practically equivalent when NO is present.

NO$_2$ photodissociates extremely rapidly:

$$NO_2 + h\nu \, (\lambda \leqslant 400 \text{ nm}) \longrightarrow NO + O(^3P) \qquad (3.14)$$

$$O + O_2 \, (+ M) \longrightarrow O_3 \, (+ M) \qquad (3.15)$$

$$O_3 + NO \longrightarrow O_2 + NO_2 \qquad (3.16)$$

Reactions 3.15 and 3.16 show that the photolysis of NO$_2$ is not a source of OH or HO$_2$. However, it is extremely important as a source of ozone in the troposphere, where other chemical sources of ozone are inexistent. Equally important, though often ignored, is the fact that the photolysis of NO_2 maintains a non-zero steady-state concentration of NO, even in an excess of ozone, which is indispensable for the conversion of HO_2 to OH via 3.13. Competitive reactions of the HO_2 radical, which become important at extremely low NO mixing ratios, result in chain termination and net destruction of ozone.

Some important photolysis frequencies have been measured in the field (refs. 18,19,20). The solar zenith angle dependence is depicted in figure 5. Typical photolysis frequencies and radical source strengths in moderately polluted air (e.g. downwind of an urban area) are presented in TABLE 2. The source strengths vary in proportion with the parent molecule concentrations, which may of course differ from the values assumed in TABLE 2. Note that the rate of OH formation by the photolysis of ozone is directly affected by the water vapor mixing ratio (equation 3.8).

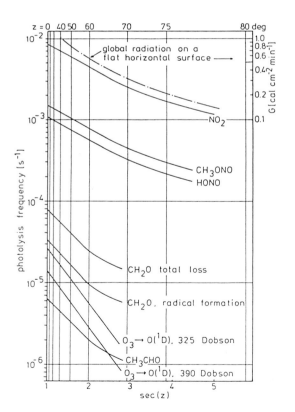

Figure 5. Important atmospheric photolysis frequencies at ground level, based on field measurements (refs. 18,19,20)

TABLE 2

Photochemical source strengths of OH and HO_2 radicals in moderately polluted air, e.g. downwind of an urban area (T = 293 K, r.h. 70 %, solar zenith angle 40°, total ozone column 355 Dobson units)

parent molecule	mixing ratio (ppb)	photolysis frequency j (1/s)	relevant product species	radical source strength (radical $cm^{-3} s^{-1}$)
NO_2	20	6.9×10^{-3}	$O(^3P)$	3.4×10^9
O_3	50	1.2×10^{-5}	OH	4.5×10^6
H_2CO	5	2.3×10^{-5}	HO_2	6×10^6
CH_3CHO	1	4.5×10^{-6}	HO_2,	1.1×10^{-5}
			CH_3O_2	1.1×10^5
H_2O_2	0.5	6.7×10^{-6}	OH	1.6×10^5
total source of OH, HO_2 and CH_3O_2				1.1×10^7

4 REACTION MECHANISMS

4.1 A "backbone" mechanism

It is not the purpose of this chapter to assemble a complete set of chemical reactions suitable for incorporation in regional and long range transport models. Such a model will be presented in chapter XVI of this book. Only the "backbone" of auch a mechanism will be discussed here.

Although only suitable for models of the clean remote troposphere, the oxidation cycles of methane and CO are simple enough, and still sufficiently detailed, to highlight the key steps in tropospheric chemistry, some of which have already been discussed in previous sections. The most important reactions occurring in humid air containing methane, CO, and NO_x as primary pollutants, can be grouped as follows:

a) Photochemical initiation:

$$O_3 + h\nu \ (+ H_2O) \ - - \rightarrow \ OH \qquad (4.1)$$

b) chain initiation and propagation:

$$CH_4 + OH \longrightarrow CH_3 + H_2O \qquad (4.2)$$

$$CH_3 + O_2 \ (+ M) \longrightarrow CH_3O_2 \ (+ M) \qquad (4.3)$$

$$CH_3O_2 + NO \longrightarrow CH_3O + NO_2 \qquad (4.4)$$

$$CH_3O + O_2 \longrightarrow CH_2O + HO_2 \tag{4.5}$$

$$CO + OH \longrightarrow CO_2 + H \tag{4.6}$$

$$H + O_2 \ (+ M) \longrightarrow HO_2 \ (+ M) \tag{4.7}$$

$$2 \ (HO_2 + NO \longrightarrow OH + NO_2) \tag{4.8}$$

+ reactions 3.9 - 3.12

c) Chain termination by radical annihilation:

$$OH + NO_2 \ (+ M) \longrightarrow HNO_3 \ (+ M) \tag{4.9}$$

$$HO_2 + HO_2 \ (+ CH_3O_2) \longrightarrow O_2 + H_2O_2 \ (+ CH_3OOH) \tag{4.10, 4.11}$$

The effects of these reactions on primary and secondary pollutant levels in the atmosphere, in summary, are the following

1. The "fuels" CH_4 and CO are slowly oxidized by OH while photo-chemical initiation occurs;

2. chain oxidation of the "fuels" results in the formation of per-oxy radicals, which oxidize NO, thus giving rise to ozone pro-duction;

3. NO_2 is irreversibly lost from the system by conversion to nitric acid, which is removed from the system by wet and dry deposition.

While in the absence of peroxy radicals ozone production by NO_2 photolysis is in exact balance with loss of ozone by reaction with NO,

$$NO_2 + h\nu \longrightarrow NO + O \tag{4.12}$$

$$O + O_2 \ (+ M) \longrightarrow O_3 \ (+ M) \tag{4.13}$$

$$NO + O_3 \longrightarrow NO_2 + O_2 \ , \tag{4.14}$$

the effect of the peroxy radicals in reactions 4.4 and 4.8 is to by-pass the loss of ozone. This is a simplified but basically correct explanation of photochemical ozone production in the troposphere, which on a global scale is limited by the availability of NO_x.

4.2 Supplementing the "backbone" mechanism

The rate constants of the reactions involved in the $CO/CH_4/NO_x/$ clean humid air mechanism are fairly well established (ref. 8), and integration of the coupled rate equations by numerical methods is feasible. Let us consider the perturbations arising from the addi-tion of anthropogenic pollutants:

An obvious effect is a tremendous increase in the complexity of the system, which arises from the large number and structural complexity of the anthropogenic hydrocarbons, and of other volatile organic pollutants. Another effect is a large increase in the NO_x mixing ratio, which can give rise to the formation of photooxidants under suitable conditions.

Although some of the steps of methane oxidation - initiation by OH attack, generation of peroxy radicals which oxidize NO - are basically similar for other hydrocarbons as well, the rate constants of the thousands of elementary reactions involved in the degradation of complex molecular structures are impossible to measure, and estimates and simplifications are inevitable in reaction mechanisms of manageable size. Also the possible existence in the atmosphere of poorly characterized short-living intermediates, particularly peroxy compounds of the nitrogen oxides,

$$RO_2 + NO_2 \rightleftharpoons ROONO_2 \ , \qquad\qquad (4.15)$$

gives rise to considerable uncertainties in the models, mainly with respect to the conversion and removal of the nitrogen oxides.

Environmental chemists who study the effects of a particular pollutant or class of pollutants, are often more interested in certain reaction subsets involving that particular pollutant. A convenient way of visualizing the reaction pathways in a sub-system is a block-flow-diagram, which is shown for the nitrogen oxides in figure 6 (ref. 21). Heterogeneous reactions have been taken into account, because they are important in the real world.

The subset of reactions summarizing the fate of NO_x emissions in the atmosphere is still fairly complex, owing to the large number of homogeneous reactions involving the nitrogen oxides. SO_2 presents a much simpler example of a chemical sub-set, figure 7. It can be shown (ref. 21) that of all potential loss reactions of SO_2 in the gas phase, only the addition of OH is really important. This relatively slow reaction leads to a homogeneous oxidation rate up to a few %/h during daytime. Note that each time an OH radical is removed by SO_2, a HO_2 radical is created:

$$OH + SO_2 \ (+ \ M) \longrightarrow HOSO_2 \ (+ \ M) \qquad\qquad (4.16)$$

$$HOSO_2 + O_2 \longrightarrow HO_2 + SO_3 \qquad\qquad (4.17)$$

$$SO_3 + H_2O \longrightarrow H_2SO_4 \ (aerosol) \qquad\qquad (4.18)$$

$$HO_2 + NO \longrightarrow OH + NO_2 \qquad\qquad (4.8)$$

Clearly the effect of SO_2 on other trace gases and their reactions in the atmosphere is minimal. Model calculations of the homogeneous oxidation of SO_2 in the atmosphere are insensitive to details of the chemical model, provided that the OH radical concentration is modelled with reasonable accuracy.

Figure 7 emphasizes that heterogeneous reactions of SO_2 are extremely important in the atmosphere, and that its reactions in the liquid phase are much more complex than the gas phase chemistry.

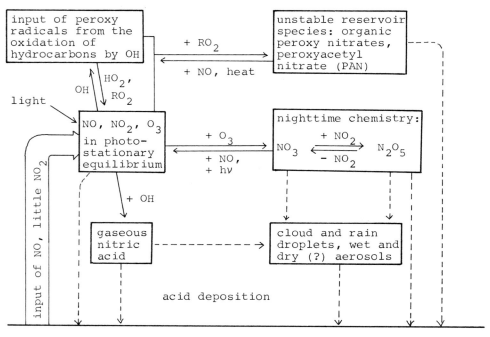

surface sources and sinks

Figure 6. Simplified block-flow diagram summarizing the homogeneous and heterogeneous reactions of the nitrogen oxides in the atmosphere. Dashed lines indicate heterogeneous paths.

4.3 Testing a reaction mechanism

Owing to the difficulties outlined in the previous section, "exact" chemical models of the polluted troposphere are out of reach. Existing reaction mechanisms are based on estimates of rate parameters and simplifications in the organic reaction scheme. A reasonable way of testing the validity of a reaction mechanism is to prove or disprove its completeness:

A mechanism starts out from N species (moecules, radicals, atoms) which can give rise, at least theoretically, to $N(N + 1)/2$ distinguishable kinds of bimolecular collisions. Many of these pairings

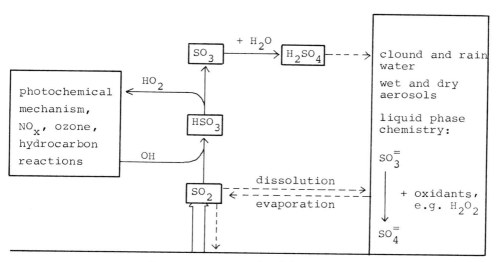

surface sources and sinks

Figure 7. Simplified block-flow diagram summarizing the homogeneous and heterogeneous reactions of sulfur dioxide in the atmosphere. Dashed lines indicate heterogeneous paths.

are strongly endothermic and thus cannot lead to products, others have been shown experimentally to be unreactive. A considerable number of reactions can be omitted from the mechanism because the corresponding rates are very much lower than the rates of other competitive reactions. The reactions which cannot be elimited on one of these grounds constitute the reaction mechanism or chemical model. The model is complete if the products of its reactions are either stable (CO_2, H_2O), or figure already among the N species initially considered. However, should one of the products be a novel reactive species, the number N of potential reactants has to be increased and, if necessary, additional reactions must be included in the model. The procedure is continued until a self-consistent set of reactions is obtained.

An empirical way of testing a model, or a sub-set of a model, is by comparison with smog chamber experiments (ref. 22). This method suffers from the possibility of undetected smog chamber effects, which invalidate the comparison. Alternatively, chemical models have been tested for consistency by applying several models to the same scenario, and intercomparing the results, or using a very detailed reaction mechanism as a reference model. The results of such intercomparisons are not completely satisfactory (refs. 1,23).

5 SUMMARY

Chemical change in the atmosphere results from radical chain reactions on the molecular level (heterogeneous reactions and reactions in solution are not considered here). Photochemical radical sources are needed for chain inititation. A chemical model is a complete set of such elementary reactions. Reaction kinetics provide simple guidelines for setting up the rate equations, which constitute a set of coupled non-linear first order differential equations (equation 1.4 in section 1). These can be integrated by numerical methods if the rate constants of the elementary steps are available from laboratory studies. Uncertainties in existing models arise (a) from the limited knowledge of reaction mechanisms involving complex organic molecules and radicals, and (b) from inaccuracies of measured or estimated rate constants.

REFERENCES

1 J.A. Leone, J.H. Seinfeld, Comparative analysis of chemical reaction mechanisms for photochemical smog, Atmos. Environ. 19 (1985) 437-464 - H. Meinel, P. Builtjes, Photochemical oxidant and acid deposition model application (PHOXA). Umweltbundesamt, Berlin 1984
2 J.S. Chang, Project Director, The NCAR Eulerian Regional Acid Deposition Model. NCAR/TN-256+STR, National Center for Atmospheric Research, Boulder Colorado June 1985
3 J.H. Seinfeld, Mathematical modeling of the polluted troposphere, in: J.S. Levine and D.R. Schryer (Eds.) Man's impact on the troposphere, NASA Ref. Publ. 1022, NASA Scientific Technical Information Office, 1978
4 R.G. Lamb, J.H. Seinfeld, Mathematical modeling of urban air Pollution, general theory. Environ. Sci. Technol. 7 (1973) 253-261
5 W.P.L. Carter, R. Atkinson, A.M. Winer, J.N. Pitts, Experimental investigation of chamber dependent radical sources. Int. J. Chem. Kinet. 14 (1982) 1071-1103
6 F. Sakamaki, S. Hatakeyama, H. Akimoto, Formation of nitrous acid and nitric oxide in the heterogeneous dark reaction of nitrogen dioxide and water vapor in a smog chamber. Int. J. Chem. Kinet. 15 (1983) 1013-1029
7 E.M. Chance, A.R. Curtis, I.P. Jones, C.R. Kirby, Facsimile: a computer program for flow and chemistry simulation, and general initial value problems. AERE-R 8775, H.M. Stationary Office, London 1977
8 R. Atkinson, A.C. Lloyd, Evaluation of kinetic and mechanistic data for modeling of photochemical smog. J. Phys. Chem. Ref. Data 13 (1984) 315-439
9 J.A. Logan, M.J. Prather, S.C. Wofsy, M.B. McElroy, Tropospheric chemistry: a global perspective. J. Geophys. Res. 86 (1981) 7210-7254
10 V.N. Kondratiev, E.E. Nikitin, Gas-Phase Reactions . Kinetics and Mechanisms. Springer, Berlin 1981; W.C. Gardiner, Rates and Mechanisms of Chemical Reactions. W.A. Benjamin, Inc., Menlo Park, California 1972

11 U. Schurath, Grundlagen der chemischen Kinetik, in: Atmosphärische Spurenstoffe und ihr physikalisch-chemisches Verhalten,
   K.H. Becker und J. Löbel (Eds.), Springer, Berlin 1985
12 R.P. Wayne, Chemsitry of the Atmospheres - an Introduction to
   the Chemistry of the Atmospheres of Earth, the Planets, and
   their Satellites. Oxford Science Publishers, Clarendon Press,
   Oxford 1985
13 U. Schurath, U. Kortmann, S. Glavas, Properties, formation, and
   detection of peroxyacetyl nitrate, in: Physico-chemical Behaviour of Atmospheric Pollutants. B. Versino, G. Angeletti (Eds.),
   D. Reidel, Dordrecht 1984
14 K.H. Becker, W. Fricke, J. Löbel and U. Schurath, Formation,
   Transport, and Control of Photochemical Oxidants. Part 1 of: Air
   Pollution by Photochemical Oxidants, R. Guderian (Ed.), Springer,
   Berlin 1985
15 A. Volz, D.H. Ehhalt, R.G. Derwent, Seasonal and latitudinal
   variation of ^{14}CO and the tropospheric concentration of OH
   radicals. J. Geophys. Res. 86 (1981) 5163-5171
16 Chapter "OH-Messkampagnen" in: Bestimmung des chemischen Abbaus
   der Stickoxide in der Atmosphäre, J. Drummond, D.H. Ehhalt, G.
   Hübler, F.J. Johnen, W. Junkermann, C. Kessler, A. Khedim, L.
   Kins, F. Meixner, D. Mihelčić, K. Müller, D. Perner, U. Platt,
   B. Rudolph, J. Rudolph, R. Schubert, A. Volz, BPT-Bericht 3/86,
   GSF München 1986
17 J.T. Peterson, Calculated actinic fluxes (290-700 nm) for air
   pollution photochemistry applications. EPA-600/4-76-025, U.S.
   EPA, Research Triangle Park, NC 27711, June 1976
18 F.C. Bahe, U. Schurath, K.H. Becker, The frequency of NO_2 photolysis at ground level, as recorded by a continuous actinometer.
   Atmos. Environ. 14 (1980) 711-718
19 F.C. Bahe, W.N. Marx, U. Schurath, E.P. Röth, Determination of
   the absolute photolysis rate of ozone by sunlight, $O_3 + h\nu \rightarrow$
   $O(^1D) + O_2(^1\Delta g)$, at ground level. Atmos. Environ. 13 (1979),
   1515-1522
20 R. Müller and U. Schurath, Entwicklung eines Geräts zur kontinu-
   ierlichen Messung der Photodissoziations-Geschwindigkeiten von
   Aldehyden in der Atmosphäre durch Nachweis des erzeugten Kohlen-
   monoxids. Final Report UC - KBF 53, GSF München, July 1985
21 U. Schurath, Chemische Reaktionen von SO_2, NO_x und organischen
   Verbindungen, in: Atmosphärische Spurenstoffe und ihr physikalisch-chemisches Verhalten, K.H. Becker, J. Löbel (Eds.),
   Springer, Berlin 1985
22 J.A. Leone, J.H. Seinfeld, Updated chemical mechanism for atmospheric photooxidation of toluene. Int. J. Chem. Kinet. 16 (1984)
   159-193
23 A.M. Dunker, S. Kumar, P.H. Berzins, A comparison of chemical
   mechanisms used in atomspheric models. Atmos. Environ. 18 (1984)
   311-321
24 W.B. DeMore, M.J. Molina, R.T. Watson, R.F. Hampson, M. Kurylo,
   D.M- Golden, C.J. Howard, A.R. Ravishankara, Chemical kinetics
   and photochemical data for use in stratospheric modeling, Evaluation No. 6, JPL Publication 83-62, California Institute of
   Technology, Pasadena, California 1983
25 J.C. Brock, R.T. Watson, Chem. Phys. 46 (1980) 477-484

Regional and Long-range Transport of Air Pollution,
Lectures of a course held at the Joint Research Centre, Ispra, Italy,
15–19 September 1986, S. Sandroni (Ed.), pp. 95–126
© Elsevier Science Publishers B.V., Amsterdam — Printed in The Netherlands

THE TRANSFER OF AIR POLLUTANTS TO THE GROUND BY WET AND DRY DEPOSITION

D. FOWLER

1. INTRODUCTION

The lifetime of individual pollutants in the atmosphere is determined by chemical transformation and/or removal processes. The chemical mechanisms have been considered in detail in the preceding lecture; this lecture is restricted to the processes of removal of pollutants from the atmosphere.

The pollutants may be present as the gases, e.g. SO_2, NO, NO_2, HNO_3, NH_3, HCl, HF, O_3, or as particles. The particles suspended in air termed aerosols may contain a complex mixture of oxidized products of some of the primary and secondary gaseous pollutants listed above. However, the aerosols over Europe and N.America generally contain $(NH_4)_2SO_4$, H_2SO_4, NH_4HSO_4, NH_4NO_3 as well as aerosols of natural origin from sea spray. The chemical properties of individual gases strongly influence components of the removal mechanisms and will be considered later, whereas for particulate pollutants, although chemical properties are important, the size of the particles has profound effects on removal mechanisms. It is, therefore, important to show the size range in which most of these particulate pollutants are found.

The size distribution of sulphate aerosol, shown for example in Fig. 1 (ref.1), provides an aerodynamic median diameter for the particles of 0.48 ± 0.1 µm, and a range of roughly 0.1 to 1.0 µm. Many other investigations of aerosol sulphate, nitrate and ammonium have provided similar size ranges.

A complication with the size distribution of aerosols is the effect of humidity. Many of the pollutant derived aerosols are hygroscopic and at the humidities and temperatures that are common over Western Europe, these particulate pollutants are present as droplets. The aerosols increase in size with increasing hymidity in a manner described by Garland (ref.2) and by Charlson et al. (ref.3).

2. DEFINITIONS

Aerosols and gases may be transferred from the atmosphere to the earth's surface by a variety of processes which may be conveniently divided into:
- wet deposition: a group of indirect processes in which the pollutants are incorporated into cloud, rain, snow or hail, and transferred to the ground by precipitation;

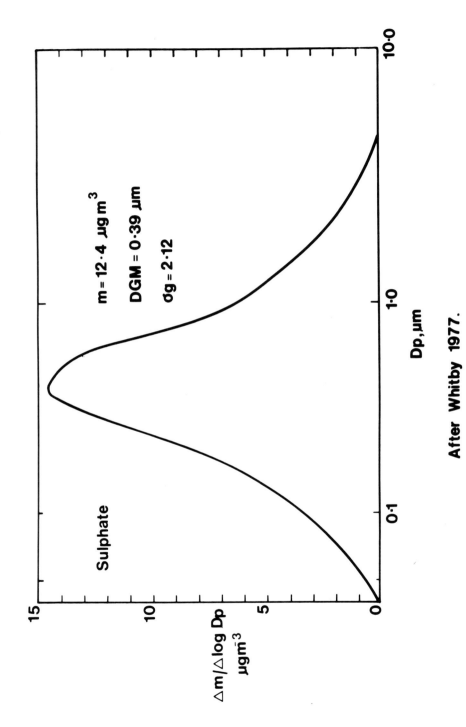

Fig. 1 – Atmospheric particle size distribution for SO_4^{2-}.

- dry deposition: processes by which gases or particles are deposited directly onto terrestrial surfaces. This includes the sorption of reactive gases at the ground and the gravitational and turbulent deposition of particles.

Within the last four or five years, interest in deposition processes on mountains has led to a number of studies of the turbulent deposition of cloud water onto vegetation. The interception of cloud water by vegetation is analogous to the transfer of large particles (a dry deposition mechanism) but as the pollutant is first incorporated in cloud water, it could also be considered a wet deposition process. The direct deposition of cloud water onto vegetation will, therefore, be treated separately.

3. WET DEPOSITION

In this section, the mechanisms by which pollutants are transferred to the surface by precipitation are described. This is followed by a short section outlining some of the principle properties of wet deposition measurements and some of the difficulties with measurement techniques.

The mechanisms by which particulate and gaseous pollutants enter cloud or precipitation droplets are very different and will be considered separately.

3.1 Particles

Considering mechanisms in temporal order, we begin with the nucleation of cloud droplets. The $(NH_4)_2SO_4$, H_2SO_4 containing aerosols are deliquescent and provide ideal condensation nuclei (CN) for the production of cloud droplets. In maritime clean air with much smaller number concentrations of particles suitable to act as CN, the clouds are generally characterized by smaller number concentrations of cloud droplets relative to cloud developing in polluted continental air, in otherwise similar conditions.

Several different investigations of scavenging processes within the last decade have demonstrated that the nucleation scavenging process provides the major pathway by which particulate pollutants enter cloud droplets (refs.2,4). There are, however, a number of cloud microphysical processes by which particles may enter cloud or rain droplets.

3.2 Brownian diffusion

This diffusion mechanism, arising from the collision of atmospheric gas molecules with particles, is very particle size dependent, becoming significant only for particle sizes smaller than 0.1 μm. Rates of Brownian diffusion for 0.01 μm diameter particles are of the order 1.4×10^{-4} $cm^2 s^{-1}$ and decrease by two orders of magnitude for particles 0.1 μm in diameter, as discussed by Dennis (ref.5).

Scavenging of particles within clouds by Brownian diffusion may, according to Goldsmith et al. (ref.6) be approximated by

$$S_B = \frac{3D_B t\, C_a}{r^2\, C_w} \qquad (1)$$

S_B = dimensionless scavenging ratio due to Brownian diffusion

t = time (s)

D_B = Brownian diffusion coefficient ($m^2 s^{-1}$)

C_a = air density (kg m^{-3})

C_w = water density (kg m^{-3})

r = drop radius (m)

The scavenging ratios produced are in the range 10 to 50 for aerosols in the size range 0.1 to 1.0 µm diameter and for droplets of 10 to 20 µm diameter after 10 hours. This mechanism is likely to contribute only a few percent of the nitrate and sulphate in precipitation since typical scavenging ratios for these substances are generally in the range 800 to 1500. For falling rain drops, their very short residence time provides even less opportunity for Brownian diffusion to supply particles to the falling droplets, thus this mechanism can largely be ignored as a contributor to the major ions in precipitation.

3.3 Impact

When atmospheric particles fail to follow the streamlines of airflow around droplets, impact (or inertial attachment) occurs, a principle used widely for the size-resolved particle measurement techniques. In the atmosphere, the efficiency of collection of the sub-micron particles (Fig. 1) by cloud or rain drops is very low, and this mechanism of particle scavenging only becomes efficient for particles larger than about 5 µm. For 5 µm particles, the efficiency of collection by small rain droplets (100 µm) reaches 0.18, and increases with rain drop size to 0.5 for rain drops of 1 mm diameter (ref.7). From the arguments of Berg (ref.8) and Garland (ref.2) that impact processes contribute between 5 and 10% of the average SO_4^{2-} in precipitation.

3.4 Phoretic mechanisms

In addition to Brownian diffusion, electrical, thermal, photo, acoustic effects and the presence of concentration gradients of other gases, all influence the movement of small particles in air (ref.9). In the presence of concentration gradients of gases in air, for example close to an evaporating or growing cloud drop, particles are swept in the direction of the mean vapour flux. The mechanism termed diffusiophoresis provides the means by which

cloud droplets may scavenge particles. Goldsmith et al. (ref.6) determined
rates of diffusiophoresis of 0.06 μm particles in concentration gradients of
water vapour. They showed that efficiency of particle capture by growing
cloud droplets was given by:

$$E = \frac{4r \ K_d sr^3 C_w}{3D_w}$$

(2)

K_d = diffusiophoresis coefficient ($m^2s^{-1}mb^{-1}$)

s = droplet concentration (m^{-3})

r = final radius of droplet (m)

C_w = density of droplets (kg m^{-3})

D_w = diffusion coefficient of water in air (m^2s^{-1})

For droplets growing to 20 μm radius and a 0.1 μm diameter, aerosol E was
3.5×10^{-3} so, thus, only a small fraction of particulate matter is captured by
this mechanism. For larger particles of the size range appropriate for most
sulphate, nitrate and ammonium (0.2 to 0.8 μm), this removal mechanism provides only 1 to 2% of the observed concentration in precipitation (ref.2).

Thermo-, photo- and electrophoretic processes are discussed in the literature (see, for example, Dennis (ref.5), Slinn and Hales (ref.10), Goldsmith
et al. (ref.6), Pruppacher and Klett (ref.11)), though none of the authors
considers these mechanisms to provide a significant proportion of the major
ions observed in precipitation.

3.5 Nucleation Scavenging

The only mechanism that appears capable of providing the observed ion concentration in cloud or rain water appears to be nucleation scavenging, or the
uptake of oxidation of gaseous sulphur and nitrogen containing gases.

Recent measurements of particle sulphate and nitrate concentrations in air
below cloud base and in droplets in an orographic cloud in northern England
have shown that the entire SO_4^{2-} and NO_3^{-} measured as small particles below
cloud base may be found in the cloud droplets, further up the hill. Such measurements have also shown that this particulate material produces cloud water
droplets very efficiently. Garland (ref.2) argues on the basis of observed
scavenging ratios for sulphate aerosol where the scavenging ratio may be expressed as concentration per unit mass of rain water/concentration per unit
mass of air that nucleation scavenging provides the majority (> 70%) of the
sulphate ions in precipitation and since the same aerosol contains nitrate
and ammonium, the same arguments would apply. Similar arguments were presented
much earlier by Junge (ref.12) who predicted scavenging efficiencies due to

nucleation scavenging in the range 50 to 100%, and recent observations
(ref.13) support this estimate with measurements suggesting a value of about
65%. Table 2 summarizes the relative contributions of the different scavenging
processes to the sulphate in precipitation.

TABLE 1

Washout of SO_2 by rain falling through a polluted layer 1 km deep assuming reversible absorption (after Garland, 1978).

Raindrop diameter	Air concentration of SO_2	Washout ratio[*]
mm	$\mu g\ m^{-3}$	
0.2	1	200
	100	20
4.0	1	20
	100	15

[*]Washout ratio for SO_2 = $\dfrac{\text{Concentration of dissolved } SO_2 \text{ per unit mass rain}}{\text{Concentration of } SO_2 \text{ per unit mass of air}}$
at ground level

3.6 Solution and oxidation of gases

In the case of SO_2, the solution and oxidation in cloud and rain water has
been the subject of considerable research effort during the last 20 years. Solution and oxidation processes can occur in the atmosphere in clouds, fogs, rain
or in aerosol droplets. The different particles providing a liquid phase occur
in very different size ranges, each having characteristic residence times.

Chamberlain (ref.14) studied the exchange of SO_2 between rain drops and the
atmosphere assuming irreversible absorption. Rates of precipitation scavenging
for rain fall rates of 1 mm h^{-1} were estimated to be 10^{-4} s^{-1}, expressing the
removal as a proportion of the gas present. Introducing the solution chemistry
of SO_2, Garland (ref.2) demonstrated that the drops equilibrate with air concentrations of SO_2 in distances shown in Table 1, which are, of course, a
function of drop size and air concentration for SO_2. In general, the removal of
SO_2 by falling rain, expressed as a washout scavenging coefficient

$$W_S = \frac{\text{concentration of dissolved } SO_2 \text{ per unit mass rain}}{\text{concentration of } SO_2 \text{ per unit mass of air}}$$

leads to values of W_S of the order 10 to 30 and only contributes by an order of
10% of the sulphur in precipitation. However, the uptake and oxidation of SO_2
in cloud water by H_2O_2 has been demonstrated to be a very rapid reaction

TABLE 2

Processes contributing to the sulphur and nitrogen in rain

PROCESS	SULPHUR		NITROGEN	
	Range of concentration in rain at ground level $\mu g\ g^{-1}\ SO_4$	Note (1) Average contribution to wet-deposited sulphur %	Range of concentration in rain at ground level $\mu g\ g^{-1}\ NO_3^-$	Note (2) Average contribution to wet-deposited nitrogen %
Diffusiophoresis	$10^{-2}-10^{-1}$	2.5	$10^{-3}-10^{-2}$	2.5
Brownian diffusion	$10^{-2}-10^{-1}$	2.5	$10^{-3}-10^{-2}$	2.5
Impact and interception	$10^{-1}-1.0$	10	$10^{-2}-10^{-1}$	10
Solution and oxidation of gaseous "species"	$0.5-3.0$	20	$10^{-2}-0.4$	15-25 (3)
Cloud condensation nucleus pathway (5)	$2.0-20.0$ (4)	65	$10^{-1}-5.0$ (4)	60-70

Notes: (1) Considering rain with geometric mean SO_4^{2-} concentration of 3.5 $\mu g\ g^{-1}$ (weighted for rain quantity).
(2) Considering rain with geometric mean NO_3^- concentration of 0.5 $\mu g\ g^{-1}$ (weighted for rain quantity).
(3) Uncertainty in this component necessarily leads to uncertainty in other components.
(4) Lower limits of range deduced from average contribution and range of concentrations measured.
(5) This table considers the whole wet deposition pathway, no distinction between RAINOUT and WASHOUT.

(ref.15) and recent field measurements (refs.16,17) have shown that in orographic cloud this mechanism may provide an efficient oxidation pathway for SO_2 scavenging. The process may be limited by either of the reactants SO_2 or H_2O_2 depending on the ambient concentrations for each. However, measurements to date show a very wide range of H_2O_2 concentrations in air (refs.18,16). In spring and summer, and in clean air, concentrations are generally much larger than in winter and/or in polluted air, the range of concentrations reported by Dollard being 0.1 to 50 $\mu m\ H_2O_2$. A major uncertainty for estimates of the importance of this oxidation mechanism is the lack of measured H_2O_2 concentrations in the gas phase and in cloud water. Measurements are in progress in continental Europe, the U.K. and N.America, which should provide regionally and seasonally representative values within the next few years.

Similar processes for NO_2 oxidation in droplets are unlikely since, at atmospheric concentrations, it is not very soluble. There is evidence, however, that mechanisms exist for droplet production of NO_3, possibly from HNO_2 and NO_3 in the gas phase (ref.15). The uncertainties are large and it is not pos-

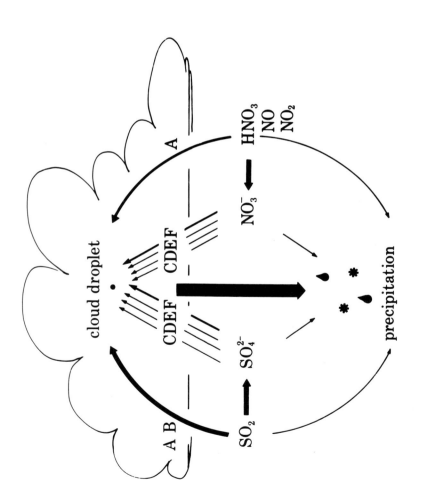

Fig. 2 – Pathways and mechanisms for the transfer of sulphur and nitrogen containing compounds to the ground in precipitation. A: dissolution; B: oxidation; C: diffusiophoresis; D: Brownian diffusion; E: impaction; F: cloud condensation nuclei pathway. Bold arrows indicate the major routes for wet removal of S- and N-containing compounds from the atmosphere.

sible to show the proportion of the NO_3^- in rain formed through droplet chemistry. The evidence for nucleation scavenging of aerosol NO_3^- suggests that like SO_4^{2-} a large proportion of the wet deposited NO_3^- is incorporated into cloud and precipitation this way. Fig. 2 summarizes the different processes leading to wet deposition.

3.7 Wet deposition measurements

The collection of rain while complicated by sampling problems arising from spatial variability and the aerodynamics of collecting equipment (ref.19) is rather straightforward. The majority of the precipitation chemistry data for Europe and N.America between 1960 and 1980 has been obtained using simple funnel type bulk collectors. These collecting methods are subject to a number of errors (ref.20). In particular, the continuously open collectors receive some material via dry deposition. The sorption of gases, especially SO_2 and HNO_3 onto the collecting equipment leads to overestimates of the wet deposited SO_4^{2-} and NO_3^-. This systematic error for SO_4^{2-} has been investigated by Fowler and Cape (ref.21) who showed that at a site with an arithmetic mean SO_2 concentration of 12 μg m^{-3} SO_2, dry deposition onto the rainfall collector over a year contributed 30% of the total SO_4^{2-}. A technique for reducing this error is to use wet-only collectors which open only during rain, but there are significant problems with detection of the beginning and end of a precipitation event and it has yet to be demonstrated by measurement that a fully automatic wet-only precipitation collector provides a more reliable and accurate measure of annual wet deposited SO_4^{2-}, NO_3^- and NH_4^+ than carefully operated daily bulk collectors!

3.8 Properties of wet deposition measurements

The variability in rainfall composition from one episode to the next is very large; at most monitoring stations in Europe, H^+ concentrations vary between 0.1 and 1000 μeq litre^{-1} for individual episodes. The variability in concentrations are similar for the major ions SO_4^{2-}, NO_3^- and NH_4^+, and the variability in precipitation amount between episodes further complicates the picture.

Rainfall amounts show a log-normal distribution (Fig. 3), rainfall occurring on about 50% of days at a site in southern Scotland. The number of rain days decreases towards the S.E. of the U.K. and across continental Europe, but the variability between episodes is rather uniform. The upper tail of the distribution shows that 30% of rain is deposited in just 6% of events, a measure of the episodicity of the process. For acidity, the episodicity is much larger: 30% of the acidity being deposited in just 2% of episodes (Fig. 4). As an extreme, 50% of annual wet deposited acidity in 1984 deposited at this site occurred on a single day. The episodicity is an important characteristic of the wet deposition measurements, the most important events occur on a small (< 20)

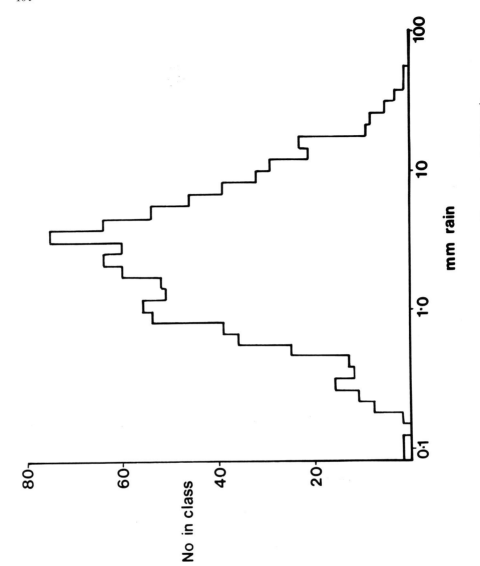

Fig. 3 – The frequency distributions of rainfall events at ITE Bush.

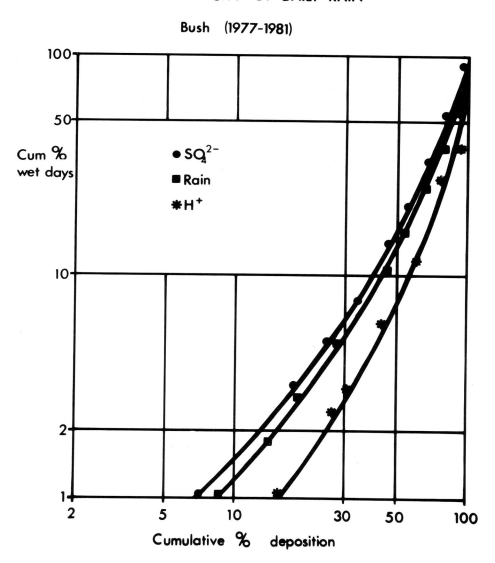

Fig. 4 – Episodicity of daily rain.

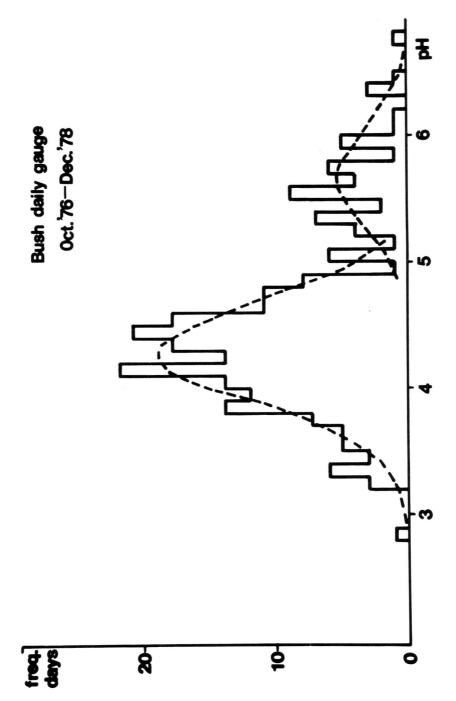

Fig. 5 - Frequency distribution of rainfall acidity at Bush.

number of days each year when very polluted air and large amounts of precipitation coincide. This leads at some sites to a bimodality in the frequency
distribution of episodes. In Fig. 5 the two modes of the distribution have
median pH's of 4.29 (75% of events) and 5.6 (25% of events). At continental
locations, where major sources of air pollutants occur upwind in most directions, there is little opportunity to associate the sample with clean
N.Atlantic air.

Despite the very large variability in composition and deposition between
episodes, there are consistent associations between many of the rainfall chemistry measurements and local meteorological and air chemistry measurements.
Such relationships may be seen by classifying all of the precipitation episodes by pH (Fig. 6).

It is clear, for example, that the most acid episodes are associated with
large particle SO_4^{2-} concentrations, poor visibility, low windspeed and small
precipitation amount. Conversely, the days with small acidity values are windy,
good visibility days with small particle SO_4^{2-} and SO_2 concentrations. The
meteorological influence over these associations may best be seen from the trajectory analyses of the two tails of the distribution (polluted and clean
precipitation, Fig. 7).

4. DRY DEPOSITION

Our knowledge of dry deposition to natural surfaces is based on limited set
of measurements over vegetation and soil. The rates of dry deposition to the
countryside, unlike wet deposition, are not monitored anywhere by anyone. The
vegetation over which measurements have been made is limited to only five different plant species (Lolium perenne, Triticum aestivum, Zea mays, Glycine max
and Pinus sylvestris), and the different soils over which field measurements
have been made are limited to about four... a small sample of the available
soil types. Nevertheless, it will be argued here that with an understanding of
the mechanism of transfer and a framework for parameterizing each stage of the
process, it is possible to provide regional inputs of SO_2 and some other pollutant gases to the countryside.

Direct transfer of gases and particles from the atmosphere to the ground
comprises three stages: first the materials must be transported through the
free atmosphere to within a few millimetres of the surface; second there is
transport through a viscous sub-layer of air close to the surface, and last,
the gas or particle must be captured by the surface. These stages in the transfer to vegetation are illustrated in Fig. 8, in which the main transfer processes are indicated.

Particles that are large enough to sediment out of the atmosphere under the
influence of gravity represent a small fraction of the particles in the atmos-

108

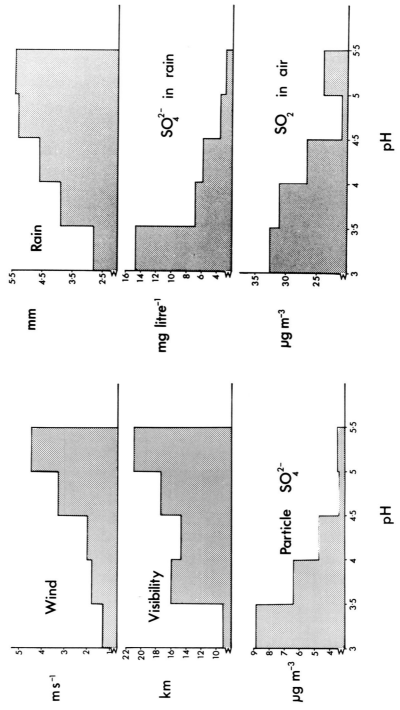

Fig. 6 – The association between rain acidity and other measured variables; data classified by pH (3 years daily data).

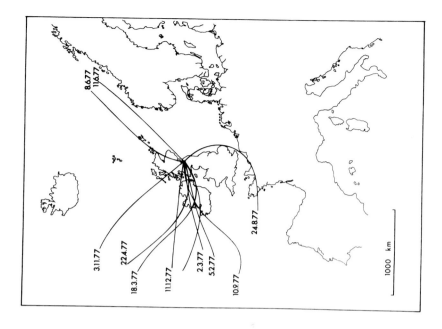

Fig. 7

Back-trajectories over 48 h leading to rainfall at
Bush with pH > 5.

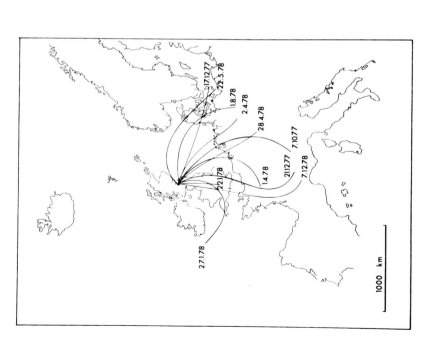

Back-trajectories over 48 h leading to rainfall at
Bush with pH < 4.

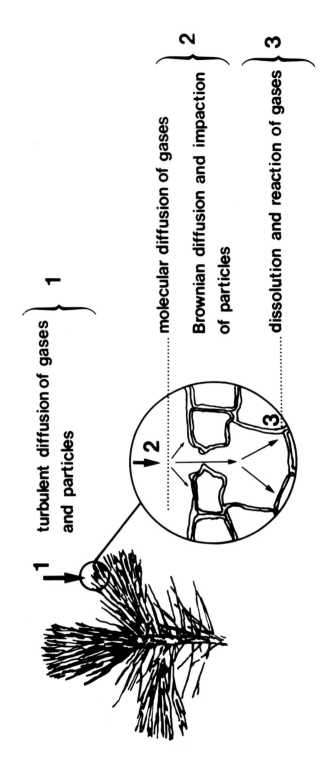

turbulent diffusion of gases and particles } 1

molecular diffusion of gases
Brownian diffusion and impaction of particles } 2

dissolution and reaction of gases } 3

Fig. 8 - Dry deposition mechanisms for gases.

phere (Fig. 1), so that the following discussion is restricted to the particle size range appropriate for atmospheric aerosols, 0.1 to 1.0 μm.

For particles smaller than 1.0 μm in diameter, transfer in the free atmosphere is effected by turbulent diffusion and depends on windspeed, surface roughness and temperature stratification in the atmosphere. Transfer through the viscous sub-layer is very particle-size dependent. The inertial forces are sufficient for large particles (> 1 μm) to penetrate this still layer of air and deposit at the surface by impact. For smaller particles, this mechanism does not work, the inertial forces are insufficient to overcome the viscosity of the air. For small particles (< 0.1 μm), Brownian diffusion and phoretic effects provide the necessary vehicle for transport, but for aerosols in the size 0.1 to 1.0 μm, more of the transport mechanism provides an efficient means of transport to the surface. The consequence is that dry deposition rates for these particles are very small. A consequence of this is that in the absence of rain (the only efficient means of removing small particles), the particle concentration increases with time and long, calm, dry periods are generally associated with large particle concentrations and poor visibility (ref.22).

Field measurements of particle dry deposition show conflicting results, with some groups obtaining consistently small deposition rates (ref.23), consistent with the detailed laboratory studies (ref.24) (Fig. 9), while other groups obtain large and very variable deposition rates (ref.25). It has been widely suggested that the groups measuring large deposition rates may be measuring in an atmosphere in which larger particles (> 2 μm) represent a significant proportion of the aerosol mass present, but this has not been confirmed by measurement.

Most information on dry deposition of gases has been obtained for SO_2, but a substantial number of measurements have also been made of O_3, NO_2 and HNO_3 deposition. The range of gases for which there are some measurements has grown rapidly and now includes NH_3, NO, HCl, HF, PAN, N_2O, CH_4, COS and H_2S among others. In this section, the principles of the exchange process will be discussed, but as the number of gases has increased, the significance of fluxes in both directions (towards and away from the surface) has gradually become appreciated. It is, therefore, misleading to think exclusively about "deposition" of many of the above gases because upward fluxes are not uncommon.

The most direct measurement of gas deposition is the net flux per unit ground area, and for most pollutants the fluxes are typically of the order 0.1 μg $m^{-2}s^{-1}$ (areas are ground area rather than specific areas m^2m^{-2}).

The vertical flux may be obtained by a variety of techniques discussed later, but over horizontal uniform vegetation, the one-dimensional (vertical) flux (F_S) is proportional to the vertical gradient in concentration, the constant of

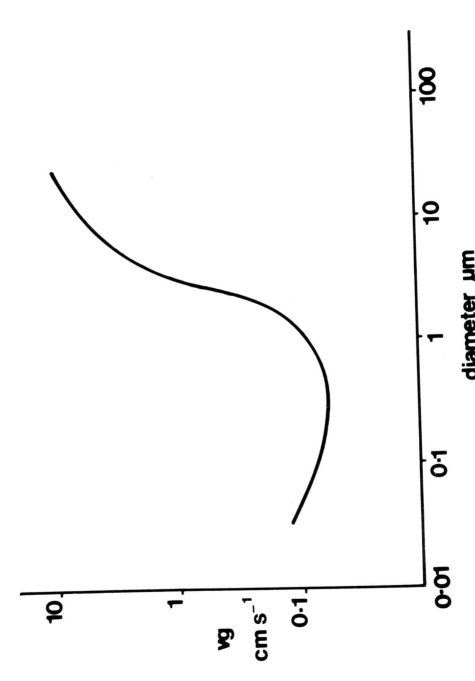

Fig. 9 – Variation of particle deposition velocity with particle size (after Chamberlain).

proportionality being termed the eddy diffusivity (K_S)

$$F_S = -K_{s(2)} \, S_\chi / S_Z \tag{3}$$

where χ is the concentration of gas, Z is height.

Integrating (4) with respect to height between Z_1 and Z_2 yields

$$F_s = \cfrac{\chi_s(Z_2) - \chi_s(Z_1)}{\displaystyle\int_{Z_1}^{Z_2} \frac{dZ}{K}} \tag{4}$$

a form analogous to Ohm's law (flux = potential difference/resistance).

If the lowest limit of the integral is the surface and assuming surface concentration as zero, Eq.(5) shows the total resistance (r_t)

$$r_{t(Z)} = \chi_{s(Z)} / F_s \tag{5}$$

The reciprocal of total resistance has dimensions of velocity and is identical to the velocity of deposition (v_g).

The main virtue in the use of a resistance analogy is that the total resistance (which may be obtained by measurement) may be sub-divided into the atmospheric and surface components shown in Fig. 10. The atmospheric transfer term r_{am} may be obtained directly from measurements of wind velocity and temperature gradients over flat homogeneous terrain.

$$r_a = U_{(Z)} / U_*^2 \tag{6}$$

where U is wind velocity (ms^{-1}) and U_* is friction velocity (ms^{-1}), but because bluff body forces augment the transfer of momentum by molecular diffusion at the surface, the atmospheric resistance to momentum transfer (r_{am}) is smaller than that for a gas (SO_2, NO_2, etc), which is limited to molecular diffusion for transfer through the viscous sub-layer. An additional resistance term r_{bg} is added in series with r_{am}, where

$$r_{bg} = (BU_*)^{-1} \tag{7}$$

where B is the dimensionless sub-layer Stanton number (ref.26). Values for B^{-1} appropriate for pollutant gases have been determined by Chamberlain (ref.24).

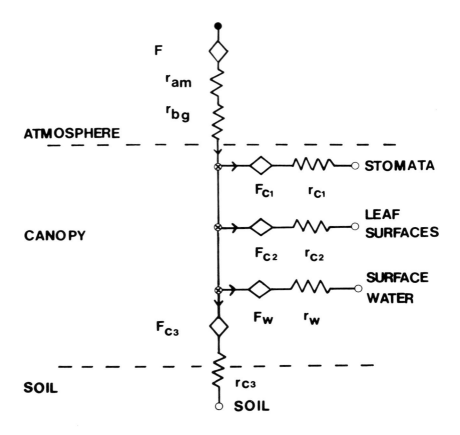

Fig. 10 – A resistance analogue for pollutant uptake by crop canopies. The total resistance r_t is the sum of atmospheric resistances $r_{am}+r_{bg}$ and canopy resistance r_c. Canopy resistance is determined by up to four parallel paths, stomata r_{c1}, external plant surfaces r_{c2}, surface water r_w and soil r_{c3}; corresponding fluxes are denoted by the prefix F.

The atmospheric resistances $r_{am}+r_{bg}$ may, therefore, be subtracted from the total resistance r_t, and the remaining term represents the bulk surface or canopy resistance r_c. The surface resistance may be thought of as the consequence of uptake at all of the different sites of absorption acting in parallel. For vegetation this may be simplified into a term representing absorption onto the external surfaces of plants (r_{c2}), a term representing the absorption by stomata (r_{c1}) and a term representing absorption by the ground beneath the canopy (r_{c3}). This simplification shown in Fig. 10 may be further developed so that the canopy is subdivided into a number of layers, a common practice in canopy gas exchange models. However, the data on which our understanding of SO_2 and O_3 deposition is based, are not generally of adequate precision to define the way in which r_{c1} and r_{c2} would change through a canopy. Many more detailed field and laboratory measurements would be necessary to test the properties of such models. Fig. 10 shows the resistances commonly used in this form of analysis. If the results of dry deposition measurements from the different experiments to date were listed, there would be no simple technique for weighting the values to provide appropriate long-term average values appropriate for modelling deposition to the countryside. The technique that is used for this exercise is to extract the values of the resistance components and observe the way in which each of them varies with changing environmental conditions.

The understanding that this exercise provides may then be used to provide a model to produce the "regional average" values necessary. As an example, the results of measurements over a wheat crop show the variation of canopy resistance with total resistance. These show that surface resistance dominates the transfer process and that for this surface there is generally little effect of windspeed or atmospheric stability on rates of deposition, the atmospheric transfer represents an average only 30% of the total resistance. Thus, variation in properties of the absorbing surface causes most of the variation in rates of deposition.

Fig. 11 shows a diurnal cycle in deposition velocity for SO_2 uptake by wheat. Over the period there was little effect of atmospheric variables, but the observed change responded very strongly to the opening and closing of stomata. This effect provides the means of separating surface resistance components for some gases. It is generally argued that uptake onto external surfaces of plants and uptake at the soil may be estimated by examining the nighttime values of canopy resistance on nights when there is no clear SO_2 and O_3.

The rates of SO_2, O_3 uptake onto the external surfaces of plants in the field are small and rather constant. In contrast, stomatal uptake is the major sink for both gases in vegetation. This causes rapid changes in deposition velocity as stomata open or close. The diurnal cycles so often seen in the literature provide a means of showing which processes control the rate of uptake and how important the different sites of uptake are. Figs. 10, 11 and 12 are used for illustration of these effects.

From large data sets of this type, the average values for r_{c3} and r_{c2}, the components of canopy resistance for soil and leaf surface uptake are estimated. In reality, it is unlikely that these are constant, but the resistances involved are large for SO_2 and O_3 and no significant variability in their values has been demonstrated except in the presence of dew.

Dew-wetted vegetation may be an excellent surface for SO_2 uptake while the rate of clear formation is large; in these conditions, the surface may have as a perfect sink ($r_c \simeq 0$) (Fig. 13). However, when dew formation stops, the S^{IV} species in solution equilibrate with air concentrations and a canopy resistance

116

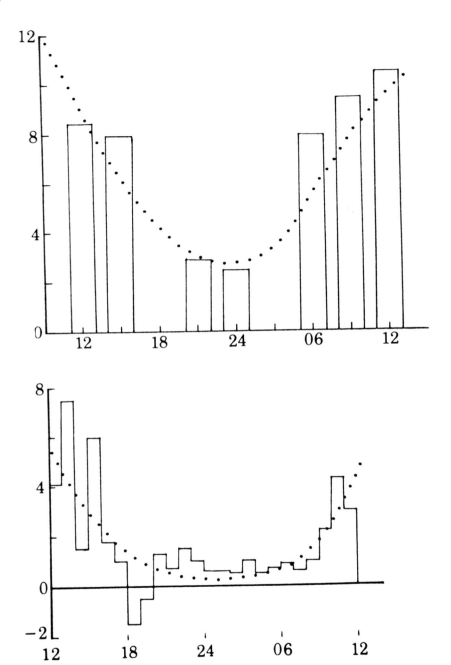

Fig. 11 — (a) SO_2 deposition on wheat; (b) SO_2 deposition on a Scots pine
forest.

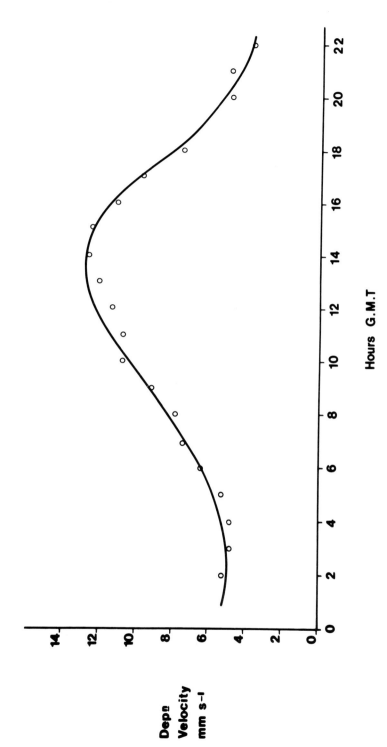

Fig. 12 - O$_3$ deposition on barley.

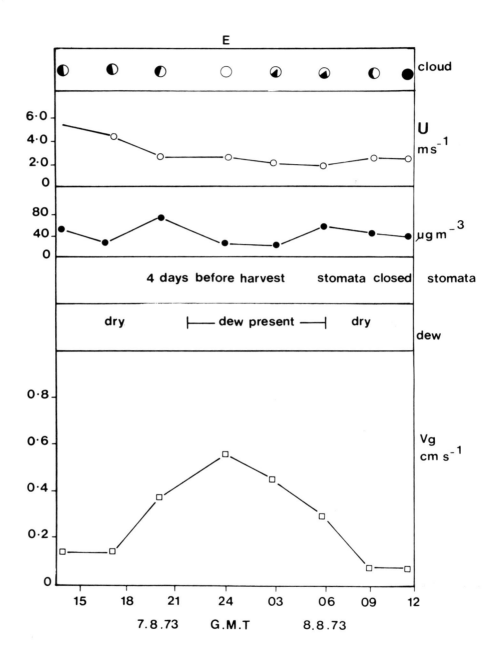

Fig. 13 - SO$_2$ deposition on senescent wheat in the presence of dew.

reappears. The combined values of r_{c2} and r_{c3} for grasses, agricultural crops and forests are generally of the order 300 to 1000 sm^{-1} for SO_2 and O_3, so that the maximum nocturnal deposition velocities (with an atmospheric resistance of $r_{am}+r_{bg} = 40$ sm^{-1}) would be in the range 1 to 3 mms^{-1} (simply adding $r_{am}+r_{bg}$ to the value of r_c and taking the reciprocal).

During the day with stomata open, the value of r_{c1} for SO_2 and O_3 may, for some crops, be as small as 50 to 100 sm^{-1}, which, with atmospheric resistances of 20 sm^{-1}, would permit deposition velocities in the range 7 to 15 mms^{-1}.

Treating the non-stomatal uptake at the surface as a constant, it is necessary to be able to calculate appropriate values for the stomatal component of canopy resistance in order to obtain the bulk surface or canopy resistance r_c ($r_c^{-1} = r_{c1}^{-1} + r_{c2}^{-1} + r_{c3}^{-1}$).

If it may be shown that the stomatal resistance to SO_2 or O_3 is equivalent to that for water vapour (modified only by the differences in molecular diffusivities), then the very extensive literature on water loss from vegetation may be used (see, for example, Baldochi et al. (ref.27)). This requires there to be no internal resistance to SO_2 or O_3 uptake within plants, or that the gas concentration of SO_2 and O_3 at the lining of the sub-stomatal cavity is zero. For SO_2 uptake by grasses and cereals, this appears to be a good approximation but for forests there is considerable doubt (refs.21,28).

With values of bulk canopy resistance from the above reasoning, all that is needed to obtain an appropriate velocity of deposition are the atmospheric terms $r_{am}+r_{bg}$. Fortunately, these are quite straightforward and may be obtained using the wind profile equation

$$U_Z = \frac{U_*}{k} \ln(Z-d/Z_o) \tag{8}$$

where:

U = windspeed

U_* = friction velocity

d = zero plane displacement

Z_O = roughness length

k = Von Karmans constant (0.4)

to obtain U_* from a known windspeed at a defined height above the surface, and Eqs.(6) and (7). The values of d and Z_O may, in a first approximation, be taken as 0.7 h and 0.13 h, where h is the height of a uniform canopy of vegetation (ref.29). In reality, such values should be modified to take account of the effects of atmospheric stability on U_*, but uncertainties in estimating appropriate values for canopy resistances are too large to justify such a refinement. Table 3 shows typical values for the atmospheric resistances for a range of

TABLE 3

Atmospheric resistance $(r_{am}+r_{bg})$ for transfer of pollutant gases onto vegetation

| Vegetation | Crop height/m | Atmospheric resistance $(r_{am}+r_{bg})/sm^{-1}$ windspeed | |
		1 ms^{-1}	4 ms^{-1}
Grass	0.4	130	30
Cereals	1.0	70	20
Maize	2.0	50	12
Forest	10.0	25	10

vegetation heights and for two windspeeds.

The use of data for SO_2 and O_3 are purely illustrative and are, of course, limited in their applicability. Other gases such as HNO_3 behave very different- ly. There is evidence from field measurements that the rate of deposition of HNO_3 vapour is limited only by atmospheric transfer, and that canopy resistances are effectively zero (ref.30). This rather simple case permits rates of HNO_3 deposition to be calculated quite simply since in this case

$$V_g(HNO_3) \cong (r_{am}+r_{bg})^{-1}$$

Dry deposition of NO_2 has proved to be quite simple to measure in cuvettes and laboratory experiments (ref.31), but several recent field micrometeorologi- cal measurements of NO_2 fluxes have provided very variable results (refs.32,33), including both upward and downward fluxes. With the current state of uncertainty over the net NO_2 deposition rates, it is not possible to provide reliable in- formation on the relative source and sink distribution for NO_2 in vegetation or soil.

The closely related flux of NO out of soil has been extensively measured (refs.34,35) and has been shown to provide fluxes in the range 5 to 200 $ng\,m^{-2}s^{-1}$ NO, with typical values of 10 $ng\,m^{-2}s^{-1}$. An example of recent results, taken from our own research group, is shown in Fig. 14.

As with nitric oxide and NO_2, the flux of NH_3 is bidirectional (ref.36). Ammonia may be released from or absorbed by the surface. In those cases where fluxes are both upward and downward, the framework for their interpretation may require more complexity than the simple resistance model outlined earlier; although such development will require more detailed field measurements to pro- vide the information for such an analysis.

121

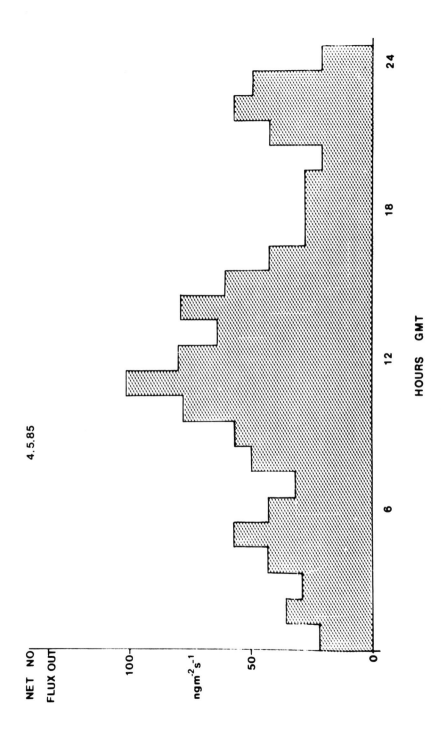

Fig. 14 – Nitric oxide release from a soil + barley crop.

4.1 Dry deposition measurement techniques

(a) Micro meteorological : - flux gradient methods
                             - Eddy correlation
(b) Box methods   : differences between inlet and outlet concentra-
                             tions (Fig. 15)
(c) Tracer   : radioactive tracer ^{35}S or dual tracer SF$_6$/SO$_2$ for
                             example
(d) Mass budget techniques: - catchment scale, long-term integrating type
                               experiments
                             - canopy-budget methods have been used.

Vegetation intercepts wind-blown cloud or fog droplets directly by impact. Such droplets lie in the size range 10-50 μm and are efficiently captured by vegetation. The process described as occult precipitation (ref.37) has been reviewed by Kerfoot (ref.38) and in areas where low cloud is common (coastal areas and high altitudes), it may represent a significant hydrological input. The process may also be important as a solute transfer process for the same areas because cloud water may show very large concentrations of the major ions (ref.39) (Fig. 16). For a site in the north of England, Dollard et al. (ref.40) showed that inputs by this mechanism may be of the order 10-20% of wet deposition for SO_4^{2-}, NO_3^- and H^+. However, the very large concentrations reported for cloud water of 2500 μeq SO_4^{2-} l^{-1}, 2400 μeq NO_3^- l^{-1} and 400 μeq H^+ l^{-1} (ref.40) may make effects of this form of deposition proportionally greater than its contribution to total deposition.

The capture of fine rain by vegetation is similar to that of cloud capture and may lead to larger inputs of water to forests than to a rain guage. Miller and Miller (ref.41) discussed the capture of fine rain by trees as a component of "filtering" and showed that concentrations of most ions collected in a rain gauge with a wire grid above the funnel were greater than those in rain collected in a conventional gauge. The degree of enhancement of concentration varied with the ionic species, but was of the order of a factor of 2 for S and total N.

The mechanism of transfer of cloud droplets to vegetation is limited only by atmospheric transfer (ref.40) with deposition velocity typically 2-3 times the bulk sedimentation velocity. Over grassland, overall deposition velocities between 30 and 60 mms^{-1} were obtained; for forests, values between 50 and 150 mms^{-1} should be expected.

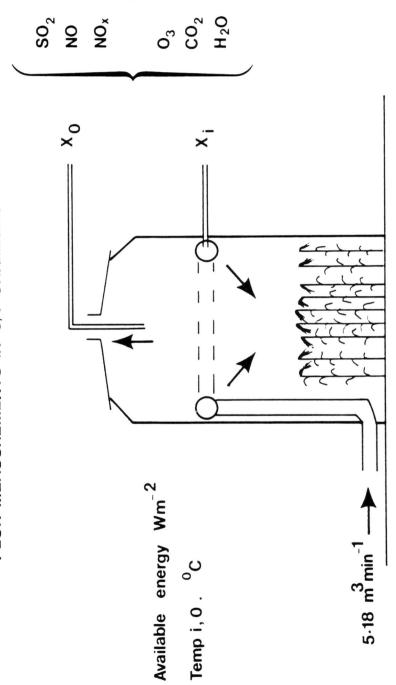

Fig. 15 – Flux measurements in O/T chambers.

124

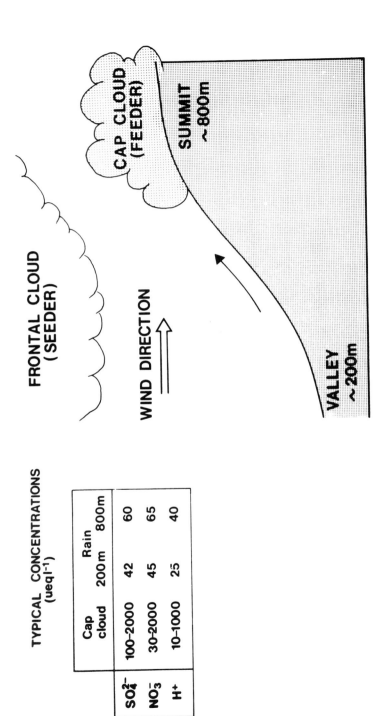

Fig. 16 – Feeder-seeder mechanism for enhanced rainfall concentrations of major ions in rain.

REFERENCES

1 K.T. Whitby, The physical characteristics of sulfur aerosols, Atmos. Environ., 12 (1978) 135-160.
2 J.A. Garland, Condensation on ammonium sulphate particles and its effect on visibility, Atmos. Environ., 3 (1969) 347-354.
3 R.J. Charlson, D.S. Covert, T.V. Larson, A.P. Waggoner, Chemical properties of tropospheric sulfur aerosols, Atmos. Environ., 12 (1978) 39-53.
4 J.M. Hales, Precipitation chemistry: its behaviour and its calculation. In: Air Pollutants and their Effects on the Terrestrial Ecosystem. Ed. S.V. Krupa & A.H. Legge, New York: Wiley (1983).
5 R. Dennis, Handbook on aerosols, Publ. Technical Information Centre, ERDA Springfield, Virginia (USA) (1976).
6 P. Goldsmith, H.J. Delafield, L.C. Cox, The role of diffusiophoresis in the scavenging of radioactive particles from the atmosphere, Q. Jl R. Met. Soc., 89 (1963) 43-61.
7 B.J. Mason, Physics of clouds, Clarendon Press, Oxford (1971).
8 T.G.O. Berg, Collection efficiency in washout by rain. Precipitation Scavenging A.E.C. Symposium Service 22, WTIS Springfield Virginia (USA) (1970).
9 C.E. Billings, R.A. Gussman, Dynamic behaviour of aerosols. Handbook of aerosols. Publ. Technical Information Centre ERDA, Springfield Virginia (USA) (1976) 40-65.
10 W.G.N. Slinn, J.M. Hales, A re-evaluation of the role of thermophoresis as a mechanism of in and below cloud scavenging, J. Atmos. Sc., 28 (1971) 1465-1471.
11 H.R. Pruppacher, J.K. Klett, Microphysics of clouds and precipitation, Reidel Publishers (1978).
12 C.E. Junge, Air chemistry and radioactivity, Academic Press, New York (1963).
13 D.A. Hegg, P.V. Hobbs, Preliminary measurements on the scavenging of sulfate and nitrate by clouds. In: Precipitation Scavenging, dry deposition and resuspension. Ed. H.R. Pruppacher, P.G. Semonin & W.G.N. Slinn, New York: Elsevier (1983) 79-89.
14 A.C. Chamberlain, The movement of particles in plant communities. In: Vegetation and the atmosphere. Ed. J.L. Monteith, Vol.1, London: Academic Press (1975) 115-201.
15 R.A. Cox, S.A. Penkett, Formation of atmospheric acidity, Proc. of the CEC Workshop, Berlin 1982. Ed. S. Beilke & A.J. Elshout, Commission of the European Communities (1983) 58-83.
16 G.J. Dollard, B. Jones, A. Chandler, M.J. Gay, Measurements of ambient SO_2 and H_2O_2 at Great Dun Fell and their evidence in cloud, Proc. of NATO ARI Acid Deposition Processes at High Elevation Sites. Ed. M.H. Unsworth & D. Fowler (1986).
17 G.P. Gervat, P.A. Clark, A.R.W. Marsh, T.W. Choularton, M.J. Gay, Controlled chemical kinetic experiments in cloud: oxidation of SO_2, Proc. of NATO ARI Acid Deposition Processes at High Elevation Sites. Ed. M.H. Unsworth & D. Fowler (1986).
18 J. Kadlecek, S. McLaren, N. Camorota, V. Mohnen, J. Wilson, Cloud water chemistry at Whiteface Mountain. In: Precipitation scavenging, dry deposition and resuspension. Ed. Pruppacher et al., Elsevier Science Publishing Co., Inc. (1983).
19 J.C. Rodda, On the questions of rainfall measurement and representativeness, Symposium on World Water Balance, IASH, Publ. 92, 1 (1971) 173-186.
20 J. Slanina, F.G. Römer, W.A.H. Asman, Investigation of the source region for the acid deposition in the Netherlands. In: Acid deposition. Ed. S. Beilke & A.J. Elshout, E.L. Shout, Publ. CEC (1982) 131-137.
21 D. Fowler, J.N. Cape, Dry deposition of SO_2 onto a Scots pine forest. In: Precipitation scavenging, dry deposition and resuspension. Ed. H.R. Pruppacher, R.G. Samonin & W.G.N. Slinn, New York: Elsevier (1983) 763-774.

22 D.M. Whelpdale, Large-scale sulfur studies in Canada, Atmos. Environ., 12 (1978) 661-670.

23 J.A. Garland, Dry deposition of small particles to crass in field conditions. In: Precipitation scavenging, dry deposition and resuspension. Ed. H.R. Pruppacher, R.G. Semonin & W.G.N. Slinn, New York: Elsevier (1983) 849-858.

24 A.C. Chamberlain, Aspects of the deposition of radioactive and other gases and particles, Int. J. Air Pollut., 3 (1960) 63-88.

25 B.B. Hicks, M.L. Wesely, R.L. Coulter, R.L. Hart, J.L. Durham, R.E. Speer, D.H. Stedman, An experimental study of sulphur deposition to grassland. In: Precipitation scavenging, dry deposition and resuspension. Ed. H.R. Pruppacher, R.G. Semonin & W.G.N. Slinn, New York: Elsevier (1983) 933-942.

26 O.R. Owen, W.R. Thompson, J. Fluid Mech., 15 (1963) 321-334.

27 D. Baldochi, B. Hicks, C. Palela, Atmos. Environ. in press

28 J.E. Hallgren, S. Linder, A. Richter, E. Troeng, L. Granat, Uptake of SO_2 in shoots of Scots pine: field measurements of new fluxes of sulphur in relation to stomatal conductance, Pl. Cell Environ., 5 (1982) 75-83.

29 J.L. Monteith, Principles of environmental physics, Edward Arnold (1973).

30 B.J. Huebert, Measurements of the dry deposition flux of nitric acis vapour to grasslands and forests. In: Precipitation scavenging, dry deposition and resuspension. Ed. H.R. Pruppacher, R.G. Semonin & W.G.N. Slinn, New York: Elsevier (1983) 785-794.

31 P. Grennfelt, C. Bengston, L. Skarby, Dry deposition of nitrogen dioxide to Scots pine needles. In: Precipitation scavenging, dry deposition and resuspension. Ed. H.R. Pruppacher, R.G. Semonin & W.G.N. Slinn, New York: Elsevier (1983) 753-762.

32 Duyzer, personal communication (1986).

33 M.L. Wesely, J.A. Eastman, D.H. Stedman, E.D. Yelvac, Eddy correlation measurement of N_O flux to vegetation and comparison to O_3 flux, Atmos. Environ., 16 (1982) 815-820.

34 C. Johansson, L. Granat, Emission of nitric oxide from arable land, Tellus 36B (1984) 25-37.

35 I.E. Galbally, C.R. Roy, Loss of fixed nitrogen from soils by nitric oxide exhalation, Nature (London), 275 (1978) 734-735.

36 J.R. Freney, J.R. Simpson, Gaseous loss of nitrogen from plants: soil systems, Dr. W. Junk, Publ., The Hague (1983).

37 A.H. Rutter, Hydrological cycle in vegetation. In: Vegetation and the atmosphere. Ed. J.L. Monteith (1975).

38 O. Kerfoot, Mist precipitation on vegetation, Forestry Ahs., Vol.29 (1968) 8-20.

39 G.H. Tomlinson, P.J.P. Brouzes, R.A.N. McLean, J. Kadlececk, Role of clouds in atmospheric transport of mercury and other pollutants. In: Ecological impact of acid precipitation. Ed. D. Drabløs & A. Tollan, Oslo, SNSF (1980) 134-137.

40 G.J. Dollard, M.H. Unsworth, M.J. Harvey, Pollutant transfer in upland regions by occult precipitation, Nature 302 (1983) 241-243.

41 H.G. Millar, J.D. Millar, Collection and retension of atmospheric pollutants by vegetation. In: Ecological impact of acid precipitation. Ed. D. Drabløs & A. Tollan, Oslo, SNSF (1980) 33-40.

Regional and Long-range Transport of Air Pollution,
Lectures of a course held at the Joint Research Centre, Ispra, Italy,
15–19 September 1986, S. Sandroni (Ed.), pp. 127–149
© Elsevier Science Publishers B.V., Amsterdam — Printed in The Netherlands

CONVENTIONAL AND REMOTE TECHNIQUES IN METEOROLOGY

P. THOMAS

.1 INTRODUCTION

In most models simulating the transport and diffusion of airborne substances a series of difference equations have to be solved at the points of a
3-dimensional grid. To start the simulation one must know
- the initial conditions at the grid points and
- the temporal development at the boundary of variables such as the
    - components of wind speed,
    - variances of these components,
    - temperature,
    - concentrations of airborne material.

The result of the simulation is a time series of some of these variables
at the grid points. To check the performance of the model the variables must
be measured continuously in terms of space and time for specific scenarios.

These variables are measured mostly by the following techniques:
- continuously in time at ground level or at a tower using conventional instruments,
- discontinuously above ground level using a tethered or a free-floating
  balloon or by radiosounding,
- continuously in time and (sometimes) in space remotely by the scattering of
  electromagnetic and acoustic waves in the atmosphere. The remote sounders are
- RADAR (radiation detection and ranging),
- LIDAR or LASER RADAR (light detection and ranging),
- SODAR (sonic detection and ranging),
- RASS (radio acoustic sounding system).

Some of the techniques and instruments listed above will be presented and
described in this paper. It is a selection of techniques that are used or will
be used in the near future at the Kernforschungszentrum Karlsruhe (KfK, Karlsruhe Nuclear Research Center). The feasibility of these techniques has been
studied extensively at KfK.

.2 CONVENTIONAL INSTRUMENTS

At KfK a Meteorological Information System (MIS) is operated (refs. 1,2).
It includes
- a 200 m high meteorological tower equipped with 37 instruments (Fig. .2.1),
- 9 ground based instruments, and
- a data acquisition and processing system comprising 2 PDP-11 computers.

.2.1 Meteorological data

The meteorological data measured by the system are the
- wind velocity at 10 levels; it is measured by 2 cup anemometers at each
  level to eliminate the influence of the tower;
- wind direction at 6 levels, measured by 1-dimensional vanes;
- wind vector at 3 levels, measured by 3-dimensional vector vanes (Fig. .2.2),
  these vanes also furnish the standard deviations σ_ϕ and σ_θ of the vertical
  and horizontal wind directions;
- temperature at 7 levels;
- dew point at 4 levels.

Near ground measurements of
- precipitation,
- radiation data, and
- atmospheric pressure
are performed.

.2.2 Data acquisition

The meteorological instruments are scanned at a
- 1-s-rhythm (vector vanes), and
- 4-s-rhythm (other instruments)
by a PDP-11/34 computer. The computer has a core memory of 256 kB and 2 magnetic disks of 10.4 MB each. A special bootstrap initializes the computer and
starts the data acquisition automatically after the power has been switched on
either manually or after breakdown of the line. Synchronization is performed
by a radio clock that communicates the date and time to the computer. The
system performs the following tasks:
- scanning the meteorological instruments,
- measuring and digitizing the electrical signals of the instruments,
- calculating physicsal values,
- checking if these values are within meaningful limits,
- compression to 10-min mean values,

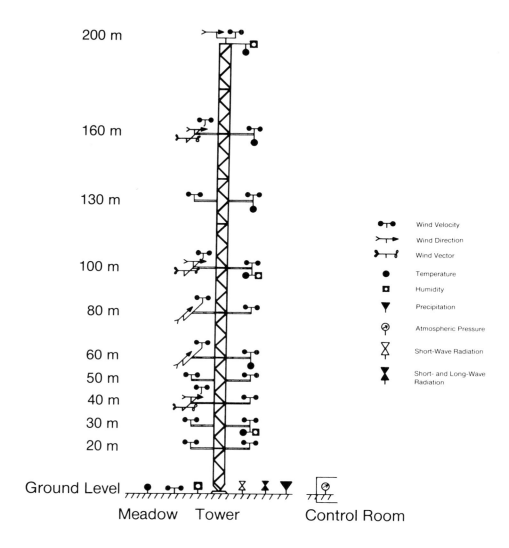

Fig. .2.1. Instrumentation of the Meteorological Informationsystem

Fig. .2.2. Vector vane.

- calculating meteorological values, which are not measured directly (e. g.
  standard deviations σ_θ and σ_ϕ of the horizontal and vertical wind direction,
  temperature gradient, radiation balance),
- attributing the date and time to the mean values,
- transferring the mean values to a PDP-11/44 computer.

.2.3 Data evaluation

   The PDP-11/44 computer has a core memory of 256 kB and the following
periphery:
- 4 magnetic disks of 10.4 MB each. On two of these disks the meteorological
  data of the last 2 months are stored for direct access. The data are stored
  on 2 disks for data security;
- 1 magnetic tape system for off-line data evaluation;
- 2 printing terminals;
- 6 remote colored displays with keybords. These displays are installed at
  those services of KfK, which are responsible for environmental monitoring,
  safeguards, public relations, etc.

- modem for remote communication with the MIS via a telephone line.

The meteorological data stored on the magnetic disks can be visualized on displays by
- tables,
- vertical profiles (Fig. .2.3), and
- time series (Fig. .2.4).

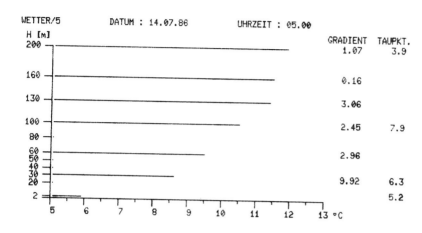

Fig. .2.3. Vertical profile of the temperature.

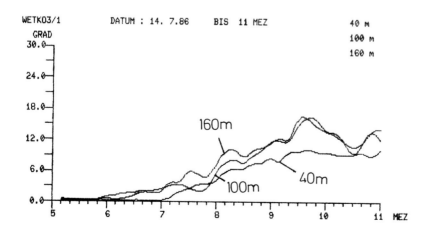

Fig. .2.4. Time series of the standard deviation σ_ϕ of the vertical wind direction.

.3 RADIOSOUNDING AND OMEGA NAVIGATION

The Vaisala company offers the so-called DigiCora System which is a combination of navigation and radiosounding.

A world-wide network of eight Omega stations is operated for marine navigation purposes. The Omega stations emit pulses of radiowaves at a frequency of 13.6 kHz in a rhythm of 10 s always in the same sequence. According to Fig. .3.1 a sonde carried by a balloon

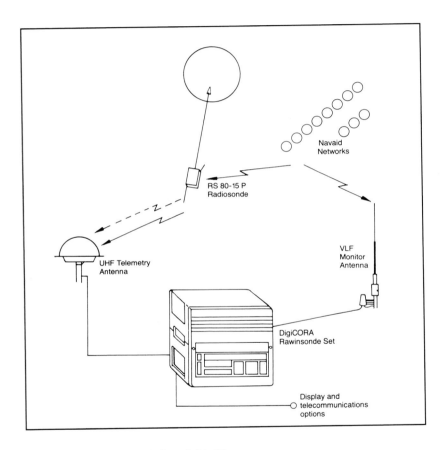

Fig. .3.1. Working principle of DigiCora.

- receives the Omega pulses,
- emits continuously radiowaves at a frequency of 400 MHz,
- modulates these radiowaves with the Omega pulses,
- measures
  - air pressure,

- temperature,

- relative humidity, and

- modulates the 400 MHz radiowaves with these data.

A ground based DigiCora station receives the 400 MHz radiowaves and determines the phases of the Omega signals from the different Omega stations against the signal of a distinct Omega station. From the well known geometry of the Omega stations and the measured phases (time lag or distance) the position of the balloon is determined in degrees of longitude and latitude or relative to the ground based DigiCora station (distance and azimuth).

This is the geometric principle of the method: The difference between the two distances from the two focuses of a hyperbola to every point on the hyperbola is constant. If the DigiCora station receives the signals of only two Omega stations via the sonde, the DigiCora station determines a hyperbola on which the sonde is situated somewhere. The two Omega stations are the focus of this hyperbola. To determine the position of the sonde on this hyperbola the signals of at least a third Omega station are needed to construct a second hyperbola. The intersection of the two hyperbolas defines the position of the sonde.

Normally the signals from all Omega stations are received continuously. Based on the signal quality (signal to noise ratio, variances of the phases) the best combination of the Omega stations is selected and their signals are weighted.

The height of the sonde above ground is calculated from the air pressure. By subsequent measurements the trajectory of the sonde is calculated and the wind speed and direction are determined.

.4 TETROONS TRACKED BY A RADAR

At KfK tetroons (tetrahedral balloons, constant level balloons, Fig. .4.1) are tracked by a RADAR to measure plume trajectories in a Lagrangian manner (ref. 3). The tetroon is manufactured from polyester dyed red film with a skin thicknes of 51 µm. Its mass and volume are 470 g and 1 m^3, respectively. It is inflated with helium to approximately 10 hPa overpressure and ballasted for the desired altitude. Once a tetroon has been launched it will rise until its buoyancy equilibrium is reached. Then its mean flight level will be on a surface of constant air density.

To facilitate the tracking of the tetroon by a RADAR it is equipped with an octrahedral corner reflector manufactured from aluminium bars and aluminium coated plastic foil.

To get rid of the ground clutter a transponder (transmitter-responder) can be attached to a tetroon instead of the corner reflector. The technical details of a transponder are listed in Tab. .4.1. The transponder receives

134

Fig. .4.1. Launching a tetroon.

TABLE .4.1
Technical details of radar and transponder.

	RADAR	Transponder
Diameter of antenna	1.2 m	
Dimensions	5.85x2.08x3.05 m^3	15x8x8 cm^3
Weight	6.3 t	450 g
Gain of antenna	38 dB	5 dB
Polarization	linear, vertical	
Frequency	9375 MHz	9212 MHz
Peak power	60 kW	180 mW
Pulse repet. frequency	800 Hz	800 Hz
Pulse length	0.25/1.0 μs	1.0 μs
Range	> 50 km	
Operating time	> 5 h	

the RADAR pulses and responds at a slightly different frequency. To track the transponder the receiving chain of the RADAR is tuned to the frequency of the transponder. Now the RADAR is able to discriminate the ground clutter which has the same frequency as the RADAR pulses emitted.

A mobile wind finding RADAR is used to track the tetroons (Tab. .4.1, Fig. .4.2). After the tetroon-reflector- or tetroon-transponder-unit has been released it is tracked manually by the aid of a video camera, which is mounted on the antenna of the RADAR, and a television monitor at the control panel of the RADAR. As soon as the echo of the reflector or transponder is received on the R/A-scope (range/amplitude) at the control panel, the RADAR is switched to the automatic tracking mode.

Fig. .4.2. Mobile RADAR for the tracking of tetroons.

During automatic tracking the following data are printed and punched on paper tape every 10 s:
- time after release of the tetroon,
- slant range between RADAR and tetroon,

136

- elevation angle,
- azimuth.

After a smoothing procedure applied off-line to these data while taking into account the earth's curvature and the refraction of the RADAR beam, trajectories of the tetroons are calculated as
- a projection of the flight path onto a topographic map (Fig. .4.3), or
- a height profile of the tetroon as a function of travel time or travel distance (Fig. .4.4). In Fig. 4.4 the profile of the terrain passed by the tetroon is indicated as a cross-hatched area.

Fig. .4.3. Projection of tetroon trajectories measured on March 29, 1984

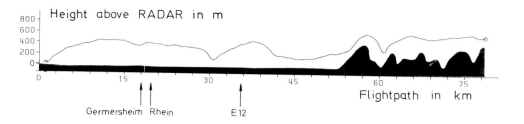

Fig. .4.4. Profile of a tetroon trajectory measured on March 29, 1984.

Besides trajectories and profiles
- vertical and horizontal velocities,
- horizontal directions,
- oscillation periods, and
- turbulence intensities
can be determined from the tetroon data.

.5 PRINCIPLES OF REMOTE SENSING TECHNIQUES
RADAR, LIDAR and SODAR emit into the atmosphere short pulses of
- electromagnetic waves (wavelength $\lambda \approx$ 8 mm - 6 m),
- light (visible and infrared, $\lambda \approx$ 0,5 µm - 11 µm),
- acoustic waves ($\lambda \approx$ 10 cm - 2 m)

as a well collimated beam. A small fraction of the transmitted energy is
backscattered by scatterers like
- atoms and molecules,
- aerosols,
- hydrometers,
- insects,
- chaff, aluminized needles of mylar (∅ 1 µm, length $\lambda/2$, deposition velocity
  0.3 m/s) which is released artificially as a tracer for a RADAR from cars,
  aircrafts or stacks,
- inhomogeneities of the refractive index of the air generated by turbulent
  fluctuations and by differences in humidity and temperature of the air
  (clear air turbulence).

Backscattering is strong if scatterers are encountered at scales of half
the wavelength of the transmitted waves (Bragg resonance).

.5.1 Doppler effect

Due to the Doppler effect the frequency of the backscattered signal, as measured by the remote sounder, is shifted, if the scatterer is moving. The frequency shift is

$$\Delta f = \pm \, 2 \, f \, \frac{v}{c} \tag{5.1}$$

f: frequency of the transmitted wave,
c: velocity of the transmitted wave (velocity of sound or light),
v: radial component of the velocity of the scatterer (line-of-sight velocity).

The frequency of the backscattered signal is increased (+ sign) or descreased (- sign), if the scatterer moves towards or away from the remote sounder. From the frequency shift Δf the radial component v of the velocity of the scatterer can be determined which corresponds to the velocity of the air carrying the scatterer.

.5.2 Range gates

The running time t of the transmitted and backscattered signal determines the distance d between the sounder and the scatterer via the velocity c of the transmitted wave:

$$d = c \cdot \frac{t}{2} \tag{5.2}$$

According to Fig. .5.1, this distance is defined along the beam by so-called range gates. In Fig. .5.1 the velocity c of the transmitted wave corresponds to the slope. The range gates are defined by the time intervals of reception and emission.

Figure .5.1 demonstrates clearly that by remote sensing no "point measurements" are performed but "volume measurements". The measured meteorological variable is a value averaged within a volume which is defined by
- the time intervals of emission and reception (range gate), and
- the solid angles of the transmitter (width of transmitted beam) and receiver.

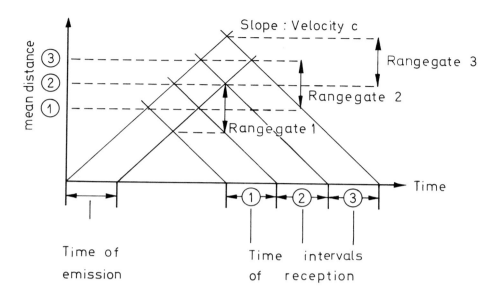

Fig. .5.1. Running time and range gates.

.5.3 Doppler moments

The backscattered signals are received and processed in such a manner that the Doppler spectrum (frequency spectrum) is obtained for each range gate (Fig. .5.2). This can be performed by a Fast Fourier Transform, e. g. as for a single emission the signal to noise ratio normally is small in most sounders, the Doppler spectra of many emissions are summed up coherently. Therefore, the meteorological parameters measured by remote sensing techniques are mean values corresponding to time intervals of about 10 s up to several minutes. From the Doppler spectrum the noise level is subtracted and the first three Doppler moments are determined.

.5.3.1 Zeroth moment

The zeroth moment is a measure of the backscattered intensity or reflectivity. The following information can be obtained:
- vertical beam:
  - height of inversion,
  - height of tropopause,
  - mixing depth,

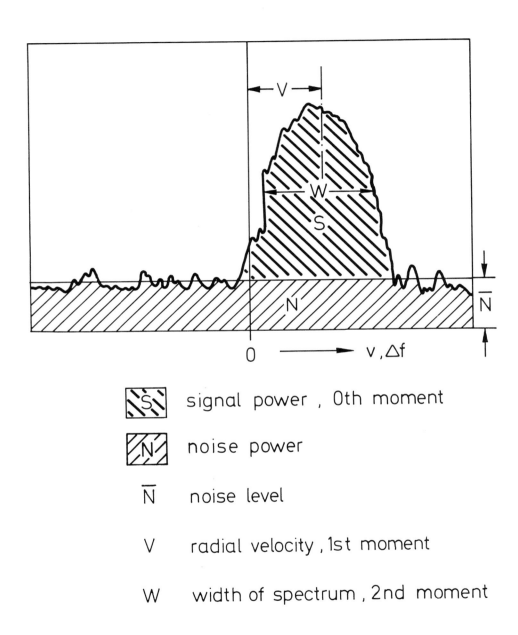

Fig. .5.2. Doppler spectrum

- scanning beam:
  - cloud dimensions,
  - surface integrated precipitation,
- profile along the beam:
  - concentration of
    - droplets,
    - aerosols,
    - molecules (humidity, pollutants),
  - temperature (from the air density via the ideal gas equation).

.5.3.2 First moment

The first moment furnishes the component of the wind velocity parallel to the beam. As the wind vector is 3-dimensional, 3 antennas with different orientations are used which emit the same frequency sequentially or emit slightly different frequencies simultaneously, or a scanning beam is used.

.5.3.3 Second moment

The second moment is the width of the Doppler spectrum. From the width the variance of the component of the wind velocity parallel to the beam can be determined. This is a measure of the turbulence intensity of the air. With a LIDAR, based on Reyleigh scattering (elastic scattering by molecules), the temperature of the air can be measured.

.6. MONOSTATIC DOPPLER-SODAR

At KfK two different 3-dimensional Doppler-SODARs have been operated (refs. 4-6):
- DS 108 manufactured by Rosenhagen, operated in 1982,
- AO manufactured by REMTECH, operated since 1983.

Figure .6.1 and Tab. .6.1 show the sounder AO, its technical details and its typical operating parameters, respectively. 3 antennas are mounted on a trailer. The base of the antenna has a parabolic shape. In the focus of the parabola a compression chamber is installed. A so-called transceiver switches the compression driver in an alternating sequence to
- emission (the chamber is a loudspeaker) and then to
- reception (the chamber is a microphone).

After a pulse repetition time of typically about 4 s the compression chamber of the next antenna is emitting and receiving, and so on. This method of operation is called monostatic contrary to bistatic. In the latter case one antenna is emitting and another or several other antennas are receiving.

Fig. .6.1. Acoustic sounder operated at KfK.

A time series of the horizontal wind speed corresponding to 100 m height is plotted in Fig. .6.2. The 10-min mean values are compared to the wind speed measured at the tower at the same level. The SODAR was located only about 200 m north of the tower. The comparison shows that the data furnished by both instruments agree well. The wind speed measured by the SODAR is somewhat smaller on the average. Only in some situations precipitation leads to a dropout of information.

As demonstrated by Fig. .6.3 the availability of measured data decreases with increasing height. This availability is a function of
- meteorological conditions (inhomogeneities of the refractive index of the air), and of
- site conditions (ambient noise, fixed echoes).

TABLE .6.1

Technical characteristics and operating parameters of the Doppler SODAR AO. The operating parameters can be changed by key board commands.

Frequency	1600 Hz
Wavelength	20 cm
Peak power	300 W_{el}, 60 W_{ac}
Pulse length	100 ms
Beam width	5° at 3 dB
Inclination of 2 antennas	18°
Lowest mean range	40 m
Range gate	20 m
Number of gates	20

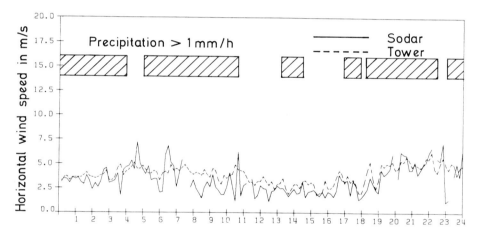

Fig. .6.2. Time series of horizontal wind speed measured at 100 m height on May 24, 1983.

.7 DOPPLER-RADAR

An important receiver component of each Doppler RADAR is the STALO (stabilized local oscillator), which oscillates at the transmitted frequency f. A portion of its signal is mixed (summed up) with the echo signal, whose strength is weak relative to the STALOs. The result of this mixing is a modulation of the carrier (transmitted) frequency with the frequency shift Δf

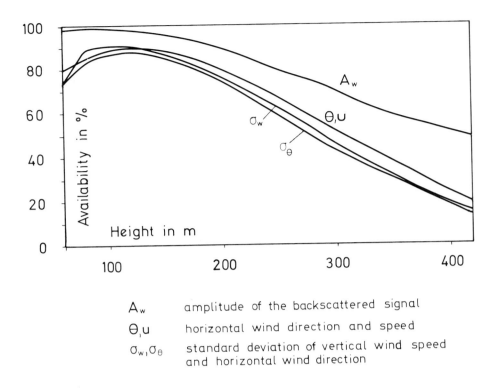

A_w	amplitude of the backscattered signal
Θ, u	horizontal wind direction and speed
σ_w, σ_Θ	standard deviation of vertical wind speed and horizontal wind direction

Fig. .6.3. Availability of SODAR data.

of the echo signal, which is produced by the Doppler effect. With this so-called heterodyne technique only the modulation envelope, whose frequency is Δf, has to be analyzed ($\Delta f \ll f$).

In the following paragraphs some Doppler-Radars will be briefly presented. Many papers and books on RADAR meteorolgy have been published, e. g. in refs. 7-9.

.7.1 The Norman RADAR

The RADAR is operated by the National Severe Storms Laboratory near Norman, Oklahoma, USA, to detect tornados. Its technical and operating parameters are summarized in Tab. .7.1.

TABLE .7.1

Technical characteristics and operating parameters of the Norman RADAR.

Frequency	2.9 GHz
Wavelength	10.5 cm
Peakpower	750 kW
Pulse length	1; 3 µs
Pulse repetition time	768; 2304 µs
Beam width	0.8° at 3 dB
Diameter of Cassgrain parabolic antennas	10 m
Scan rate	1 rev./min
Range gate	100 m, 200 m
Number of pulses for one profile of radial velocity	64

Typical color coded outputs of backscattered intensity, radial velocity and standard deviation of radial velocity on displays are in the
- RHI-(range height indicator) mode, a vertical cross section; the RADAR beam scans at different elevation angles and constant azimuth;
- PPI-(planar position indicator) mode, a horizontal cross section; the RADAR beam sweeps azimuthally at a constant elevation angle.

To determine the vertical profile of horizontal wind speed and direction the VAD-technique is applied (velocity azimuth display): The RADAR beam sweeps azimuthally a 360° revolution at constant elevation angle. The measured radial velocity is plotted versus the azimuth. This plot is fitted by a sine curve. From the sine curve the following data can be determined:
- amplitude: horizontal wind speed,
- vertical offset: vertical wind speed,
- azimuthal position of maximum: horizontal wind direction.

This procedure is performed for each range gate (height gate). The meteor-ological data are averages corresponding to a cylinder, whose radius increases with the height.

.7.2 SOUSY

The SOUSY-VHF (sounding system-very high frequency) RADAR is operated by the Max Planck Institut für Aeronomie at Katlenburg-Lindau, Federal Republic of Germany in the Harz mountains. Its technical and operating parameters are summarized in Tab. .7.2.

TABLE .7.2
Technical characteristics and operating parameters of the SOUSY-VHF-RADAR.

Frequency	53.5	MHz
Wavelength	5.6	m
Peak power	600	kW
Pulse length	0.8-100	µs
Duty cycle	4	%
Half power beam width	5°	

Lowest range	750	m
Range gate	> 120	m
Number of gates	1024	
Mean values	typically 5	min

SOUSY is a so-called MST (mesosphere, stratosphere, troposphere) RADAR. It consists of a phased array of 176 four element Yagi antennas covering a surface of 2440 m^2. The beam is swung to any position within 30° off the zenith by computer controlled electrical phase shifting.

.7.3 The Colorada Wind Profiling Network
    A network of six wind profiling RADARs has been established by WPL, NOAA (Wave Propogation Laboratory, National Oceanic and Atmospheric Administration) near Boulder, USA. The network is composed of:
- 4 VHF-RADARs (50 MHz): These RADARs transmit simultaneously in two
   directions (15° off-zenith to north and east) by a fixed phased array of
   colinear-coaxial dipoles covering an area of 50 m x 50 m.
- 1 UHF- (ultra high frequency, 405 MHz) RADAR: The technical and operating
   parameters are summarized in Tab. .7.3. It is a phased array of 100 five-
   element Yagi antennas covering an area of 9 m x 9 m. The RADAR has three
   electronically selectable bandwidths (modes of operation): 1,3 and 9 µs
   pules are transmitted. One transmitter is sequentially switched to two
   different pointing directions (15° off-zenith to north and east).
- 1 UHF-RADAR (915 MHz): One transmitter and an offset paraboloidal antenna
   with three offset feeds produce the zenith and two oblique pointing
   directions sequentially. The operating parameters are similar to those
   compiled in Tab. .7.3.

TABLE .7.3

Technical and operating parameters ot the Colorada UHF-Profiler.

Frequency				405	MHz
Wavelength				74	cm
Peak power				30	kW
Two way beam width				4.3°	
Resolution of radial velocity				0.5	m/s
Pulse length	1	3	9		μs
Pulse repetition time	100	150	300		μs
Lowest range	400	2400	4000		m
Range gate	100	290	870		m
Number of gates	24	24	14		

In the normal sequence of operation, hourly wind profiles are measured using 12 profiles with each pulse width. The profilers alternate between the different modes of operation and finish the data acquisition cycle in about 45 min. The last 15 min are idle to allow the Central Profiler Computer at WPL to poll all the outlying sites and to allow for changes of parameters or to obtain diagnostic outputs. Communication is via telephone transmission.

.8 LIDAR

The WPL, NOAA also operates a LIDAR for sensing atmospheric winds (ref. 10). Some technical and operating parameters of the LIDAR are summarized in Tab. .8.1. The source of this coherent Doppler LIDAR is a CO_2-LASER producing infrared (eye-safe) light pulses that are scattered by aerosols. The scattered and received energy is mixed on an infrared detector with energy from a reference source (another infrared LASER) to produce a modulation with the Doppler frequency as described in Ch. .7. A hard-wired data processor digitizes the Doppler signal of a 200 μs long burst at a rate of 15 MHz and determines the first three Doppler moments. The 200 μs burst, containing the energy scattered from one transmitted pulse, is divided into 100 gates (each gate is 2 μs long).

The LIDAR generally can measure the radial component of wind anywhere within horizontal ranges of 20 km and vertical altitudes of 8 km. As the angular divergence of a LASER beam is extremely low (50 microradians), the probed volume at any range is a long cylinder, about 0.5 m in diameter by 150 m in length. Accuracy of radial velocity generally is $\pm$ 0.5 m/s within the boundary layer.

In order to map the detailed structure of the windfield, the LASER is operated in the RHI-, PPI-, and VAD-mode as described in Ch. .7.2. In a special raster scan a computer guides each LASER pulse through a small interval of regularly spaced elevation and azimuth angles. When the monitor is formatted to display color coded radial velocities in azimuth versus elevation angle, a picture of the flow across arbitrary cross sections is created. One cross section corresponds to one range gate. The colour of each pixel on the monitor indicates the radial wind component sensed for a single transmitted pulse.

TABLE .8.1.
Technical and operating parameters of the WPL LIDAR.

Wavelength	10.6	μm
Pulse length	2	μs
Pulse energy	< 0.15	J
Pulse repetition time	100	ms
Minimum range	> 1000	m
Range gate	> 150	m
Number of gates	100	

.9 RASS

With RASS the vertical profile of the temperature of the air can be measured up to heights of about 1000 m above ground level (ref. 11). RASS consists of
- a SODAR beaming short pulses of acoustic sine-waves vertically toward the zenith, and
- a RADAR emitting and receiving continuously electromagnetic waves in a bistatic manner (two antennas for emission and reception, respectively).

The electromagnetic waves are backscattered by the inhomogeneities of the refractive index of the air which is intensively compressed by the acoustic pulse traveling upward at the velocity of sound. Due to this velocity the frequency of the backscattered electromgnetic waves are shifted by the Doppler effect. From this frequency shift the velocity of sound is determined which is proportional at every height to the square root of the local absolute temperature. The height at which the electromagnetic waves are scattered and at which the temperature is determined is a function of the running time of the acoustic pulse.

The faint echo of the backscattered electromagnetic waves is maximized by choosing an acoustic wave in Bragg resonance with the electromagnetic wave. Therefore, the electromagnetic wavelength should be twice as long as the acoustic wavelength, e. g.

- RADAR: λ = 42 cm, 700 MHz,
- SODAR: λ = 21 cm, 1600 MHz.

Normally, a 3-dimensional SODAR is used together with the RADAR to measure, besides the temperature profile, that of wind data as described in Chap. 6.

REFERENCES

1 R. von Holleuffer-Kypke, W. Hübschmann, F. Süss, P. Thomas, MIS-Meteorologisches Informationssystem des Kernforschungszentrums, Atomkernenergie Kerntechnik 44(4) (1984) 300-304.
2 P. Thomas, Meteorological Information System - A Tool in Accident Consequence Assessment, Workshop on Real-time Computing of the Environmental Consequences of an Accidental Release to Atmosphere from a Nuclear Installation, Luxembourg, 1985.
3 S. Vogt, P. Thomas, Untersuchung mesoskaliger Luftströmungen in der Umgebung des Kernforschungszentrums Karlsruhe mit radarverfolgten Tetroons, KfK 3565 (1984).
4 P. Thomas, R. von Holleuffer-Kypke, W. Hübschmann, Doppler Acoustic Sounding Performance Test, 2nd Int. Symp. on Acoustic Remote Sensing of the Atmosphere and Oceans, Rome, 1983.
5 R. von Holleuffer-Kypke, W. Hübschmann, P. Thomas, Testbericht über das monostatische Doppler SODAR B, KfK 3928 (1985).
6 R. von Holleuffer-Kypke, W. Hübschmann, P. Thomas, Testbericht über das monostatische Doppler SODAR R, KfK 3929 (1985).
7 R.J. Doviak, D.S. Zrnic, Doppler Radar and Weather Observations, Academic Press Inc., Orlando, 1984.
8 M.F. Larsen, J. Röttger, VHF and UHF Doppler RADARs as Tools for Synoptic Research, Bull. Am. Meteor. Soc. 63(9) (1982) 996-1008.
9 R.G. Strauch et al., The Colorado Wind-Profiling Network, Journal Atm. and Oceanic Techn. 1(1) (1984) 37-48.
10 M. J. Post, W. D. Neff, Doppler Lidar Measurements of Winds in a Narrow Mountain Valley, Bull. Am. Met. Soc. 67(3) (1986) 274-281.
11 G. Bonino, P. Trivero, Automatic Tuning of Bragg Condition in a Radio-Acoustic System for PBL Temperature Profile Measurement, Atm. Env. 19(6) (1985) 973-978.

Regional and Long-range Transport of Air Pollution,
Lectures of a course held at the Joint Research Centre, Ispra, Italy,
15–19 September 1986, S. Sandroni (Ed.), pp. 151–193
© Elsevier Science Publishers B.V., Amsterdam — Printed in The Netherlands

CONVENTIONAL AND REMOTE MONITORING TECHNIQUES*

STRATEGY AND PERFORMANCE

R.H. VAREY

1.   INTRODUCTION

There are several approaches to the study of regional and long range
transport of air pollution. One is to examine long term trends and
establish a data base for annual average values of emissions, air
concentrations and deposition of gaseous species and material in
precipitation. Such studies are generally accompanied by a compilation of
meteorological data which is used to determine the statistics of factors
such as precipitation and air trajectories. Another approach is to
concentrate instrumentation and analysis on episodes of special
meteorological interest or on the transport of air masses along particular
trajectories.

Such investigations cover distances ranging from a few tens of kilometres
up to a thousand or more. The range of measurements required and the
suitability of different types of instrumentation vary with the type of
study being undertaken. This chapter reviews the species of most
importance in studies of acid deposition from the atmosphere and the
availability of instruments to measure them. Both gas and liquid phase
processes are reviewed briefly and simple examples are given to illustrate
the contribution of hydrocarbons and oxides of nitrogen to the chemistry
of photo-oxidants . The present capability of remote sensing techniques,
correlation spectroscopy and lidar, are outlined. The application of
different measurement techniques and strategies are illustrated with
descriptions of surveys of NO_2, SO_2 and O_3 in London and with an account
of recent experiments on the oxidation of SO_2 in cloud.

2. CHEMICAL SPECIES TYPICAL CONCENTRATIONS AND METHODS OF MEASUREMENT

2.1 Gases

2.1.1. Primary pollutants

The main gaseous species of interest are listed in Table 1. First there are the primary pollutants, SO_2 and NO_x. For the purposes of regional studies concentrations of SO_2 in plumes from industrial sources reach concentrations sometimes exceeding 100 ppb and close to large sources can be considerably higher. Over greater distances where plumes are more

TABLE 1
Species and Concentrations

SPECIES	CONCENTRATION (ppb)		
	Background	Rural	Industrial Urban Plumes
SO_2	0.1 - 1	2 - 20	20 - 300
NO	0.1 - 1	1 - 20	20 - 200
NO_2	0.1 - 1	1 - 20	20 - 100
NH_3	0.1 - 5	5 - 15	10 - 100
CO	100	200	400 - 1000
CH_4	1000-1500	1000-1500	1000-1500
PAN	<0.1	1	3 - 30
HCl	0.1 - 3	0.1 - 3	3 - 30
HNO_3	<0.1	0.1 - 1	3 - 10
O_3	20 - 50	50 - 100	10 - 200
OH	1×10^{-4}	1×10^{-4}	5×10^{-4}
H_2O_2	0.1 - 1	0.1 - 1	0.1 - 3
HO_2	5×10^{-3}	5×10^{-3}	5×10^{-2}
NO_3	<0.01	<0.05	0.1-0.3

dispersed concentrations are reduced and are generally in the range 10 –
50 ppb. It is also of interest to determine concentrations in background
air, that is in air which has traversed large distances over areas with no
anthropogenic sources, such as the oceans. Then concentrations can be at
the sub-ppb level. Fortunately there are well developed methods of
measuring SO_2 in this range of concentration. If longer term average
values are required there is the standard bubbler technique where air is
pumped through an H_2O_2 solution. SO_2 is rapidly oxidised to H_2SO_4 and is
determined by light absorption in a barium-thoranal complex. This
technique is usually employed to give 24 hour average concentrations and
has a lower detection limit of about 1 ppb. For reference to well
established methods the reader is referred to standard texts such as that
by Harrison and Perry, (ref. 1). For shorter term measurements flame
photometric or ultra-violet fluorescence instruments are available which
are capable of giving reliable results in the range from a few ppb up to
several hundred ppb. At the lower limit care has to be taken in checking
for base line drift which with some instruments is temperature and
humidity sensitive. For background measurements concentrations down to
0.2 ppb can be made with the Severn Sciences instrument which depends for
its operation on the disproportionation of mercury ion in solution to give
mercury vapour which is determined spectroscopically, the time constant
being 1-2 minutes. With this range of instrumentation, provided that care
is taken with calibration, SO_2 can be measured reliably at the
concentrations of interest with time averages ranging from a few minutes
to several days.

The situation is similar for NO_x. Concentrations of interest for regional
and long range transport studies are generally less than 100 ppb and
although higher concentrations do occur they are generally close to
sources of industrial or vehicular emissions. Long term measurement of
NO_2 can be made with diffusion tubes. These are passive devices
consisting of small tubes open at one end with stainless steel discs
coated with triethanolamine at the other, closed end. The tube is placed
vertically, open end downwards. NO_2 traverses the tube by molecular
diffusion, is absorbed on the discs and is subsequently measured
colorimetrically using the Greiss/Saltzmann method (ref. 2). Diffusion
tubes are inexpensive and can be deployed easily in the field. They are

not suitable for shorter term measurements such as hourly averages but are
adequate for measurements over a range from a few days to several weeks.
The most commonly used instrument for shorter term measurements is the
chemiluminescence analyser which depends for its operation on photons
emitted in the reaction of NO with O_3. NO can be detected directly and
total NO - NO_2 by first reducing NO_2 to NO by passing through a heated
stainless steel tube. Concentrations down to a few ppb can be measured
with a time constant of a minute or so.

CO arises from combustion and vehicular emissions. As shown in Table 1
concentrations range from about 100 ppb in background air to above 1000
ppb in urban plumes. Concentrations in this range can be measured using
infra-red absorption techniques to give an accuracy of 10 ppb or so with a
time constant of about a minute.

HCl is not as widespread in industrial emissions as SO_2 and NO_x. It
arises from the burning of coal with a chlorine content and from the
incineration of refuse with chlorine containing materials such as PVC. In
background air HCl is found at concentrations around a ppb or so
originating from sea salt aerosol. The presence of HCl is generally
related to specific sources and concentrations are very variable. Although
remaining in the gas phase up to high relative humidities (ref. 3) it is
highly soluble in water and is rapidly removed in cloud or rain. Unlike
SO_2 and NO_x there is no readily available method of measuring HCl at the
concentrations required, a few ppb, with a time constant of a minute or
so. At higher concentrations close to incinerator plant, or oddly enough
in the vicinity of space rocket launches, where concentrations reach the
ppm level, HCl has been measured with lidar (ref. 4) and with a
chemiluminescence method based on the oxidation of luminol (ref 5). At
lower concentrations in power station plumes some success has been
achieved with diffusion/denuder tubes (ref 6). The technique here is to
draw ambient air through a tube coated on the inside with potassium
fluoride. Laminar flow is maintained and HCl diffuses to the wall where
it is absorbed. Any particulate matter, in particular salt sea aerosol,
is carried through the tube in the laminar flow and does not reach the
wall.

The tube coating is subsequently dissolved in water and analysed by
selective ion electrode or ion chromatography for chloride. Recent
experiments using tubes 70 cm long, 2 cm diameter with a flow rate of one
litre/min indicate that an HCl concentration of 100 ppb can be measured
with a sampling time of a few minutes and a lower practicable limit of a
few ppb can be achieved with a sampling time of about an hour. This
technique may provide some useful information but for detailed
investigation of the contribution of emissions of HCl to regional and long
range transport studies a continuous monitor such as those available for
SO_2 and NO_x would be valuable.

H_2SO_4 does not occur in the atmosphere in the gas phase, any produced by
the reactions outlined in Section 2.1.2 below appearing as aerosols.
However, HNO_3 does appear in gaseous form in concentrations up to 10 ppb
or so. A filter method using NaCl impregnated teflon filters for 24 hours
sampling has been reported (ref. 7) but the most promising technique at
present is infra-red absorption spectroscopy. Hanst et al (ref. 8) have
reported a detection limit of 10 ppb and more recent measurements (ref. 9)
have shown the diurnal variation in HNO_3, Figure 1, with peak
concentrations reaching 40 ppb in urban smog and
a detection limit of about 5 ppb. Such methods are not however generally
available.

Ammonia plays an important role in the oxidation of SO_2 in cloud water,
see below. It arises from the bio-degradation of organic matter and is
found in concentrations ranging from about 0.1 - 5 ppb in background air
up to 100 ppb in industrial/urban plumes. The standard method of
measuring NH_3 in ambient air is based on the indophenol reaction, the
formation of a blue dye when NH_3 reacts in a phenol-hypochorite
solution. In principle this provides a detection of a few ppb in a
sampling time of 30 minutes but interferences from aerosols containing
ammonium compounds introduce some uncertainty at low concentrations and in
industrial and urban plumes. More recent developments include the use of
diffusion/denuder tubes (refs. 10 and 11) and long path infra-red
spectroscopy (ref. 9).

156

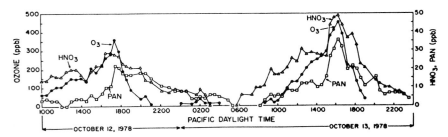

Fig. 1.   Time-concentration profiles of ozone and other gaseous pollutants measured by kilometre-pathlength FT-IR spectroscopy in photochemical smog at Claremont, CA, 12 and 13 October 1978 (Tuazon et al., 1981).

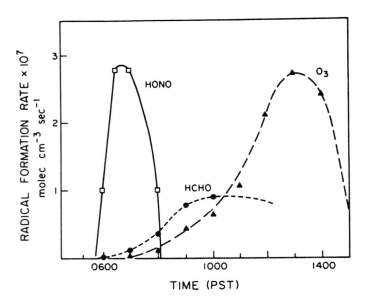

Fig. 2.   Calculated rates of radical formation initiated by photolysis of HONO, and O_3 as a function of time for typical Los Angeles atmospheric conditions. (Winer, 1986).

2.1.2 Oxidants

The formation of acid species from the primary pollutants SO_2 and NO_x is by oxidation, either in the gas or liquid phase. The primary source of oxidants is ozone. That is not to say that ozone is the only oxidant involved in the reactions. In some cases ozone reacts directly but in others it is the primary source of other oxidants involved in the reactions.

The main route for oxidation of SO_2 in the gas phase is the reaction with the OH radical

$$SO_2 + OH \longrightarrow HSO_3. \tag{1}$$

The second step, to produce H_2SO_4, is less certain but is believed to involve the HO_2 radical (ref. 12). A second route is the reaction with NO_3

$$SO_2 + NO_3 + H_2O \longrightarrow H_2SO_4 + NO_2. \tag{2}$$

In the liquid phase, cloud or rainwater, oxidation involves ozone and H_2O_2 with some contribution in certain circumstances from reactions with O_2 catalysed by the presence of particulate containing metallic species such as molybdenum, iron and manganese.

The oxidation of NO occurs principally in the gas phase in reactions involving ozone and the radicals OH and NO_3

$$NO + O_3 \longrightarrow NO_2 + O_2$$
$$NO_2 + OH \longrightarrow HNO_3. \tag{3}$$
$$NO_2 + NO_3 \longrightarrow N_2O_5$$
$$N_2O_5 + H_2O \longrightarrow 2HNO_3. \tag{4}$$

It can be seen therefore that the oxidants of interest in the formation of acid species are O_3, H_2O_2, OH, and NO_3. The occurance of these species in the atmosphere originates from O_3 formed in the stratosphere via the reaction

$$O_2 + h\nu \longrightarrow 2O \qquad \lambda < 242 \text{ nm}$$
$$O + O_2 \longrightarrow O_3. \tag{5}$$

158

Stratospheric ozone is transported to lower regions of the atmosphere where concentrations are generally about 20-30 ppb. Episodes of much higher concentrations occur, generally in summer anticyclonic conditions when ozone is generated photochemically from NO and hydrocarbons, see below. There are three routes for the production of OH

$$O_3 + h\nu \rightarrow O_2 + O^* \qquad \lambda < 319 \text{ nm} \qquad (6)$$
$$O^* + H_2O \rightarrow 2OH$$
$$HONO + h\nu \rightarrow OH + NO \qquad \lambda < 400 \text{ nm} \qquad (7)$$
$$HO_2 + NO \rightarrow OH + NO_2 \qquad (8)$$

Production from O_3 builds up during the day as ozone concentrations and photon flux increase. Production from nitrous acid HONO is concentrated in the morning as the overnight accumulation of HONO is photolysed and production from HO_2 continues through the day following the production of HO_2 as described below in Section 2.1.3. The relative importance of these reactions depends upon solar radiation and the relative concentrations of species in the atmosphere. An example is given in Figure 2, taken from Winer (ref. 13).

H_2O_2 arises from the reactions

$$CO + OH + O_2 \rightarrow CO_2 + HO_2$$
$$2HO_2 \rightarrow H_2O_2 + O_2, \qquad (9)$$

other sources of HO_2 radicals being reactions involving hydrocarbons, Section 2.1.3. NO_3 is produced by

$$NO_2 + O_3 \rightarrow NO_3 + O_2. \qquad (10)$$

The concentrations of these oxidants in the atmosphere at ground level are shown in Table 1. Instruments are available commercially to measure ozone at the required concentrations, depending on the chemiluminescent reaction with ethylene or absorption of light in the ultra-violet. Small radiosonde sensors, based on electrochemical detection of iodine released when ozone is bubbled through a potassium iodide solution, are available and are suitable for attachment to small balloons for the measurement of concentration profiles through the atmosphere.

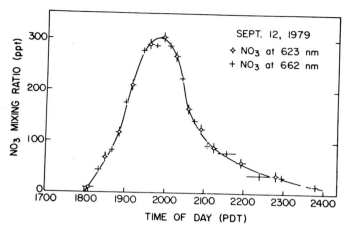

Fig. 3.   NO₃ time-concentration profile observed with DUVVS system at
Riverside, CA with 970 m pathlength (Platt et al., 1980).

TABLE 2
Hydrocarbons important in
photochemical episodes

ETHENE	C_2H_4
O-XYLENE	$C_6H_4(CH_3)_2$
N-BUTANE	$CH_3(CH_2)_2CH_3$
TOLUENE	$C_6H_5CH_3$
I-PENTANE	$CH_3CH_2CH(CH_3)_2$
PROPENE	$CH_3CH\ CH_2$
N-PENTANE	$CH_3(CH_3)_2CH_3$
M-XYLENE	$C_6H_4(CH_3)_2$
P-XYLENE	$C_6H_4(CH_3)_2$
ACETYLENE	C_2H_2

Concentrations of H_2O_2 are of interest in the range from about 0.1 ppb up to several ppb. This is difficult to measure since H_2O_2 is not very stable, particularly in the presence of metal surfaces, and is highly soluble in water. However, a chemiluminescence method based upon reaction with luminol in the presence of a copper catalyst has achieved some success (ref. 14) in which a detection limit of 0.4 ppb has been reported for a 15-20 minute air sample. More recent developments of the same technique (ref. 15) promise to improve both the sensitivity and time constant of this method. Measurements have also been made with Fourier transform infra-red spectroscopy (ref. 16) but the apparatus required for such measurements make its application difficult for field work.

Concentrations of NO_3 range from less than 0.01 ppb up to 0.1 ppb or so in polluted atmospheres at night. There is no readily available technique for measuring such concentrations. The only successful method so far is that of long path differential absorption spectroscopy (refs. 17 and 18). Such a system is claimed to have a detection limit in the region of 0.001 ppb (ref. 19), and in Los Angeles, using a path length of several kilometers, Platt and co-workers have measured diurnal variations in NO_3 concentration with maxima reaching 0.3 ppb, Figure 3. Some success has been achieved in the field with a mobile van mounted system (ref. 20) but clearly such systems are not in general use.

In spite of its importance in the gas phase chemistry of atmospheric pollution the OH radical occurs at only very low concentrations, typically around 10^{-4} ppb. In view of this low concentration and the high reactivity of the radical it is not surprising that in spite of considerable effort there has so far been little success in obtaining reliable measurements of its presence in the atmosphere. The most promising developments appear to be those using optical absorption in the ultra-violet (ref. 21) and induced fluorescence (ref. 22) but systematic measurements in polluted atmospheres have not yet been achieved.

2.1.3 Hydrocarbons

Hydrocarbons play an important part in the chemistry of atmospheric pollution. Their role in photochemical smog episodes of the Los Angeles type has been widely studied and there is growing interest in the occurrence of episodes of high photochemical activity in Europe. The most widely found hydrocarbon in the atmosphere is methane. It originates principally from the bio-degradation of organic matter and occurs at a concentration of 1.3-1.6 ppm. A very large and diverse range of other hydrocarbons - alkanes, alkenes, aromatics, aldehydes etc. - arises from industrial processes, vehicle exhausts and evaporation of petrochemical products. These are generally found at concentrations of a few ppb. In polluted atmospheres peroxyacetyl nitrate, PAN, is found at concentrations up to a few tens of ppb. In the study of the production of acid species in the atmosphere from industrial emissions an important feature of the hydrocarbon chemistry is the production of the oxidants described in the previous section.

Not only is hydrocarbon chemistry important it is also complicated. Models of atmospheric chemistry may contain several hundred reactions involving hydrocarbons. It may be useful here to illustrate schematically the process of photochemical oxidant production which occurs in mixtures of hydrocarbons and nitric oxide, particularly in urban conurbations where traffic emissions are a source of both components.

Take for example ethane C_2H_6. The first stage is reaction with OH

$$C_2H_6 + OH \longrightarrow C_2H_5 + H_2O \qquad (11)$$

This is followed by

$$C_2H_5 + O_2 \longrightarrow C_2H_5O_2$$
$$C_2H_5O_2 + NO \longrightarrow C_2H_5O + NO_2 \qquad (12)$$

Where NO has entered the reaction scheme and NO_2 has been produced. NO_2 is a potential source of ozone via the reaction,

$$NO_2 + h\nu \longrightarrow NO + O \qquad 295 < \lambda < 430 \text{ nm}$$
$$O + O_2 \longrightarrow O_3 \qquad (13)$$

Thus the hydrocarbons together with NO provide a source of ozone in polluted air, in fact reactions such as (11-13) above are the only significant non-stratospheric source of ozone. After (13) the reaction proceeds further

$$C_2H_5O + O_2 \longrightarrow CH_3CHO + HO_2$$
$$CH_3CHO + h\nu \longrightarrow CH_3 + CHO$$
$$CHO + O_2 \longrightarrow CO + HO_2 \tag{14}$$

Here photons have entered the scheme and two HO_2 radicals have been produced. The process is not finished since there remains CH_3 which can react further producing more NO_2 and HO_2.

$$CH_3 + O_2 \longrightarrow CH_3O_2$$
$$CH_3O_2 + NO \longrightarrow CH_3O + NO_2$$
$$CH_3O + O_2 \longrightarrow HCHO + HO_2$$
$$HCHO + h\nu \longrightarrow H + CHO$$
$$H + O_2 \longrightarrow HO_2$$
$$CHO + O_2 \longrightarrow CO + HO_2 \tag{15}$$

The example used here is C_2H_6 but this represents just one of a range of hydrocarbons which are available for a similar set of reactions which are usually generalised in the form

$$RH + OH \longrightarrow R + H_2O$$
$$R + O_2 \longrightarrow RO_2$$
$$RO_2 + NO \longrightarrow RO + NO_2$$
$$RO + O_2 \longrightarrow R'CHO + HO_2$$
$$R'CHO + h\nu \longrightarrow R' + CHO$$
$$CHO + O_2 \longrightarrow CO + HO_2$$
$$R' + O_2 \longrightarrow R'O_2 \tag{16}$$

Where R is a general hydrocarbon (in the case of ethane R = C_2H_5) and R = R' + HCH and R' can reprocess through the reaction scheme. This is a highly simplified schematic sketch of photo-oxidant reactions. What happens in practice is strongly dependent on the widely different reaction rates for different hydrocarbons, the relative concentrations of the various species including NO, competing reactions for OH radicals and of course, on solar radiation, the driving force for the photochemistry. Nevertheless equations (9-14) provide an example of the processes involved. A more detailed but concise description of the chemistry is given by Winer (ref. 13) which contains an extensive bibliography. An example of a more comprehensive description of the atmospheric chemistry is Calvert and Stockwell (ref. 23).

Hydrocarbons can be measured at the ppb level in the atmosphere directly using gas chromatography and mass spectroscopy and indirectly by taking air samples for subsequent analysis. The relative importance of the numerous species depends upon their concentration and reactivity. An example of those considered most important for photo-oxidant formation would include those shown in Table 2 in addition to which PAN should also be measured.

To sum up then, of the species listed in Table 1 the primary pollutants SO_2, NO, NO_2, CO, CH_4 and the oxidant O_3 can in the main be measured at the required concentrations with commercially available instruments. The exception being low concentrations of NO_x, less than a few ppb. Hydrocarbons can be measured by taking samples for laboratory analysis. Some success has been achieved in measuring HCl and NH_3 using diffusion/denuder tubes but more direct methods with greater sensitivity and shorter time constants would be valuable. The same is true for H_2O_2 in the gas phase since although success has been achieved with the luminol technique it is difficult to handle and is subject to interference from other species. Spectroscopic methods have been successful in measuring NO_3, HNO_3 and other species, and have provided fundamental information on atmospheric chemistry. In general however the instrumentation required for such measurements is not appropriate for normal field work. As for the radicals OH and HO_2, in spite of their central role in the chemistry of air pollution, and some progress with absorption and fluorescence techniques, they remain the most elusive of the species to be measured in the atmosphere.

2.2 Aerosol

The principal aerosols of interest are those containing sulphate, nitrate and ammonium. They arise as a result of gas phase reactions and from processes in cloud droplets which leave aerosol residues after evaporation. The size of the aerosol ranges up to a micron or so. Concentrations in background air are in the region of 0.1 ug m^{-3}, in rural air 1-2 ug m^{-3} and in urban/industrial plumes sulphate aerosol may be 50 ug m^{-3} or so. The usual method of measurement is to collect the aerosol by drawing air through filters with subsequent washing and analysis by ion chromatography. This method is adequate for measurements at the concentrations quoted above, the sampling times required for accurate results depending upon the rate at which the air is sampled and the concentration, with minimum times being generally several tens of minutes and regular survey sampling for twenty four hours.

2.3 Cloud and Rain Water

Liquid phase processes are not significant for the oxidation of NO_x, principally because of the low solubility of NO and NO_2. For SO_2 the liquid phase is important and reactions with ozone and H_2O_2 provide the two main routes for oxidation while in some circumstances reactions with oxygen catalysed by Mn and Fe have a role to play. The reaction with ozone is potentially dominant since ozone concentrations are much larger than those of H_2O_2 and are equal to if not larger than those of SO_2 in many situations. However the ozone reaction is strongly pH dependent, with the result that the reaction is quickly quenched as SO_2 is oxidised and the pH of the solution falls. It is here that the presence of ammonia in the atmosphere is important since in solution it inhibits the drop of pH and allows oxidation to proceed further without quenching. The overall reaction with H_2O_2 is not pH dependent and the reaction rate is fast. Here however, with concentrations of H_2O_2 not in general being higher than 1 ppb or so the amount of SO_2 which can be oxidised is limited. Currently both reactions are believed to play a part, that which is more important depending on prevailing conditions such as droplet size distribution, temperature, H_2O_2 concentration, SO_2 concentration, the presence of ammonia and the timescale of the cloud/precipitation event.

In cloud and rain water it is necessary to measure pH, the concentration of dissolved H_2O_2 and ions such as $SO_4^=$, NO_3^-, Cl^-, Na^+, Ca^{++}, K^+, Mg^+, NH_4^+. H_2O_2 is found in concentrations up to about 100 μM and the ions listed above at concentrations in the range 10-100 μM. Mn and Fe may be found at concentrations in the range 0.01-1 μM, there generally being more Fe than Mn and in very polluted air Fe may reach 10 μM. With care pH can be measured accurately and quickly. H_2O_2 can be measured using the luminol method, Section 2.2.3. This method has a good sensitivity and with care to avoid interference from other species is capable of measurements of concentrations below 0.1 μM (ref. 24). The ions listed above can be measured at concentrations down to a few μM using methods such as ion chromatography and plasma spectroscopy.

There remains the question of collecting the rain or cloud water. The collection of rain water has been the subject of much discussion and many recommendations, the main features of which concern differentiating between wet and dry deposition and avoiding contamination from extraneous sources such as vegetation and bird droppings. The most effective systems have mechanisms which open to expose the collector only when it is raining and are designed and located to minimise any effects of local terrain. Cloud water content can be up to 2 gm^{-3} for precipitating cumulus clouds and down to 0.1 gm^{-3} for other systems. In fog and mist it is even lower, 0.01 gm^{-3}. About 10 cc or so are required for analysis and cloud water collectors come in many shapes and sizes. They can be active rotating mechanisms or passive devices extracting water as cloud passes through them. A commonly used class of detector uses an array of strings or rods on which cloud droplets impact and then run down to a collector at the base. An example of this type of collector is shown in Figure 4. As with rain water care has to be taken to avoid contamination from dry deposition and from extraneous sources. With both rain and cloud water samples are generally obtained in the field and then taken back to the laboratory for analysis. However, it is possible to measure pH in real time during the rain/cloud event. Also the luminol method can be used for continuous measurement of H_2O_2. It is in any case important to determine H_2O_2 concentration soon after collection since it is not very stable in stored

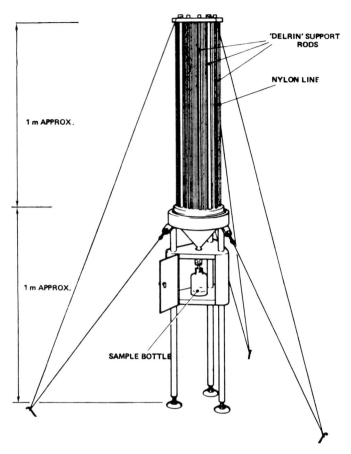

Fig. 4. Ground based cloud water collector, (after Falconer and Falconer, 1979).

samples, typically 50% can be lost in twelve hours. Some success has been achieved in continuous measurement of other species using selective ion electrodes (ref. 24).

3.   REMOTE SENSING

3.1 Introduction

The importance of remote sensing is principally in providing measurements not just at ground level but at heights through the first kilometre or so of the atmosphere. It has played an important role in studies of plume rise and dispersion close to point sources (refs. 26 and 27). Remote sensors mounted in vehicles or aircraft can be used to track plumes over considerable distances (refs. 28-30). An important use of remote sensing arises when ground level concentrations of species are dominated by meteorological factors or nearby local sources. The ability of remote sensing to make measurements unaffected by these factors is particularly valuable.

There are two types of remote sensor in common use, correlation spectrometers and lidar. Correlation spectroscopy has received wide application in the measurement of SO_2 and NO_2 and lidar can be used to measure both these species and ozone.

3.2 Correlation Spectroscopy

Correlation spectroscopy is described in detail by Millán in Chapter IX. Briefly, the principle of operation is similar to absorption spectroscopy where light from a standard source is passed through a volume of the gas being studied, the concentration of the gas being deduced from the attenuation suffered by the light in passing through the gas. Attenuation is measured at two adjacent wavelengths corresponding to weak and strong absorption in order to normalise the effects of extraneous losses in the optics and detection system. If the absorption coefficients of the gas are known at the two wavelengths then the ratio of the transmitted signals gives the number of molecules in the light path. Alternatively the system can be calibrated by introducing cells containing known concentrations of the gas of interest into the light path.

Correlation spectroscopy is similar except solar radiation is used instead
of a standard source. It is not direct solar radiation which is monitored
but scattered radiation from the bright sky background, so called
skylight, the spectrometer being generally arranged to collect radiation
from vertically above. This radiation is modified by pollutants in the
atmosphere and its spectrum reveals the total absorption produced by
molecules in the field of view of the instrument, that is the integrate
effect, or total column content of molecules in the atmosphere above the
instrument. Thus there is no discrimination between a narrow layer of
high concentration or a thicker layer of lower concentration. Nor does
the instrument give any indication of vertical distribution of the gas in
question. However, it does have the attribute of a remote sensor in that
it is sensitive to the presence of the gas being measured whether it is at
ground level adjacent to the instrument or high in the atmosphere, remote
from it. The units used in correlation spectroscopy are usually
micro-metres, μm, that is the equivalence of a concentration of one ppm
through a depth of one metre.

In practice the useful region of the spectrum is limited to the
ultraviolet and visible and the method is restricted to the measurement of
SO_2, in the region of 300 nm, and NO_2 at about 440 nm. The detection
limits and accuracy of the instrument vary with skylight radiance.
Clearly, it does not work at all at night and is less effective in the
early morning and late evening than in the middle of the day. Typically
for several hours around noon on a spring or autumn day the noise on a
carefully adjusted instrument is likely to be about 10 μm for SO_2 and 5 μm
for NO_2. This means that the instrument can for example detect changes of
around 5 ppb NO_2 distributed through a mixing layer 1000 m deep or 50 ppb
in a layer 100 m deep. Such sensitivity can be maintained for a longer
period in mid-summer and for considerably less in winter. A further
important feature of correlation spectroscopy is that measurements can be
made with a short time constant, generally in the region of one second or
so compared with typical ground level sensors which have a time constant
of about one minute. This is an important factor when a plume is being
traversed by an instrument mounted in a vehicle and particularly if an
aircraft is used.

A typical result of measurements made with correlation spectrometers is shown in Figure 5. Here instruments mounted in a van were traversed under a power station plume 10 km or so downwind of the source. Such measurements can be used to give the flux of material in the plume and to study the rate of conversion of NO to NO_2 which occurs principally from the mixing of ambient ozone into the plume. Figure 6 shows the increase in the rate of conversion from morning through to noon.

One problem with correlation spectroscopy is that both the baseline and sensitivity change with solar elevation and with cloud cover. This is not generally a serious problem for traverses such as those shown in Figure 5 which were under a well defined plume and were of only short duration. For longer traverses of less well defined plumes such as may be encountered in surveys of extended industrial areas baseline drift and changes in sensitivity can be the source of considerable inaccuracy (ref. 31) and care is required in calibration and instrument adjustment (refs. 32 and 33).

3.3 Lidar

Unlike correlation spectroscopy lidar does not depend upon skylight. Thus it is not limited to daylight operation, indeed it works more effectively at night. It is not limited to measuring total overhead burden since it can provide measurements with spatial resolution. Furthermore it does not suffer from problems of baseline drift and changing sensitivity, though it does have problems of its own arising principally from changes in the backscatter characteristics of different atmospheric aerosols (ref. 34).

The principle of lidar is that a laser pulse is fired into the atmosphere and as it proceeds along its path radiation is continuously scattered by molecules and aerosol particles back towards the laser where it is collected with a telescope and measured with a detector. The signal is analysed to provide information on the magnitude of backscatter and attenuation experienced by the pulse in its passage through the

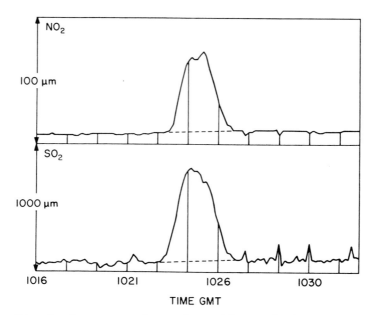

Fig. 5. Traverses under a plume with correlation spectrometers. Simultaneous measurements of SO_2 and NO_2 showing linear interpolation used to remove background. (Varey, 1986).

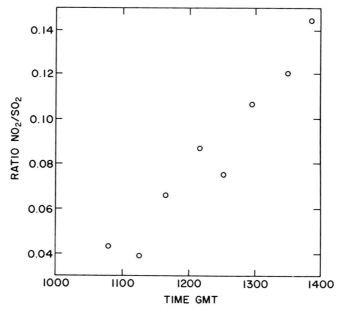

Fig. 6. Diurnal vatiation in the oxidation rate of NO in a power station plume indicated by the changing ratio of the fluxes of NO_2 and SO_2 measured with correlation spectrometers. (Varey, 1986).

atmosphere. The technique is applied in two ways. The first is to use a laser operating at a single fixed wavelength to monitor the distribution of aerosol particles. This has applications in meteorology, Chapter X, and in studies of plume rise and dispersion. The second mode of operation is to use a tuneable laser in which the wavelength can be changed. This enables the distribution of selected molecules in the atmosphere to be determined. The principles of lidar are discussed in detail elsewhere (refs. 38-38).

The accuracy and detection limits of differential lidar cannot be defined in the same way as those for a normal point sampler since they depend on factors such as atmospheric visibility, the spatial resolution required, the range at which the measurement is being made and the time resolution i.e. the number of pulses being averaged. Essentially it is a question of the signal-to-noise ratio in the measured signal which in turn depends on the number of photons received from each measurement. However, it is generally possible to measure SO_2 and O_3 with a range resolution of 50 m and a sensitivity of about 10 ppb out to a range of a few hundred metres and with a sensitivity of 20 ppb or so at a kilometre. Sensitivity for NO_2 is in principle somewhat lower but to date fewer measurements have been made (refs. 27, 29, 40), most effort being concentrated on SO_2 (refs. 41-43), and O_3 (refs. 34, 44, 45). A typical result is shown in Figure 7 where a vehicle carrying a lidar system made a traverse under four power station plumes.

The mode of operation of a lidar system is quite different to that of a point monitor such as a flame photometric SO_2 analyser or a chemiluminescent NO_x analyser since lidar cannot in general be left to operate for long periods. It requires periodic tuning of its lasers and if tunable dye lasers are used, as is most common, regular dye changes. Consequently lidar systems are usually employed in short term studies rather than in establishing longer term average values of SO_2, and NO_2 or ozone. However, the reliability of lidar systems is improving. Figure 8 shows the results of continuous operation for a period of 24 hours measuring ozone. More recently the same system has been operated for a period of ten days with attention only twice a day, providing continuous results with only minor interruptions.

172

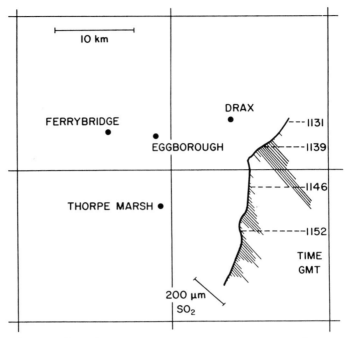

Fig. 7. Differential lidar measurements of four power station plumes
showing the route followed by the vehicle and the overhead burden of SO_2.
(Sutton, Central Electricity Generating Board, private communication).

4.   APPLICATION OF TECHNIQUES IN FIELD SURVEYS

4.1   Introduction

The objective of this section is to illustrate the way some of the
techniques described above can be used in field surveys. Three examples
are taken from surveys in London, each with different objectives and
methods of measurement. The fourth example is one of the few which can be
termed an 'experiment' in that some control was exercised over the
conditions under which the measurements were made.

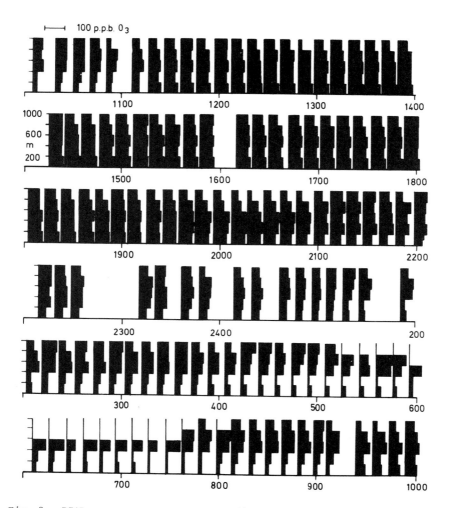

Fig. 8. DIAL ozone measurements over 24 h continuous measurement. (Sutton, Central Electricity Generating Board, private communication).

174

4.2 Survey of Nitrogen Dioxide in London Using Diffusion Tubes

A one year survey of NO_2 was carried out in London during 1984/85 by the Greater London Council in collaboration with local Boroughs and AERE

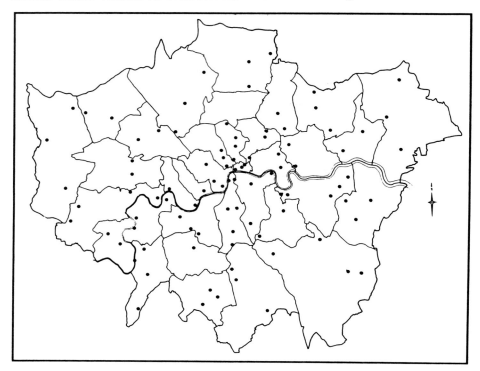

Fig. 9. Location of 'background' sites used in the 1984/85 nitrogen dioxide survey. (London Environmental Supplement No. 14, 1985. Reproduced with permission of the Editorial Board).

Harwell (ref. 46). The primary objective was to determine annual average background levels of NO_2, that is levels away from the vicinity of local sources, principal of which in the London area are vehicle exhausts. The area to be covered is approximately 1600 Km^2 and so required monitoring at a large number of sites. Some measurements had already been made at a

175

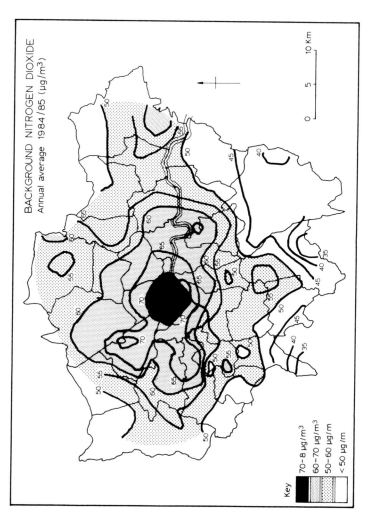

Fig. 10. Computer-generated isopleths of annual average nitrogen dioxide concentrations in Greater London based on measurements at 'Background' sites during 1984/85. (London Environmental Supplement No. 14, 1985. Reproduced with permission of the Editorial Board).

176

small number of sites with continuous chemiluminescent monitors but at a cost of about £10,000 each it was not feasible to deploy a sufficient number of such monitors to achieve the objective. Consequently diffusion tubes were used as described in Section 2.1.1. These were deployed at over 100 sites and in all about 5000 tubes were deployed in the period July 1984 to July 1985. In this survey the tubes were 70 mm long and 12 mm diameter. They were mounted vertically, open end downwards and the criteria for their deployment was that they should be at least 50 m away from any road carrying more than 20,000 vehicles a day and that they should be mounted 1.5 to 5.0 m above the ground, clear of any structure. The sampling time for each tube was one to two weeks. In addition to background measurements tubes were deployed at 21 roadside sites. Locations of the sites are shown in Figure 9. Results from the background sites are summarized in Figure 10 which shows annual average background concentrations ranging from about 40 μg m^{-3} on the outskirts of the city to 80 μg m^{-3} in the centre (94 μg m^{-3} = 50 ppb). Concentrations at

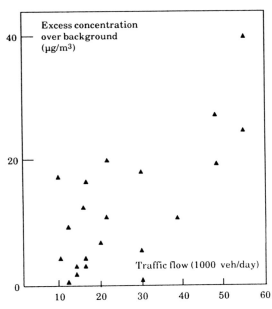

Fig. 11. Nitrogen dioxide concentrations at roadside sites in excess of local 'background' versus traffic flow. (London Environmental Supplement No. 14, 1985. Reproduced with permission of the Editorial Board).

roadside sites were higher than the local background as shown in Figure 11 where it can be seen that the excess, ranging up to 40 μg m^{-3}, appears to be related to the density of traffic flow. Other factors affecting the excess include distance of the sampler from the road, local topography and wind flow, and local traffic conditions such as speed and the proportion of diesel vehicles.

An interesting feature of these results is that although the measurements were averages over one to two weeks it appears that they can be related with reasonable accuracy to short term concentrations. This finding is based on measurements made with chemiluminescence monitors at both urban background and roadside sites. Results are shown in Table 3 where it can be seen that there is reasonable consistency between 98 and 50 percentiles of hourly averages and long term averages.

TABLE 3
Data from the Scientific Services Branch survey of nitrogen dioxide by chemiluminescence monitors, and its use in calculating the conversion factors necessary for estimating 50 and 98 percentiles from annual averages in London (London Environmental Supplement No. 14 (1985). Reproduced with permission of the Editorial Board).

Site	Period	Concentrations μg/m^3			Ratios	
		Average	50 percentile	98 percentile	50 percentile / Average	98 percentile / Average
Roadside						
Archway Rd, LB Haringey	Oct 83 - Jul 83	68	63	174	0.926	2.56
E India Dock Rd, LB Tower Hamlets	Jan 83 - Feb 83	75	67	185	0.893	2.47
Elephant & Castle, LB Southwark	Jan 83 - Sep 83	77	69	193	0.896	2.51
Manor Rd, LB Newham	Mar 81 - Sep 81	45	38	126	0.844	2.80
N Circular, City of Westminster	Oct 83 - Mar 84	90	92	187	1.022	2.08
Orchard St, City of Westminster	Apr 82 - Jun 82	138	134	272	0.971	1.97
Ripple Rd, LB Barking & Dagenham	Feb 82 - Jun 82	57	55	141	0.965	2.47
York Road, County Hall, LB Lambeth	1978	67	63	163	0.940	2.43
	1979	67	59	191	0.881	2.85
	1980	57	54	132	0.947	2.32
	1981	89	80	205	0.899	2.30
	1982	99	92	229	0.929	2.31
	1983	53	52	119	0.981	2.25
	1984	86	82	185	0.953	2.15
			Roadside	mean ± s.d.	0.93 ± 0.05	2.39 ± 0.25
Urban Background						
Brent Cross, LB Barnet	Dec 82 - Mar 83	64	63	126	0.984	1.97
Bushy Park, LB Richmond/Thames	Jul 82 - Sep 82	37	33	98	0.892	2.65
Kew Gardens, LB Richmond/Thames	Jan 80 - Dec 80	31	27	80	0.871	2.58
	Jan 81 - Apr 81	39	36	105	0.923	2.69
Maryland St, LB Newham	Nov 82 - Jul 82	51	46	128	0.902	2.51
Upminster Rd South, LB Havering	Jan 82 - Mar 82	43	42	90	0.977	2.09
County Hall, LB Lambeth	1978	50	44	130	0.880	2.60
	1979	41	40	86	0.976	2.10
	1980	45	44	98	0.978	2.18
	1981	52	46	140	0.885	2.69
	1982	53	50	132	0.943	2.49
	1983	59	57	130	0.966	2.20
	1984	58	55	126	0.948	2.17
			Backgrnd.	mean ± s.d.	0.93 ± 0.04	2.38 ± 0.26
				Overall	0.93 ± 0.04	2.38 ± 0.25

The relationships for both roadside and background values are:

98 percentile = 2.4 x annual average

50 percentile = 0.93 x annual average.

Using these factors it is possible to interpret the diffusion tube results in terms of 98 and 50 percentile hourly values. This is shown in Table 4 for the 21 roadside sites. Similar results were deduced for background sites.

TABLE 4
Annual average nitrogen dioxide concentrations as measured by diffusion tube at 'roadside sites' in London and estimated 50 and 98 percentiles. (London Environmental Supplement No. 14 (1985). Reproduced by permission of the Editorial Board).

Borough	Site	Concentration µg/m³		
		Measured annual average	Estimated	
			50 percentile	98 percentile
Barking & Dagenham	Ripple Road (BD3)	74	69	175
Brent	Acton Lane (BR2)	75	70	180
Camden	Euston Road(EN18)	96	89	230
Ealing	Chase Road (EL1)	68	63	160
	High Street (EL2)	84	78	200
Hackney	Church Street (HK2)	79	73	190
	Lower Clapham Road (HK3)	67	62	160
Hammersmith & Fulham	Harwood Road (HM3)	73	68	175
Haringey	Albert Road (HY1)	60	56	145
	St Annes Road (HY2)	65	60	155
	Stroud Green Road (HY3)	73	68	175
Harrow	Uxbridge Road (HW3)	53	49	125
Lambeth	York Road (EN15)	93	86	220
Merton	Central Road (MT2)	52	48	125
	Vestry Hall (MT3)	58	54	140
Newham	East Ham Town Hall (NH3)	65	60	155
Southwark	Evelina Road (SW1)	77	72	185
	Old Kent Road (SW2)	91	85	215
Tower Hamlets	Philpot Street (TH2)	66	61	155
	East India Dock Road (EN17)	103	96	245
Waltham Forest	The Ridgeway (WF1)	56	52	135

This survey is a good illustration of the use of a relatively simple method of pollution measurement to obtain results at a large number of sites covering a considerable area. It enabled the investigators to identify sites and areas where concentrations were high with a view to more detailed measurements using instruments with shorter time constants and better sensitivity.

4.3 Measurements of SO_2 in London with Mobile Surveys

Another approach to surveying pollution in an extended area is to use instruments mounted in vehicles and traverse the area under investigation. A survey of this type was carried out in London by the Central Electricity Generating Board during the winter months of 1982-83 and 1983-84 (ref. 47) One objective of the study was to determine the contribution to SO_2 levels in the city of power stations sited along the Thames to the east. A route running north/south through the centre of the city was chosen for the survey since this would traverse the station plumes in easterly winds. In addition to making measurements in easterly winds regularly weekly surveys were made irrespective of wind direction to establish SO_2 concentrations along the route in the absence of power station contributions. Two vehicles, MAPSU and DIAL, were used in the survey each equipped with a variety of instrumentation. In each, SO_2 was measured with a Meloy flame photometric instrument. Measurements were made on the move as the vehicles traversed the route, the sampler inputs being at the top of the vehicles. The route followed by the vehicles is shown in Figure 12. The procedure was generally for both vehicles to drive into central London and then separate. One going north along the A10 and the other south along the A23. Each would then carry out as many complete traverses as possible in a working day stopping on each traverse in the centre to calibrate the instruments. Usually it was possible to achieve two complete traverses with each vehicle each day. The largest data set was obtained between October 1983 and March 1984 when the vehicles operated on a total of 26 days.

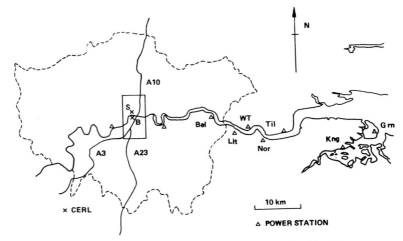

Fig. 12. London Survey Area. S - Sudbury House, B - Bankside House,
Lit - Littlebrook, WT - West Thurrock,
Nor - Northfleet, Til - Tilbury, Kng - Kingsnorth
Grn - Grain (Bennett et al., 1986)

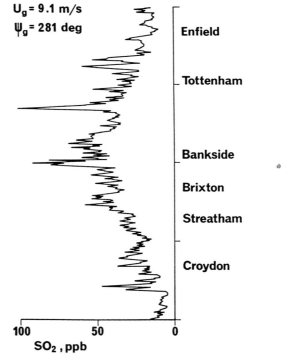

Fig. 13. London Pollution Profile on 2.2.83 between 950 and 1200,
vehicle MAPSU. (Bennett et al., 1986)

TABLE 5
Meteorological parameters and mean SO_2 concentrations (ppb) measured during mobile surveys in Central London (Bennett et al., 1986).

Date	u_g $(m\,s^{-1})$	ψ_g (deg)	T_0 (°C)	MAPSU concn.		DIAL concn.	
				a.m	p.m.	a.m.	p.m.
12/10/83	18.7	248	15.0	17.7	20.8		
19/10/83	14.6	281	14.0	21.5	21.5	25.7	18.5
26/10/83	7.8	286	10.5	28.9	28.5	45.3	36.3
2/11/83	4.6	140	14.5			164.5	206.6
9/11/83	9.0	116	15.5	60.2	43.3		
10/11/83	7.4	112	14.6	137.5	59.6		
14/11/83	16.1	104	6.0	31.6	27.0		
22/11/83	7.0	42	3.2			80.5	52.9
23/11/83	2.5	231	−1.9	109.4		134.4	129.7
7/12/83	4.1	317	1.7	124.1	100.9	126.9	84.0
14/12/83	20.3	213	5.8	19.3	22.9	18.2	15.0
4/1/84	14.8	333	6.3			39.0	39.7
11/1/84	24.2	255	11.3	12.9	11.4	13.4	13.1
19/1/84	11.7	48	3.2			65.0	30.4
25/1/84	9.0	267	4.4			51.1	26.6
1/2/84	13.0	210	6.1	29.6	17.0	24.4	10.4
8/2/84	20.4	342	7.9	12.5	26.2	21.2	25.8
13/2/84	6.6	104	4.8	55.2	214.9		
14/2/84	8.1	86	5.5	52.4		62.9	
15/2/84	11.7	63	−0.6	48.3		45.7	39.5
29/2/84	3.7	339	4.3	24.7	41.3		
7/3/84	4.6	37	9.6			39.1	26.4
13/3/84	13.4	61	4.5		22.3		
14/3/84	11.7	100	8.6	46.3	65.2		
15/3/84	18.7	69	5.3	24.1	24.7	35.0	36.0
21/3/84	0.9	274	5.9			59.7	49.0

The result of a typical traverse is shown in Figure 13. SO_2 concentrations vary widely, with peaks on this occasion reaching 100 ppb or so, and an underlying trend showing a rise in concentration towards the centre of the city. The large variations in concentration indicate the contributions of local sources along the route. Table 5 shows the results for the central 10 km of the survey. The SO_2 concentration is the mean measured over that section of the route and is separated into readings taken before and after 1200 GMT. The table also shows midday surface temperature T_o taken at Heathrow and surface geostrophic wind speed u_g and direction ψ_g (ref. 47). Even when averaged in this way concentrations show considerable variation. It might be expected that they would depend upon wind speed and Figure 14 shows a plot of morning SO_2 concentration, against wind speed. Occasions when power stations may have been making a contribution, wind directions 080° - 120°, have been omitted in order not to mask the wind speed dependence. It can be seen that although there is considerable scatter the inverse relationship between wind speed and concentration is reasonably well defined. It was useful to define a quantity the 'horizontal surface flux', χu_g. Since this should be more or less independent of wind speed any dependence on other variables such as wind direction or temperature should become apparent. Figure 15 shows $u_g \chi$ plotted against temperature and although again there is considerable scatter in the results an inverse relationship is indicated.

In one respect this data set is the converse of that represented by that obtained with diffusion tubes for NO_2 as described in the previous section. There, long term averages were measured and conversion factors were derived to estimate shorter term values. In the SO_2 survey the traverses on single days are just samples of the complete data set for the winter. It is not practicable to make measurements with the vehicles every day and it is useful to devise a method of estimating winter average values. First, from the results shown in Figures 14 and 15 a relationship was established between SO_2 concentration and wind speed and temperature

$$\chi = \begin{cases} 1370/u_g & \mu g\ m^{-3} \quad T_o < 7.5^\circ C \\ \\ 950/u_g & \mu g\ m^{-3} \quad T_o > 7.5^\circ C \end{cases} \qquad (17)$$

where the temperature dependence has been approximated into just two levels. In spite of the approximations involved the relationship (17) is quite a good description of the results obtained on the traverses as shown in Figure 16. Since values of wind speed and temperature are available for every day through the winter equation (17) can be used to estimate the SO_2 concentration on each day and thus derive a winter mean. The value obtained was 107 $\mu g\ m^{-3}$ an interesting outcome of which was that a careful

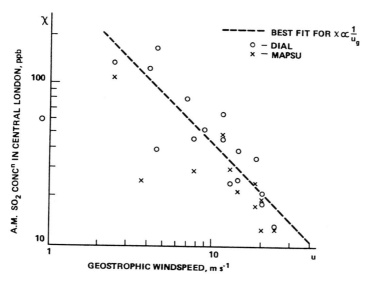

Fig. 14. Morning mean SO_2 concentrations in Central London vs wind speed. (Bennett et al., 1986).

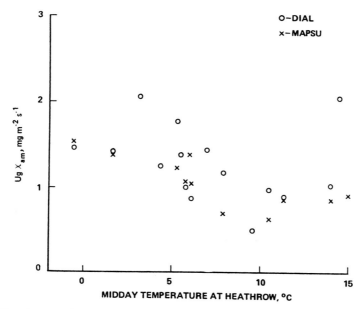

Fig. 15. Morning surface SO_2 flux in Central London vs midday temperature at Heathrow. (Bennett et al., 1986).

184

comparison of this figure with others obtained from static point samplers
led to the conclusion that at the roadside diesel vehicle emissions
contribute substantially, about 50%, to the concentration of SO_2.

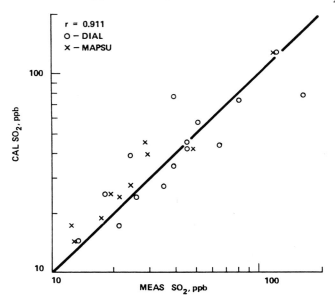

Fig. 16. Calculated and measured morning SO_2 concentrations in Central
London. (Bennett et al., 1986).

4.4   Measurement of Ozone in the London Urban Plume

As outlined in Section 2.1, ozone and related oxidants play a central role
in producing acid species from primary industrial emissions. Ozone itself
can cause damage to trees and crops at high concentrations. In general,
the main source of ozone is the stratosphere but during photochemical
episodes when ozone concentrations are many times higher than normal,
generation from urban and industrial emissions of NO and hydrocarbons
becomes the dominant source. Measurement of the rate of ozone production
in the urban plume as it drifts downwind from the city presents problems
different to those encountered in the NO_2 and SO_2 surveys described above.
Long averaging times are of little use because significant generation can
occur in a few hours. A minimum time resolution of one hour is required

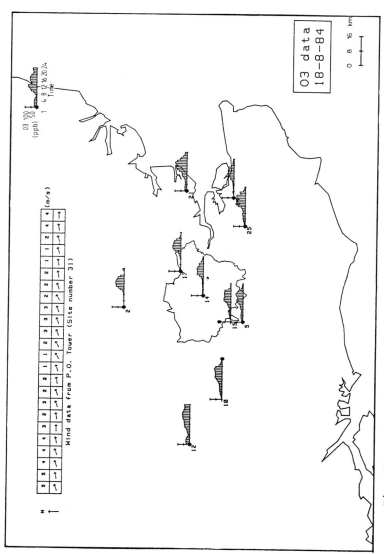

Fig. 17. Hourly mean ozone concentrations around London, 18 August 1984.
(London Ozone Survey 1982-5).

and important features in ozone concentration can occur on an even shorter
time scale. Regular surveys such as those used to measure SO_2 cannot be
planned since episodes of high ozone concentration depend on the occurance
of anticyclonic relatively calm, sunny weather. Ozone is different also
in that it has no localised point sources, the generation of ozone does
not have strong small scale spatial variation. However, there are strong
sinks for ozone. Principal among these is NO, which does have intense
local sources and thus produces large local reductions in ozone
concentration. Indeed at ground level in urban areas ozone concentration
can be virtually zero even during periods when outside the city
concentrations are very high. Averaging over several monitors or using a
mobile sensor, as with SO_2, does not solve the problem, since the effect
of NO is always to reduce the ozone concentration. The only solution is
to monitor NO_x as well as ozone or to be certain that the sites chosen for
measurements are free of NO interference, a difficult task in an urban
plume. Another major sink for ozone is the ground. This is particularly
apparent on clear, calm nights when the atmosphere becomes stable. Then
ozone is destroyed at the ground and not replenished from above. Thus
leads to the strong diurnal variation in ozone concentration measured at
ground level which is not necessarily representative of concentrations
above ground level. An example of this is shown in Figure 8 where it can
be seen that during the night ozone concentrations near the ground fall to
much lower levels than at a height of a few hundred metres. A similar
effect can occur in early morning when emissions of NO_x from road vehicles
are trapped below a low level inversion. This produces high
concentrations of NO_x until the ground is heated by solar radiation and
the inversion lifts. This onset of growth in the mixing layer can be
clearly marked and lead to a rapid reduction in the ground level
concentration of NO_x. At the same time air mixing down from above brings
with it ozone which has remained aloft overnight. Of course in many
situations it is the ground level concentration of pollutants which is of
most importance since it is there that damage can occur. However, in the
study of photochemical production of ozone it is important to distinguish
between ozone generated photochemically and that appearing at the ground
from above. It is in situations of this sort that remote sensing is
useful and where results from ground level sensors can be misleading.

Nevertheless, ground level measurements can be useful. In a survey in London supported by the CEC and involving several laboratories the strategy was to use a network of ground level chemiluminescence and ultra-violet absorption monitors which were as far as possible operated continuously through the summer months. This provides results whenever photochemical episodes occur and the data base can be augumented by lidar measurements whenever possible and also sometimes by measurements from an aircraft. In order to investigate ozone generation in the urban plume analysis has been confined to a period of seven hours or so in the middle of the day when the atmosphere is well mixed and ground level concentrations of ozone are representative of those in the plume as a whole. Lidar measurements are useful to confirm the vertical uniformity of ozone concentrations. An example of the results is shown in Figure 17. On this occasion winds were light but reasonably well defined in direction. Upwind of the city concentrations in the afternoon are about 60 ppb whereas downwind they reach about 100 ppb. In order to estimate

TABLE 6
Selected results from the London ozone study, illustrating the enhancement of ozone levels downwind of London and the time for air to travel from central London to the monitoring site. (London Ozone Survey, 1982-85)

Date	Maximum hourly ozone downwind (ppb)		Approx. time downwind (h)	Wind direction
	Measured	Enhancement		
8 July 1982	150	41	7	100°
17 Sept 1982	93	17	3	80°
8 July 1983	153	72	7	100°
12 July 1983	102	16	1	50°
15 July 1983	112	27	3	250°
16 July 1983	165	72	7	200°
9 June 1984	87	27	3	30°
10 June 1984	92	16	1	280°
8 July 1984	130	40	2	160°
19 Aug 1984	134	33	6	150°

188

ozone generation produced as a result of urban emissions it is not
sufficient to compare upwind and downwind concentrations since in the time
taken to reach a downwind site generation could occur even in the absence
of urban emission, due to precursors present in upwind air or those
introduced in its passage over rural/semi-rural areas typical of south
east England. The procedure followed is to compare the downwind
concentration of sites clearly in the plume with those as far as possible
the same distance downwind but not within the plume. Examples of results
obtained in this way are shown in Table 6.

4.5 Field Experiments to Measure SO_2 Oxidation in Cloud

An experiment of a different kind is being carried out at Great Dun Fell
in Cumbria, England. The objective is to measure the oxidation rate of
SO_2 in cloud. Great Dun Fell is a ridge which rises to a height of 850 m

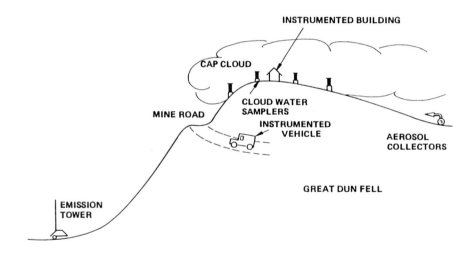

Fig. 18. Schematic diagram of Great Dunn Fell experiment.

and is cloud capped for some 200 days in the year. The experiment
involves the release of SO_2 on the upwind slope of the ridge below cloud
base and in appropriate wind conditions this is carried up the ridge and
is incorporated into the cloud system. A tracer gas, SF_6, is released
simultaneously with the SO_2 from the same point which serves to define the
trajectory of the releases over the ridge and also provides a means of
determining changes in the ratio SO_2/SF_6 during oxidation processes. It
has been found that a release rate of SO_2 of 100 kg per hour from a 10 m
tower at a range of 5 km is sufficient to produce a maximum ground level
concentration of about 10 ppb at the summit. For SF_6 a release rate of
1.6 kg per hour produces a concentration of about 100 ppt. A diagram of
the set up is shown in Figure 18. SO_2 is monitored at the summit with a
Meloy flame photometric analyser and a Severn Sciences instrument (see
Section 2.1). On the side of the ridge cross-wind traverses are made with
a vehicle equipped to measure SO_2 with a Melo analyser, and O_3 and NO_x
with chemiluminscence instruments. The tracer is detetected using a
continuous electron capture instrument (ref. 48) mounted in the vehicle
and by taking bag samles at various points on the ridge. H_2O_2 is measured
using luminol detectors (Section 2.1). Cloud water is collected on the
slope of the ridge and at the summit using several types of rod and string
instruments, such as that shown in Figure 4. Analysis of cloudwater for
H_2O_2 is carried in real time again using a luminol detector (ref. 49) and
the samples are refrigerated at about 4°C and subsequently analysed in the
laboratory for SO_4^{2-}, Cl_{2+}^{-}, NO_3^{-} and NH_4^{+} by ion chromatography and for
Na^+, K^+, Ca^{2+}, Mg^{2+} and trace metals by inductively coupled plasma
spectroscopy. At the summit wind speed and direction and wet and dry bulb
temperatures are measured together with cloud size spectra.

In principle the experiment appears straightforward. In practice it is
difficult since appropriate cloud cap conditions, with the wind in a
suitable direction to carry the SO_2 release over the summit where the
instruments are located, are relatively rare. It has been found that even
using long range weather forecasts if the instruments are deployed for a
period of three weeks it is only on one or two days that conditions are
suitable. However, some preliminary results have been obtained. Figure 19
shows the result of measurements of SF_6 and H_2O_2 in cloudwater at the
summit in winter. The SF_6 measurements show that the plume of SO_2 from

190

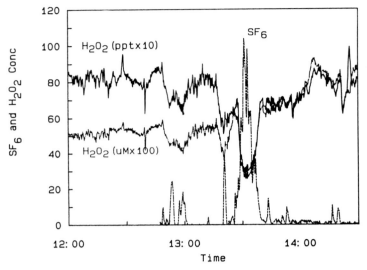

Fig. 19. SF$_6$ and cloud water H$_2$O$_2$ - Great Dunn Fell.
(G.P. Gervat and P.A. Clark, Central Electricity Generating Board,
private communication).

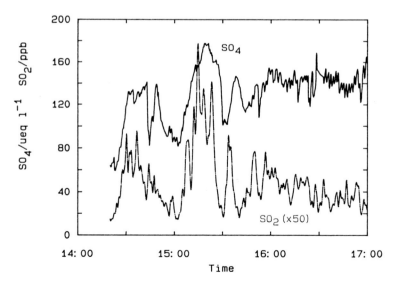

Fig. 20. SO$_2$ and cloud water sulphate - Great Dunn Fell.
(G.P. Gervat and P.A. Clark, Central Electricity Generating Board,
private communication).

the release point was reaching the summit intermittently and that in the cloud water associated with the plume the concentration of H_2O_2 is significantly depleted. Figure 20 shows the concentration of sulphate in cloud water and the concentration of SO_2 in air during a different, spring time experiment. There is an apparent base line drift in the sulphate measurement, probably a result of a decrease in cloudwater content in the course of the experiment, but there is also a clear indication of enhanced sulphate content in the presence of the plume. Results of this sort are being used to calculate oxidation rates in cloud. The experiment is so far in its early stages but the results are promising.

ACKNOWLEDGEMENTS

Many colleagues have provided advice and data for this chapter and their help is gratefully acknowledged. The work was carried out at the Central Electricity Research Laboratories and is published with permission of the Central Electricity Generating Board.

5.   REFERENCES

1.   Harrison, R.M. and Perry, R. (Eds), 1986. Handbook of Air Pollution Analysis (Second Edition), Chapman and Hall, London.
2.   Atkins, D.H.F., Healey, C. and Tarrant, J.B., 1978. AERE Report No. R 9184.
3.   Cocks, A.T. and McElroy, W.J., 1984. Atmos. Environ., 18, 1471-1483.
4.   Herrmann, W., Heinrich, H.J., Michaelis, W. and Weitkamp, C., 1980. Abstracts, 10th Int. Laser Radar Conf., Maryland.
5.   Gregory, G.L. and Mayer, 1977. Rev. Sci. Inst., 48, 1464-1468.
6.   Possanzini, M., Febo, A. and Liberti, A., 1983. Atmos. Environ., 17, 2605-2610.
7.   Harrison, R.M. and Pio, C.A., 1983, Tellus, 35B, 155-162.
8.   Hanst, P.L., Lefohn, A.S. and Gay, Jr B.W., 1973. Appl. Spectrosc, 27, 188-193.
9.   Tuazon, E.C., Winer, A.M. and Pitts, J.N. Jnr., 1981. Environ. Sci. Technol, 15, 1232-7.
10. Ferm, M., 1979. Atmos. Environ., 13, 1385-1393.
11. Dimmock, N.A. and Marshall, G.B., 1986. Central Electricity Generating Board, Report No. TPRD/L/2923/R85.
12. Stockwell, W.R. and Calvert, J.G., 1983, Atmos. Environ., 17, 2231-2235.
13. Winer, A.M., 1986. In 'Handbook of Air Pollution Analysis', (Second Edition) Eds Harrison, R.M. and Perry, R., Chapman and Hall, London.
14. Kok, G.L., Darnall, J.R. Winer, A.M. et al., 1978. Environ. Sci. Technol, 12, 1077-1083.
15. Lazarus, A.L., Kok, G.L., Lind, J.A., Gitlin, S.N., Heikes, B.G., and Shelter, R.E., 1986. Analyt. Chem., 58, 594-597.

16. Hanst, P.L., Wilson, W.E., Patterson, R.K. et al., 1985. EPA
     Publication 650/4-75-006, Environmental Protection Agency,
     Research Triangle Park, NC.
17. Platt, U., Perner, D., Winer, A.M. et al., 1980. Geophys. Res.
     Lett., 7, 89-92.
18. Noxon, J.F., 1983. J. Geophys. Res., 88, 11017-21.
19. Platt, U. and Perner, D., 1983 in 'Optical and Laser Remote Sensing'
     (Eds Killinger, D.K. and Mooradian, A.), Springer-Verlag, Berlin,
     Vol. 39, pp 97-105.
20. Platt, U.F., Winer, A.M., Biermann, H.W. et al., 1984. Environ.
     Sci. Technol., 18, 365-369.
21. Perner, D., Ehhalt, D.H., Patz, H.W., Platt, U., Roth, E.P. and
     Volz, A., 1976. Geophys. Res. Lett., 3, 466-468.
22. Davis, D.D., Heaps, W.S., Philen, D., Rodgers, M., McGee, T.,
     Nelson, A. and Moriarty, A.J., 19879. Rev. Sci. Instr., 50, 1505-1515
23. Calvert, J.G. and Stockwell, W.R., 1984. In 'SO_2, NO and NO_2:
     Oxidation Mechanisms, Atmospheric Considerations', Ed Calvert,
     J.G., Butterworth, Boston.
24. Ames, D.L., 1983. Central Electricity Generating Board, Report No.
     TPRD/L/2478/N83.
25. Falconer, R.E. and Falconer, P.D., 1979. ASRC Pulication 741.
26. Scriven, R.A., 1979. CEC Report EUR 6420 EN, Commission of the
     European Communities.
27. Bowne, N.E., Londergan, R.J., Murray, D.R. and Borenstein, H.S.,
     1983. EPRI Report EA-3074, Electric Power Research Institute,
     Palo Alto, California.
28. Hamilton, P.M., 1974. Central Electricity Generating Board, Report
     No. RD/L/N131/74.
29. Millán, M.M. and Chung, Y.S., 1977. Atmos. Environ., 11, 939-44.
30. Carras, J.N. and Williams, D.J., 1981. Atmos. Environ., 15,
     2205-17.
31. Varey, R., 1986. In 'Handobok of Air Pollution Analysis' (Second
     Edition), Eds Harrison, R.. and Perry, R., Chapman and Hall.
32. Millán, M.M. and Hoff, R.M., 1977. Atmos. Environ., 11, 857-60.
33. Millán, M.M. and Hoff, R.M., 1978. Atmos. Environ., 12, 853-64.
34. Browell, E.V. et al., 1985. Appl. Opt., 22, 522-34.
35. Byer, R.L., 1975. Opt. Quant. Electron., 7, 147-77.
36. Collis, R.T.H.and Russel, P.B., 1976 in 'Laser Monitoring of the
     Atmosphere' (Ed. E.D. Hinkley), Vol. 14, Topics in Applied
     Physics, Springer-Verlag, Berlin.
37. Brassington, D.J., 1984. In 'Optical Remote Sensing of Air
     Pollution', Eds Camagni, P. and Sandroni, S., Elsevier,
     Amsterdam.
38. Varey, R.H., 1984. In 'Optical Remote Sensing of Air Pollution',
     Eds Camagni, P. and Sandroni, S., Elsevier, Amsterdam.
     (Second Edition).
39. Rothe, K.W., Brinkman, U. and Walther, H., 1974. Appl. Phys., 4,
     181-2.
40. Fredriksson, K., Galle, B., Nystrom, K. and Svanberg, S., 1979.
     Appl. Opt., 18, 2998-3003.
41. Sugimoto, N., Takeuchi, N. and Okuda, M., 1980. Abstracts, 10th
     Int. Laser Radar Conf. Maryland.
42. Adrain, R.S., Brassington, D.J., Sutton, S. and Varey, R.H., 1979.
     Opt. Quant. Electron., 11, 253-64.
43. Fredriksson, K., Galle, B., Nystrom, K. and Svanberg, S. 1981.
     Appl. Opt. 20. 4181-9.
44. Baumgarten, R.A. et al. 1979.. EPRI Report EA-1267, Electric Power
     Research Institute, Palo Alto, California.
45. Uchino, O., Maeda, M. and Hirono, M, 1979. IEEE Trans., QE-15,
     1094-107.

46. Clark, R.G. et al., 1985. London Environmental Supplement No. 14.
47. Bennett, M., Rogers, C. and Sutton, S., 1986. Atmos. Environ, 20, 461-470.
48. Blackburn, A.J. and Dear, D.J.A., 1981. Central Electricity Generating Board, Report No. RD/L/2136/R81.
49. Ames, D.L., 1983. Central Electricity Generating Board, Report No. TPRD/L/2552/N83.

Regional and Long-range Transport of Air Pollution,
Lectures of a course held at the Joint Research Centre, Ispra, Italy,
15–19 September 1986, S. Sandroni (Ed.), pp. 195–213
© Elsevier Science Publishers B.V., Amsterdam — Printed in The Netherlands

AIRBORNE MEASUREMENTS

Strategy and Performance

A.R.MARSH

1   INTRODUCTION

There are several chemical and physical processes which govern the dispersal and removal of pollutants. Fig.1 diagramatically represents some of these processes.

Emissions of pollutants generally disperse reasonably quickly into the boundary layer at the bottom of the atmosphere. The emitted material can then be removed by several pathways. Material which comes in contact with the ground may be absorbed by different surfaces, a process known as dry deposition. This process can occur over both land and sea. In addition material can undergo chemical transformations in the atmosphere. These chemical transformations can in turn affect the removal processes and may involve a change in state, for example from gas to aerosol. The rates at which these chemical transformations occur varies enormously depending on the species involved and the character- istics of the atmosphere. Characteristic timescales for chemical transformation can range from seconds to years, with corresponding consequences on travel distances within the atmosphere. Soluable materials can be incorporated in cloud and rain systems. Clouds and rain can absorb gases with a range of solubilities, while particulate matter may either form the nucleus of a cloud drop, if it is an active cloud condensation nuclei, or particulates may be captured by cloud droplets or precipitation via a variety of physical processes.   The material incorporated into the aqueous phase by these mechanisms may then undergo further chemical transformations. All of these physical and chemical possibilities should be considered when aircraft measurements are planned and made.

Aircraft measurements of pollution are usually made for one of two purposes. The first is the characterisation of the atmosphere. In the absence of any knowledge of an airborne material, simple concentration measurements are made when a suitable detector has been developed. As knowledge increases, the purpose of the measurements usually evolves into attempts to quantify fluxes of material in the atmosphere. The second purpose of aircraft measurements is enlighten,

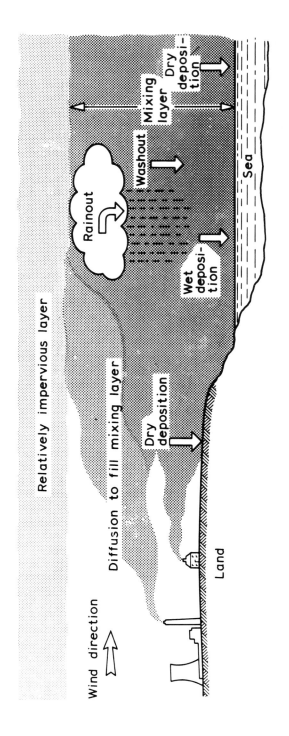

The mixing layer and deposition processes

confirm or quantify the understanding of chemical and physical processes in the atmosphere. The object being to understand the evolution of observed fluxes of the material of interest, and ultimately, to be able to make quantitative predictions of the behaviour of that material in the atmosphere.

For the purposes of this paper, these types of experiments have been divided into the following two catagories:-

Budget Studies

This type of aircraft experiment represents an attempt to characterise fluxes of material in the atmosphere. A wide range of experiments fall into this catagory depending on the specific objective of the flight. The scale of the experiment may vary from a few kilometres to hundreds of kilometres.

Process studies

This type of aircraft study represents attempts to measure chemical and physical processes of the type described above.

These catagories should not be thought of as separate, but rather at either ends of a spectrum of experiments. Flying aircraft is expensive and each flight usually has a variety of objectives. However, flights should be organised with a hierarchy of objectives and in this context, it is the primary objective that determines the flight catagory. This in turn will determine the rigidity or flexibility of the flight pattern.

This paper will examine both of these "end" catagories of flights using case studies as examples. However, before individual case studies can be examined, it is necessary to understand the tool being used, ie. the aircraft and its instrumentation.

2 INSTRUMENTATION

There are several common characteristics required of aircraft instrumentation. Installation of equipment in aircraft is not difficult but there are several mechanical pitfalls to be avoided. Two areas require special attention:- vibration and pressure effects.

2.1 Robust Equipment

Probably the most important characteristic required of equipment is robustness. An aircraft represents a harsh environment for any equipment both in terms of vibration and electrical noise. Equipment, therefore, has to be securely mounted, usually on anti-vibration mountings.

Siting of equipment within the aircraft requires careful consideration. The amplitude and frequency of vibration usually varies throughout an aircraft. The line of the propellers, on a multi-engined aircraft is usually one of the worst positions from a vibrational viewpoint, and should be avoided as a mounting position for vibrationally sensitive equipment.

Equipment should normally be fixed in quick release mountings to allow removal to the laboratory, or interchange with a different detector. Standardisation of mounts and instrument boxes allows the option of interchangeability of equipment. If equipment is not easily removable, attention to instruments will require access to the aircraft and consequently this raises the possibility of logistic problems.

A further consideration in siting equipment within the aircraft concerns the position of the actual air sampling point. Several sampling points are possible on an aircraft including the nose, the leading edge of the wings, emergency hatches and window blanks. The latter are to be preferred to holes in the fuselage as the original aircraft specification can be readily recovered. Ideally, sampling points should be in undisturbed air, and certainly clear of the air disturbed by the propellers or wings. For gases the distance between the sampling point and the detector should be kept in principle to a minimum.

Significant losses can occur in aerosol sampling in two ways. If aerosols are sampled by bringing the airstream into the aircraft, this will probably require pipework with bends. The loss of aerosols to pipe bends depends on angle, radius and flow velocity. Large diameter particles, >5um are readily lost in such systems. The second loss mechanism involves the actual probe to collect aerosols. Ideally the sampling should be iso-kinetic with respect to the aircraft velocity. If the flow velocity is too low then the fine aerosols, <1um, will follow the gas stream lines around the probe and fail to be collected. The sample will therefore be relatively enriched by coarse particles, the degree depending on the difference in velocities. Conversely, if the flow through the probe is too fast then the fine particulates will be drawn in along the gas streamlines from outside the cross-sectional area of the probe and the sample will become relatively enriched in fine material. Thus aerosol sampling is usually restricted to aircraft velocities appropriate to the probe design.

Aircraft travel at relatively high speed through the atmosphere and the response time of measuring equiment is important. The detailed planning of an experiment should allow for the sampling time delays, (due to lengths of sampling intakes etc.) and time constants of the instrumentation. The former is only really important in determining exact positions of features while the latter is important in determining the tactics of the flight. To make measurements near to a source fast instrumentation is required. At greater distances, slower reponse instrumentation can be used. Deconvolution can be used to increase the information from a relatively slow response instrument but great caution is needed with this approach since some measuring systems are not always simple linear systems, eg memory effects from sampling lines will invalidate a

Fourier transform approach. An introduction to these techniques is given by references (1-3).

2.2 Altitude effects

Most detectors and collectors used on aircraft require a pump to obtain air samples. The performance of these pumps depends on the detailed design, displacement, vane, etc. However, the performance will be affected by the change in atmospheric pressure with height. If the pump is used to draw air through a detector or collector the mass of material drawn into the detector or collector will decrease with increasing altitude because of the falling air pressure. The consequences for a detector depend on whether the detector measures total mass in some sort of cell or measures concentration. A vertically uniformly distributed species would be correctly determined by a concentration detector but not by a mass detector. An example of the fall off in performance of a chemiluminescent ozone detector is shown in Fig.2. This response curve to a constant concentration of ozone was obtained in the laboratory by throttling the intake to the instrument. The detector measures total photon flux from reaction in the detector cell and hence is a mass detector. The initial fall off in instrument performance reflects the declining intake pressure. Eventually, the pump effectively stops pumping, because it is exhausting to ground level atmospheric pressure and consequently the performance of the instrument falls

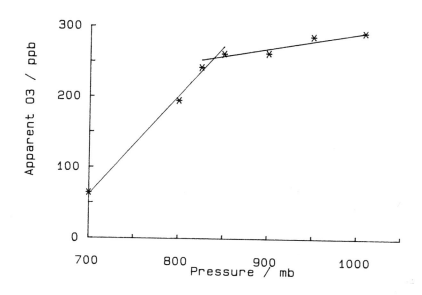

Fig. 2 Pressure response Ozone Monitor

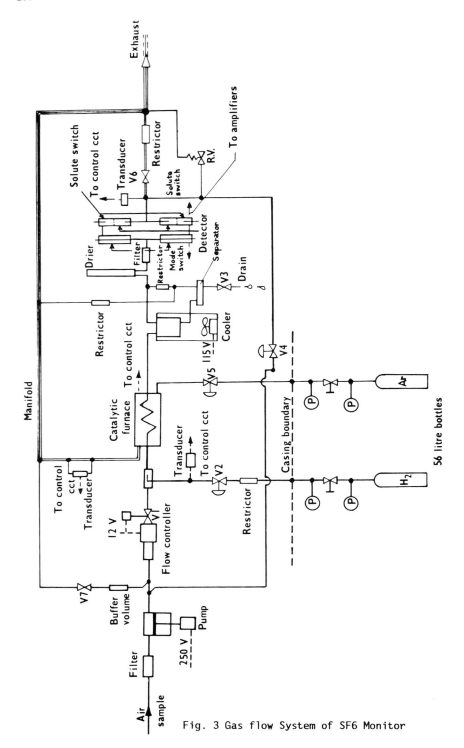

Fig. 3 Gas flow System of SF6 Monitor

dramatically as shown in Fig.2. It follows that for this type of instrumentation, its pressure characteristics must be deptermined and allowed for in the
measurement, ideally, by measuring the pressure at the instrument intake. Further, to extend the vertical range before instrument failure, it is sensible
to exhaust pumps to the outside ambient pressure in pressurised aircraft. An
alternative approach is to place a pump before the detector or collector and
maintain the detector or collector at a constant pressure. Fig.3 shows schematically the plumbing diagram for a sulphur hexafluoride detector which uses
this pressurised approach. The pressure is maintained constant by bleed air
lines and flow controllers. However, the pump will have an intake pressure that
falls with height and a relatively high exhaust pressure. Consequently, such a
system can only be used to a height that corresponds to the pump performance
in a similar manner to the suction system.

The design of individual instruments has to be taken into account when the
effect of altitude is determined. Fig.4 shows the decline of the performance
of a flame photometric detector for sulphur dioxide with a pressure reduction
at its intake. The instrument is a effectively a mass detector monitoring the
photon flux from excited sulphur dimer molecules in the flame gases. The instrument's falling behaviour with pressure reflects the fall in mass within the
detection chamber. However, the behaviour is more complicated than this because

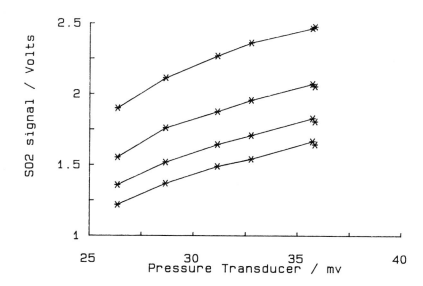

Fig. 4 Meloy signal versus pressure

the instrument relies on a hydrogen-air flame. As the air pressure falls, the air-fuel ratio of the flame changes since the hydrogen is supplied at fixed pressure from a cylinder. This in turn alters the flame temperature and hence the background emission from the flame. Consequently, a complex behaviour is observed in response to pressure changes. Further, for this instrument, the flame temperature depends on the relative humidity, since the water content of the air is important in determining its heat content. In the real atmosphere there is generally an increase in relative humidity with height, until cloud base is reached. Fortunately this proves to be a second order effect for this particular instrument.

Since it is not always possible to avoid cloud and rain when flying, it is sensible to take precautions against the ingestion of cloud and rain drops into detectors and collectors. In the case of the CERL/CIT aircraft, this is achieved by sampling downstream of a cloud water collector.

3 BUDGET STUDIES

Some general features of budget studies will be discussed followed by examples from three case studies.

The simplest type of budget study would be to measure concentrations of material along a line at right angles to the wind as illustrated in Fig.5, (ref. 4). The flux of material crossing that line is then simply the product of concentration, depth and wind speed.

It is worth examining this simple idea in more detail to illustrate the problems that arise. Assuming that the instruments give reliable estimates of concentration, there remains the problem of defining depth and wind speed.

Forecast wind speeds can be used provided they are available for the height used. However, the forecast wind field will be uniform over the grid scale of the forecast model and it is unlikely that real wind fields will be uniform over the relatively large model scales and local effects can profoundly affect the apparent flux estimate.

The illustrated concentration profile in Fig.5 was flown over the sea. This has the advantage that topography does not directly influence the wind field, but has the disadvantage that surface and ascent observations are not available. If budget studies are to be made then it is sensible to equip the aircraft with wind speed and direction instrumentation. This is no small task and usually involves an instrumented nose boom and sophisticated navigational aids.

A measurement at a single height as shown in Fig.5 yields no information on the depth of material. The assumption that material from a particular source has uniformly mixed vertically within the boundary layer cannot be made for sources close to the line of flight. A distance corresponding to at least 2-3 hours plume travel time is required depending on prevailing conditions. The

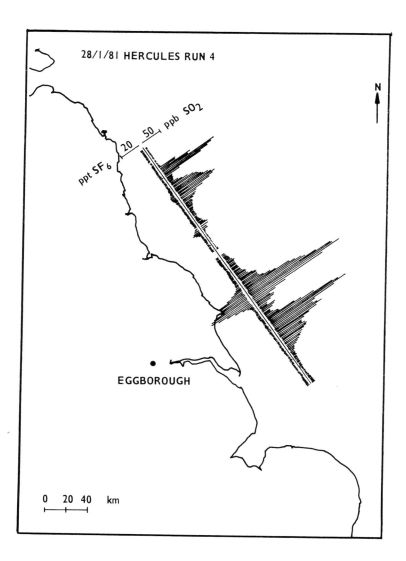

Fig. 5 Crosswind Traverse at 150m, 28/1/81

actual depth occupied by material has to be estimated in some manner. This can be deduced from meteorological information but a field measurement is desirable in order to achieve a reliable estimate of depth. At least two vertical ascents should be made at either end of the line to determine the temperature profile of the atmosphere. This will require suitable instrumentation.

A single boundary layer with one temperature inversion at the top of it as shown in Fig.1 does not occur with a very high frequency. The structure of the boundary layer is often complex with several stratified layers. Each of these layers will probably have a different concentration of material. The obvious solution to these problems is to fly several heights within the boundary layer with ascents at either end of the flight line. However, some practical problems arise. the line shown in Fig.5 is about 250km long. The aircraft flying speed was 90m/s, ie. about 45 mins flying time. To repeat this leg at say two other heights and perform two vertical ascents at either end, (180m/min), would take about 3 hours. The aircraft has to transit to the sampling area and return. Thus in planning budget studies the endurance of the aircraft becomes very important.

The structure of material observed in Fig.5 is interesting, but it is not necessarily essential to observe the detailed structure in order to estimate fluxes. For example, filters could be used for aerosols or absorbent filters or denuder tubes could be used for gases. These would integrate the concentration over the exposure time. Since it takes a finite time to change filters it would seem sensible to expose such devices for the duration of the flight leg and change filters while the aircraft turns and changes height for the next run. This is based on the assumption that one flight leg will provide enough sample for analysis. If not, as is likely to be the case with say denuder tubes, then multiple legs will have to be flown at a fixed height until sufficient exposure time has been achieved. This will have consequences on the number of legs that can be flown because of the limited aircraft endurance.

Such a postulated experiment is really only the determination of flux in a single vertical plane and more practically fluxes across areas are usually required in order to characterise the emissions from an area. A simple approach to this problem is shown by the first case study.

3.1 Case Study 1

Fig.6 shows the flight pattern close to the east coast of the UK (ref. 5), flown in summer by a Hercules aircraft equiped with chemical instrumentation. The object of the study was to determine the total loss rate of sulphur dioxide over the sea by the processes of dry deposition and chemical transformation to sulphate. Three lines were flown, each at two heights. The position of the lines were chosen on the basis of the meteorological prediction of the wind field. Plume trajectories from inland power stations were calculated and the flight lines chosen to straddle the forecast position at right angles to the forecast wind. Each run was approximately 150km long. The flight pattern, with vertical profiles and transit times took of the order of 6 hours to fly. This is of course, a significant fraction of the diurnal period and chemical transformation rates could have altered during the flight time.

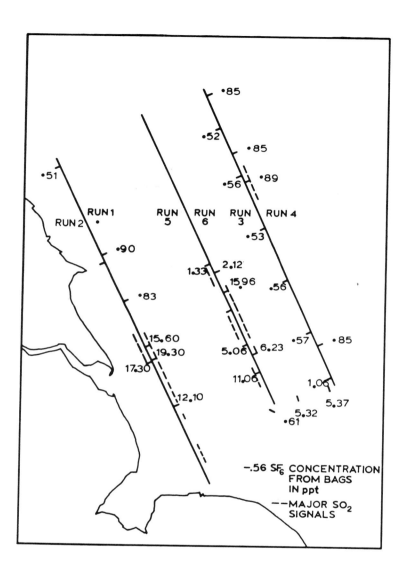

Fig. 6 Schematic Diagram of Hercules Flight

The forecast wind speed was 5m/s, consequently plume material spent about 5 hours traversing the area. Since the aircraft travelled much faster than the plume, the same air was not sampled on either side of the area.

In the event, the plume direction was not quite as forecast because of the effects of hills just inland from the coast. The plume was measured to the south of the flight area and was only partially intercepted at the southern ends of runs 5 and 6 shown in Fig.6. This illustrates the problems of a fixed flight pattern and the situation was alleviated by measurements made outside the runs described as illustrated by the sulphur hexafluoride samples shown in Fig.6. However, movement of the aircraft was restricted in this extended area by air traffic control.

It was possilbe to make an estimate of the loss of sulphur dioxide by chemical transformation into sulphate aerosol from this study. The value obtained of about 1% per hour, is in agreement with expectations for a summer day.

3.2 Case Study 2

A more elaborate budget study was mounted in January 1981 over the North Sea, (ref. 4). The objective was further study of the relatively slow chemical transformation of sulphur dioxide in the atmosphere. In order to obtain reliable estimates of transformation rates, such a study requires the plume travel time between sampling to be large so that a significant amount of conversion has occurred.

In order to locate plumes at long distances from the source, where concentrations of plume species are only slightly greater than background, it was decided to emit an inert tracer, sulphur hexafluoride from a power station. This tracer enabled the plume to be located as it could be detected on the aircraft in real time, (cf. Fig.2), and its natural background in the area was relatively lower than that of the other plume species, (ref. 6).

Since the sulphur hexafluoride tracer was inert in the troposphere, it served a second purpose of enabling estimates of the tracer flux based on the measurements to be compared with the known source emission rate of 50kg/h. Where agreement was achieved this gives confidence to the estimates of fluxes of other plume constituents, several of which were also monitored at source. A second tracer, perfluoromethylcyclohexane, PP2, was also released at 6 hour intervals for a duration of half an hour. This tracer could be detected in real time on board the aircraft, (ref. 6). The purpose of this second tracer was to mark an air parcel so that it could be positively identified at long distances from the source. The position of such a detected pulse of tracer allowed an estimate to be made of the transit time of the air parcel from the source. This could then be compared with the estimates of the integral of the wind speed from the time of release.

In the ideal experiment, the maximum possible period should elapse between measurements of a given PP2 marked air parcel, to allow the maximum possible chemical transformation. Thus, ideally, the experiment would involve measurements on two consecutive days. Fig.7 shows the flight space and time scales required in such an experiment.

Experience had shown that zig-zagging down a plume was difficult even with tracer, since if the plume were lost, it was not known in which direction to

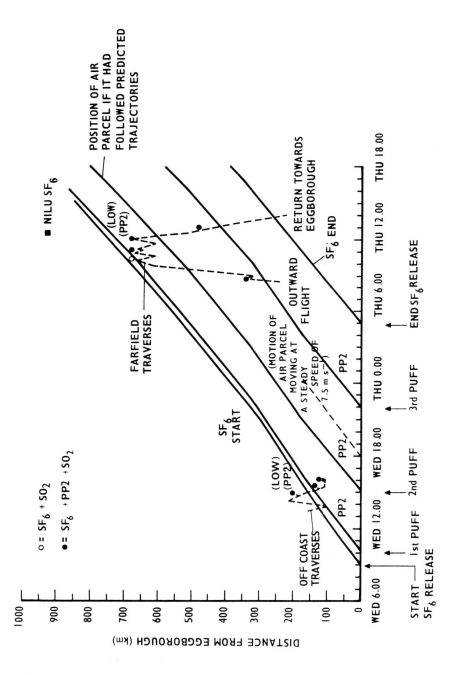

Fig. 7 Distance – Time plot for Aircraft and Air Parcel on 28–29/1/81

search - laterally or vertically? Thus, finding the plume at the far distance was based on plume trajectory forecasts. Provided the wind speed is reasonably high, >5m/s, these predictions can be reasonable accurate. A traverse at mid-distance allowed the plume position to be checked on route to the far-field and provided addition chemical data. In searching for the plume at the mid and far distances the aircraft was positioned well to one side of the predicted plume position and a cross wind traverse towards the plume was executed. This positive error in aircraft position enabled a confident prediction of the direction of the plume at the start of the cross wind traverse.

Since the opportunities for the full experiment were limited, a fall back position was anticipated. If the second days flight could not be achieved then at least some information concerning chemical transformations could be inferred from the plume transit time between the source and the first days measurements. In case the opportunities for the two day experiment proved too limited, additional plans for a single day experiment were also prepared.

In the event, a two day experiment was achieved and despite extensive cloud cover, little oxidation of sulphur dioxide to sulphate in cloud water occurred because of the relatively poor dispersion of the plume under these conditions. The consequence of the poor dispersion was the poor replenishment of oxidants into the plume from the surrounding ambient air. The overall transformation rate was of the order of 0.1% per hour.

The experiment worked well, although the meteorological constraints of the required wind speed and direction meant that the opportunity for such a study is very infrequent.

3.3 Case Study 3

A closed box type budget study was made in W.Germany in 1985 as part of the TULLA campaign, (ref. 7). The objective was the investigation of pollution dispersion over complex terrain. Fig.8 shows the concentration profile observed for ozone by the CERL/CIT aircraft on the 23/3/85. The TULLA campaign used several aircraft and the strategy was to fly up to four aircraft together at different heights in the boundary layer. All the aircraft flew one circuit of the perimeter of the budget box. Several flights were made during a two week campaign.

The flight pattern shown in Fig.8 required about 3 hours flying time. the start point was near the city of Karlsruhe, (marked with a K in Fig.8), and it should be noticed from Fig.8 that the ozone concentration was not the same on return to the start point. This reflects the increase in ozone concentration during the daytime and presents an obvious problem in making budget estimates across the area.

It should be noted in passing that the ozone record around the flight track,

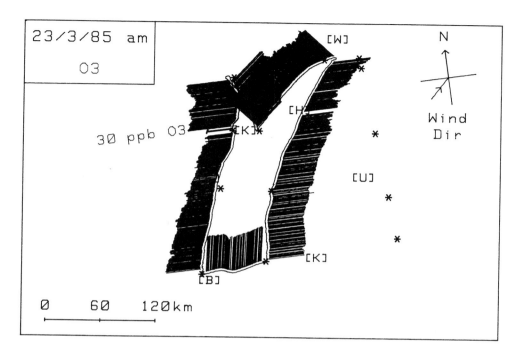

Fig. 8 Ozone record from one of the TULLA flights

as shown in Fig.8, indicates troughs in the ozone concentration to the north east of Heilbronn, (shown as H in Fig.8). This is consistent with NO emissions from that area given the nominal wind direction shown in Fig.8 based on the observations at Stuttgart

There is a further complication in making budget estimates using a box flight pattern. The flight pattern had to be well defined, especially as it involved multiple aircraft in a busy air traffic area, and no flexibility to the pattern could be allowed. Consequently, the aircraft tracks could not be made at right angles to the wind direction, (cf. Fig.8). The correction to the flux estimate when the wind is not at right angles to the aircraft heading involves the sine of the difference in the angle between the aircraft heading and the wind direction.   Consequently, if this angle is small, ie. the aircraft is almost flying along the wind direction, then the estimate of flux becomes very unreliable. In the limit, if the aircraft flyies along the wind, then no information on the flux can be obtained. This proved to be a problem on some of the TULLA flight legs. The ideal for a box type budget study would be for the legs all to be at 45 degrees to the wind direction.

The TULLA campaign results are still being evaluated, but it is clear that for the vast majority of flights reasonable estimates of the flux arriving into the box pattern and estimates of the fluxes generated within the box can be made.

4 PROCESS STUDIES

While budget studies are used to understand gross features in the atmosphere, direct study of chemical processes may require specialist instrumentation. Normally, in process studies the structure of the concentration of material is very important and this requires fast response instrumentation. In general aircraft are not a suitable platform from which to attemp chemical kinetic measurements. However, an example is given below which does not require fast response instrumentation.

4.1 Case Study 4

Fig.9 shows observations of several species around London in July 1984. These represent an attempt by the CERL/CIT aircraft to measure the build up of ozone as a consequence of photochemical activity involving hydrocarbons, (ref. 8). It is important to distinguish different hydrocarbons to understand the complex chemistry. However, at present adequate anaysis is only available at a central laboratory and it is not feasible to mount such equipment in an aircraft. Consequently, the aircraft collects grab samples of air. these require about one minute to fill a special steel air bottle. The wind direction and to a lesser extent, the ozone record, can be used to determine when bottle samples should be collected. At present only 12 bottles can be carried on any one flight. The representativeness of these samples therefore must always be questionable.

However, the success of the experiments lies in the large number of hydrocarbons that can be analysed. The predominate removal process for most simple hydrocarbons is reaction with the OH radical. The concentration of this radical is largely determined by other species such as ozone and NOx. Thus, the decay of simple hydrocarbons can be considered to be a first order process. Taking the ratio of concentrations at time tafter emission yields;-

$$CO1/CO2 = (CS1/CS2)exp(-(kf1-kf2)t)$$

where CO is the observed concentration for species 1 and 2 and CS is the source concentration for species 1 and 2. kf is the pseudo first order rate constant for species 1 and 2 given by:-

$$kf = (OH)k$$

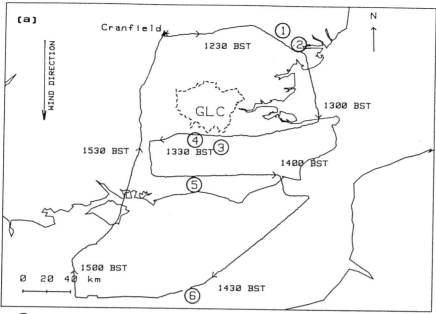

◯ – Hydrocarbon Sample

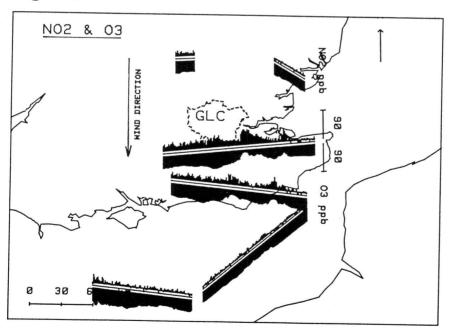

Fig. 9 Measurements on Crosswind traverses in the London area

where k is the true bimolecular reaction rate.

The same equation can be applied between two observations at different distances downwind of sources, provided there are no new sources of the particular hydrocarbons of interest between the sampling points. The implicit assumption in the application of these equations is that the hydrocarbons do not affect the OH radical concentration. This may happen indirectly if significant increases in ozone occurs between sampling points as a result of the hydrocarbon emissions.

Using information on the sources of hydrocarbons in the London area and the published rate constants for the OH radical reaction for species such as acetylene, ethane, propane and propene, estimates of the average value of the OH radical concentration could be made. Values of about 5E+6 molecules/cc were obtained for this flight.

5 CONCLUSION

Experience has eliminated some of the pitfalls that can occur when making airborne measurements. However, careful planning and clear objectives are required to ensure success.

6 REFERENCES

1. Anderson, N.A., 1963, Step-analysis method of finding time constant, Instruments and Control Systems, 36, 131.

2. Mage, D.T., and Noghrey, J., 1972, True atmospheric pollutant levels by use of transfer function for an analyzer system, JAPCA, 22, 115

3. Cooper, M.J., 1977, Deconvolution: if in doubt, don't do it, Physics Bulletin, Oct. 1977, 463.

4. Clark, P.A., Fletcher, I.S., Kallend, A.S., McElroy, W.J., Marsh, A.R.W., Webb, A.H., 1984, Observations of cloud chemistry during long-range transport of power plant plumes, Atmos. Env., 18, 1849.

5. Kallend, A.S., Clark, P.A., Cocks, A.T., Fisher, B.E.A., Glover, G.M., Marsh, A.R.W., Moore, D.J., Sloan, S.A., and Webb, A.H., 1982, The fate of atmospheric emissions along plume trajectories over the north sea; Final report to E.P.R.I. Contract RP1131-1. Central Electricity Research Laboratories Report TPRD/L/2265/R82.

6. Dear, D.J. and Laird, C.K., 1984, In pursuit of acid rain, New Scientist, 22 November 1984.

7. TULLA, 1986, Extract from Report on the CERL aircraft's participation in the TULLA measurement campaign in Baden-Wurttemburg, March 1985. in press.

8. Lightman, P., Kallend, A.S., Penkett, S.A., Jones, B.M., Glover, G.M. and
    Marsh, A.R.W., 1986, The influence of hydrocarbons on the formation of at-
    mospheric acidity; Final Report to CEC in press.

Regional and Long-range Transport of Air Pollution,
Lectures of a course held at the Joint Research Centre, Ispra, Italy,
15–19 September 1986, S. Sandroni (Ed.), pp. 215–247
© Elsevier Science Publishers B.V., Amsterdam — Printed in The Netherlands

PERFLUOROCARBON TRACER TECHNOLOGY

R.N. DIETZ

1 INTRODUCTION

Whether we want to know how air pollutants are dispersed and diluted as they travel hundreds to thousands of kilometers across the countryside, how cooled nighttime mountain air flows down into the valleys, how air leaks into buildings, or gases or liquids escape from containments, there is a significant role for the technology of tagging and tracing.

Tagging, in these cases, is the process of adding a particular gas, a tracer, to the substance or system to be traced, and then quantitatively detecting or tracing its presence with a variety of sampling and analysis tools. Subsequently, models are used to quantify or to make inferences as to the physical and perhaps chemical nature of the processes governing the particular application.

One of the problems facing the research community is the effect of pollutant sources on local and long-range receptor sites. Regardless of the nature of the pollutant emissions, that is, their physical and chemical interaction with the environment as well as their consequence on human health, the prevailing factors in source-receptor evaluations are the direction of transport of pollutants after they leave the source, how they are dispersed as the pollutants are carried downwind and, ultimately, what will be the diluted concentrations at various distant sites due to dispersion alone.

These processes of atmospheric transport, dispersion, and dilution are complex in different ways depending upon the scales of atmospheric transport (see Table 1). For some time, physical models have been applied to the forecasting of trajectories and dispersion and the prediction of downwind concentrations. Traditionally, model verification has been accomplished through the use of pollutant air quality data and other tracers-of-opportunity such as the anthropogenic use of halocarbons or through the intentional release of conservative tracers.

Such tracers can be either gases or particles, the latter not truly being conservative since they can be scavenged by rain and dry deposited on scales of transport other than short-range. For long-range transport, the required properties of conservative gaseous tracers (see Table 2) are met by a family of perfluorocarbon tracers (PFTs), which are the focus of this presentation.

TABLE 1

Atmospheric transport, dispersion and dilution

1. Scales of atmospheric transport
   - Short-range (up to 10 km) --- coastal, urban, fumigation
   - Long-range (up to 500 to 1500 km) --- acid rain precursors, storms
   - Continental (>2000 km) --- climate and weather
   - Complex terrain (up to 15 km) --- mountains, valleys
2. Modeling long-range transport (up to 1000 km)
   - Trajectories
   - Dispersion
   - Concentration predictions
3. Experimental model verification
   - By use of air quality data and tracers-of-opportunity
   - By use of intentionally released conservative tracers

TABLE 2

Conservative gaseous tracers

1. Properties
   - Non-depositing
   - Non-scavenged
   - Non-reactive
   - Low atmospheric background
   - Limited industrial use
   - Sensitively detectable
2. Perfluorocarbon tracers (PFTs)
   - PDCB (P[a] - dimethylcyclobutane)
   - PMCP (P - methylcyclopentane)
   - PMCH (P - methylcyclohexane)
   - PDCH (P - dimethylcyclohexane)

[a]P as in PDCB means "perfluoro"

Descriptions of the use of other familiar gaseous conservative tracers, for example, sulfur hexafluoride (SF_6), halocarbons, and deuterated methanes (refs. 1-2), will only be briefly covered. The four PFTs listed in Table 2, which are in the generic family of perfluoroalkylcycloalkanes, have been used in a number of tracer experiments, some of which will be described later.

Table 3 lists the units for reporting atmospheric tracer concentrations, since a variety of these may have been used in the figures or may be mentioned in the course of this presentation. Convention would have us use pL/L and fL/L.

The purpose of this presentation, as outlined in Table 4, is to describe the perfluorocarbon tracer technology developments at Brookhaven, including the latest identified as well as available PFTs and air sampling and analysis tools, to demonstrate their utility in a number of different atmospheric tracer experiments as well as in other applications, and to provide food-for-thought

TABLE 3

Units for atmospheric PFT concentrations

1. Parts-per-trillion
   - ppt
   - $X\ 10^{-12}$
   - $pp10^{12}$
   - nL/m^3 (nanoliters/cubic meter)[a]
   - pL/L (picoliters/liter)[a]

2. Parts-per-quadrillion
   - ppq
   - $X\ 10^{-15}$
   - $pp10^{15}$
   - pL/m^3
   - fL/L (femtoliters/liter)[a]

[a]nano (10^{-9}), pico (10^{-12}), femto (10^{-15})

TABLE 4

Outline of perfluorocarbon tracer technology presentation

1. Introduction

2. Description and design of PFT technology
   2.1 PFT identification and availability
   2.2 Air samplers
   2.3 Vertical atmospheric sampling cables
   2.4 Analysis tools

3. Applications of PFT technology
   3.1 Atmospheric tracer experiments
   3.2 Building air infiltration and indoor air quality
   3.3 Potential new applications

4. Conclusions

on new ways in which the PFTs can be applied in other research objectives. All of the important tools will be described, but emphasis will be given to work that has not previously been given in detail and to the latest developments in the technology.

2 DESCRIPTION AND DESIGN OF PFT TECHNOLOGY

The complete system for performing long range atmospheric transport and dispersion experiments consists of the PFTs, the air samplers, and the analytical methodology.

As will be shown shortly, availability and cost of PFTs are two key factors for consideration in the design of any long-range tracer experiment. In addition, the equipment for automatically and routinely releasing the tracer must be considered.

Air samplers are used for routine collection by adsorption, using both programmable and passive samplers, located both on the ground and aloft. The

platforms for vertical samples consist of programmable or miniature sampling
tubes in aircraft, balloon-supported vertical sampling cables with pumps on the
ground, and passive samplers connected to tethered balloon cables.

Analyses are performed routinely on adsorbent-based samplers returned to the
laboratory and also in real-time in the field using a dual-trap analyzer, which
can be mounted in aircraft.

First proposed to the U.S. National Oceanic and Atmospheric Administration
(NOAA) nearly a decade ago by James Lovelock (ref. 3), an intensive reduction
to practice has already been implemented by the U.S. Department of Energy (DOE)
at its Brookhaven National Laboratory (BNL) and Environmental Measurements
Laboratory (EML) facilities and by NOAA's Air Resources Laboratory in terms of
quantitatively and, more or less, routinely releasing, collecting, and automatically analyzing PFT air samples.

Before describing the PFT technology in detail, it will be useful to look at
a simplified picture of how the tracers are analyzed in order to understand the
advantages of the PFTs over other types of gaseous tracers. The PFTs are analyzed by gas chromatography which is shown in a simplified schematic in Fig.
1. The constituents in an air sample are thermally desorbed from the sample
tube and are injected into the carrier gas stream via the sample valve. Before
entering the chromatographic column, all the components will be present as a
slug (see the square wave in Fig. 1a). After passing through the column, the
constituents will be physically separated to an extent which depends on the
nature and conditions of the column. However, as shown in Table 5, the atmosphere contains many compounds whose concentrations exceed those of the PFTs
and which are detectable in the electron capture detector (ECD) used to measure
the PFTs. Included are O_2, nitrogen oxides, chlorofluorocarbons (Freons), SF_6,
and others, each of which could interfere with the early eluting PFTs (Fig.
1b). As will be described later, physical means (e.g., sampling onto an
adsorbent with subsequent purging) removes most of the oxygen and some of the
Freons, but a catalyst bed operating at about 200°C (Fig. 2) is needed to
destroy many of the remaining interfering compounds so that the surviving
PFTs can be detected (Fig. 1c). The importance of the catalyst bed should not
be underestimated in the successful determination of the PFTs.

Referring back to Table 2 then, it is the physical and chemical inertness of
the PFTs that not only prevents their loss in the atmosphere but also helps in
their separation and analysis from the less-stable interfering compounds and
makes them biologically inactive; thus they are perfectly safe to use (ref.
1). Because of their low solubility in H_2O and moderate vapor pressure they
are not readily scavenged nor deposited in the atmosphere, but, unlike SF_6,
their vapor pressures are low enough to allow them to be readily sampled onto
solid adsorbents. Their limited industrial use not only results in a low

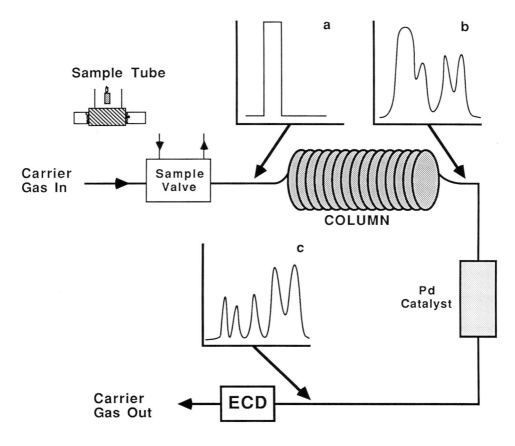

Fig. 1. Simplified schematic of a gas chromatograph (GC) system showing the function of the sample valve, the GC column, and the palladium (Pd) catalyst bed prior to electron capture detection (ECD). (a) Sample. (b) Interfering compounds and PFTs. (c) PFTs alone.

ambient background concentration but also precludes the possibility of numerous higher local concentrations which might confuse atmospheric tracer experimental results.

Lastly, the high affinity of PFTs for their reaction with electrons makes them some of the most sensitive compounds for detection on the ECD, which is a small (0.1 to 0.2 mL) reaction chamber containing an electron source. The cloud of electrons in the chamber is periodically collected, producing a current. When tracer molecules enter the cell, the reacted electrons can not be collected; this resulting reduction in current is a measure of the PFT concentration.

TABLE 5

Some components in the atmosphere

Gas	Rural Concentration, pp 10^{12}	Formula	Concentration Range
Nitrogen	780,900,000,000	N_2	
Oxygen	209,500,000,000	O_2	%
Argon	9,300,000,000	A	
Carbon dioxide	335,000,000	CO_2	
Methane	1,480,000	CH_4	ppm
Nitrous oxide	315,000	N_2O	
Ozone	35,000	O_3	ppb
Nitrogen oxides	3,000	NO,NO_2	
Methyl chloride	630	CH_3Cl	
Freon 12	305	CCl_2F_2	
Freon 11	186	CCl_3F	ppt
Carbon tetrachloride	135	CCl_4	
Chloroform	20	$CHCl_3$	
Sulfur hexafluoride	0.85	SF_6	
Bromotrifluoromethane (F13B1)	0.75	$CBrF_3$	
Perfluorodimethylcyclohexane	0.022	C_8F_{16}	
Perfluoromethylcyclohexane	0.0045	C_7F_{14}	ppq
Perfluoromethylcyclopentane	0.0032	C_6F_{12}	
Perfluorodimethylcyclobutane	0.00034	C_6F_{12}	
Deuterated methane	0.00030[a]	CD_4	sub-ppq
Deuterated methane	0.00001[a]	$^{13}CD_4$	

[a]Current limits of detection; actual background estimated at $<0.000001pp10^{12}$.

2.1 PFT identification and availability

Since many researchers have used a variety of gaseous tracers other than the PFTs (refs. 1-2), a comparison of the properties and costs of their use in long-range atmospheric tracer experiments will be made in order to demonstrate the cost-advantages of the PFTs. Also presented will be analyses of ambient air samples for currently-used PFTs to demonstrate background levels, analyses of the manufacturers' PFT samples to determine purity and to identify various isomers, examples of new tracers to be used in up-coming continental-scale tracer experiments, and lastly, a brief description of future potential tracers.

The physical properties of a few of the important compounds previously used as tracers are shown in Table 6. The tracers are listed in decreasing order with respect to the cost of the quantity released, which is based on the ability to detect the tracer concentration above its ambient background level. Many of these tracers are liquids at room temperature and must be released by either atomization or vaporization followed by dilution to prevent condensation until the vapors are sufficiently diluted below the ambient dew point

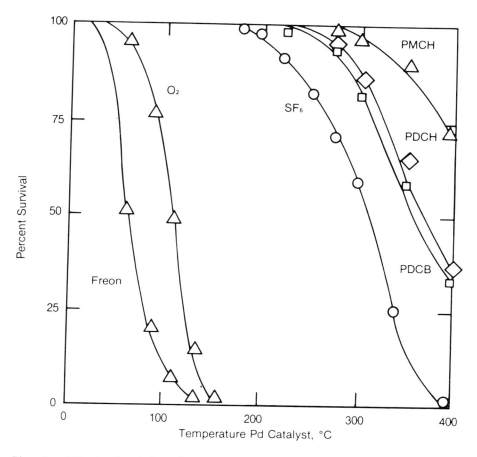

Fig. 2. Effect of catalyst bed temperature on destruction of PFTs and interfering compounds such as O_2 and Freons (chlorofluorocarbons).

TABLE 6

Properties of some gaseous conservative tracers

Symbol	Formula	Molecular weight	Phase at 20°C	Boiling point, °C	Supplied form
SF_6	SF_6	146	gas	-64	liquefied gas
F13B1	$CBrF_3$	149	gas	-58	liquefied gas
PDCH[a]	C_8F_{16}	400	liquid	102	liquid
F12B2	CBr_2F_2	210	liquid	25	liquid
PMCH[a]	C_7F_{14}	350	liquid	76	liquid
PDCB[a]	C_6F_{12}	300	liquid	45	liquid
CD_4	CD_4	20	gas	-160	gas
$^{13}CD_4$	$^{13}CD_4$	21	gas	-160	gas
ocPDCH[a]	C_8F_{16}	400	liquid	102	liquid

[a]The chemical names for these four PFTs are given in Table 8.

concentration. Thus the release apparatus is somewhat more complex than for the liquefied gaseous tracers such as SF_6 and F13B1. Descriptions of the tracer release equipment as well as the calculations to be made to estimate the release quantities have been summarized elsewhere (ref. 1). An alternative technique for the release of liquefied gases such as SF_6 was based on the release of the liquid through fuel burner nozzles (ref. 4).

The relative tracer costs for plume centerline detection of tracer at 100 times background concentrations were shown in Table 7. Even though PFTs cost

TABLE 7

Relative gaseous conservative tracer costs

Distance: 100 km
Desired Concentration: 100 times background at centerline
Release time: 3 hours

Symbol	Formula	Ambient conc.,fL/L[a]	Cost $/kg	Released qty., kg	Relative[d] tracer cost, 1000 $
SF_6	SF_6	2000[b]	10	2320	23.2
F13B1	$CBrF_3$	750	15	887	13.3
PDCH	C_8F_{16}	22	120	70	8.4
F12B2	CBr_2F_2	<20	30	<33	<1.0
PMCH	C_7F_{14}	4.5	100	12.5	1.2
PDCB	C_6F_{12}	0.34	500	0.81	0.41
CD_4	CD_4	0.60[c]	3000	0.095	0.29
$^{13}CD_4$	$^{13}CD_4$	0.02[c]	50000	0.0033	0.17
ocPDCH	C_8F_{16}	0.3	220	0.95	0.21

[a]1000 fL/L equals pL/L or 1 part-per-trillion.
[b]Near-urban SF_6 is 2000 fL/L or more in many locations because of significant use; tropospheric background is 850 fL/L.
[c]Values for deuterated methanes represent current limits of detection (S/N=2) for a 1 m^3 air sample; actual backgrounds are about 0.0005 fL/L.
[d]Cost is for required release quantity and does not reflect the cost of the analyses.

10 to 50 times that of SF_6, the significantly lower background concentrations mean less material needs to be released. Thus, for PMCH, the relative tracer cost is more than 10-fold less than if SF_6 or F13B1 had been used in a given tracer experiment, for a savings of about $10,000 to $20,000. Of course, for a 1000-km experiment, the savings would be an order of magnitude higher. For the last PFT, the ortho(cis) isomer of PDCH, the cost would be another factor of 6 lower. Only the deuterated methanes have comparable tracer costs, but their analytical costs are substantially higher, perhaps more than an order of

magnitude (ref. 2). In addition to their lower costs, the family of useful PFTs has the potential to grow to 10 to 12 tracers sampled and analyzed on the same equipment.

(i) <u>Previously used PFTs</u>. Two example chromatograms of the analysis of 14.5- and 3.8-L ambient air samples are shown in Fig. 3. Specific details on the sampling and analysis technique will be described later. Briefly it consisted of pulling the air through tubes packed with a solid, charcoal-like adsorbent which was then thermally desorbed into the gas chromatograph. The identification of the PDCB, PMCP, and PMCH peaks as well as the group of peaks representing the PDCH isomers had been previously determined (ref. 2). The definitions of the symbols for each of the named peaks are given in Table 8.

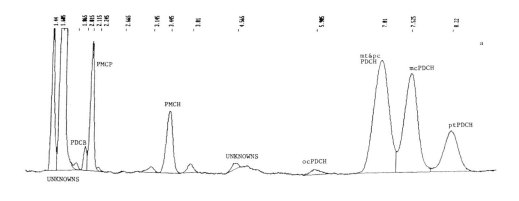

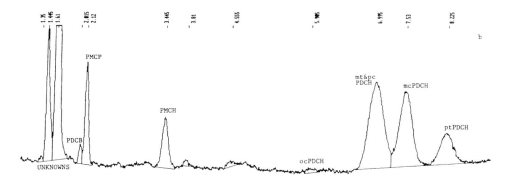

Fig. 3. Chromatograms of adsorbent-collected ambient air samples analyzed on a 6-foot (1.8-m) by 0.093-inch (2.36-mm) ID, 0.1% SP 1000 on Carbopack C (Supelco) column at 140°C with 22 mL/min of 5% H_2 in N_2 (a) 14.5-L air sample. (b) 3.8-L air sample.

TABLE 8

Identification of current and potential PFT components

No.	Symbol	Name(perfluoro-)	Elution[a] time,min	Ambient air conc., fl/l
1	PDCB	dimethylcyclobutane	1.26	0.34 ± 0.01
2	PMCP	methylcyclopentane	1.32	3.22 ± 0.03
3	PMCH	methylcyclohexane	2.08	4.46 ± 0.05
4	ocPDCH	ortho[b](cis)[c]-DCH[d]	3.43	0.3 ± 0.1
5	mtPDCH	meta(trans)-DCH	3.97	7.7
6	pcPDCH	para(cis)-DCH	4.04	2.2
7	PECH	ethylcyclohexane	4.12	<2
8	otPDCH	ortho(trans)-DCH	4.16	0.2
9	mcPDCH	meta(cis)-DCH	4.24	8.2
10	ptPDCH	para(trans)-DCH	4.62	3.4 ± 1
11	1 PI	1-indane	4.53	?
12	2 PI	2-indane	5.99	?
13	1 PTCH	1-trimethylcyclohexane	8.98	0.07
14	2 PTCH	2-trimethylcyclohexane	9.50	<0.03

[a]Mid-point of tracer peak in gas chromatogram at conditions given in Fig. 3, but with column at 160°C.
[b]ortho, meta, and para mean the 1,2-, 1,3-, and 1,4-isomers.
[c]cis and trans mean the alkyl groups (e.g., methyl) are on the same or opposite sides, respectively, of the molecular plane.
[d]DCH represents dimethylcyclohexane.

By collecting and analyzing ambient air samples of about 0.5 to 14.5 L in volume and plotting tracer quantity versus air sample volume as shown in Fig. 4, it was possible to determine the ambient air concentrations of each of the PFTs as listed in Table 8. The standard deviations on the concentrations do not reflect the uncertainty in the calibration gas standards which is another ±5%.

The actual determination of which isomers of PDCH were represented by the peaks in Fig. 3 at retention times of about 6 to 8 minutes (the numbers at the top of the figure) was only recently resolved. The elution order for the isomers of the hydrocarbon, dimethylcyclohexane (DCH), was previously determined by gas chromatography on a graphitized carbon support (ref. 5). The seven isomers of DCH (the cis and trans of each of the ortho, meta, and para isomers plus the 1,1-isomer) were available as relatively pure hydrocarbons. These were analyzed on the same support used in our PFT gas chromatograph, confirming the elution order found in ref. 5.

Starting with pure quantities of the ortho-, meta-, and para-xylenes as well as ethyl benzene, the catalytic (cobalt trifluoride) fluorination by the PFT manufacturer (ISC Chemicals Limited, Avonmouth, Bristol, Great Britain) yielded the perfluorinated versions. Subsequent analysis by thermal conductivity gas chromatography gave the chromatograms shown in Fig. 5 and the PDCH compositions shown in Table 9. Thus, starting with pure ortho-xylene, the resultant product is only 84.4% ortho-PDCH, 46.4% as the cis isomer and 38.0% as the trans

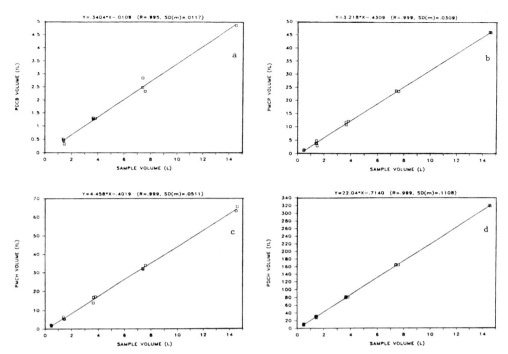

Fig. 4. Quantity of four PFTs found in the ambient air for samples ranging from 0.5 to 14.5 L. The slope of the line representing PFT quantity (fL) versus air sample volume (L) is the background ambient air concentration (fL/L). See Table 8 for tracer names. (a) PDCB. (b) PMCP. (c) PMCH. (d) PDCH is the total of all isomers.

isomer. The purities of the resultant meta- and para-PDCH, which were only 83.1 and 91.8%, respectively, as shown in Table 9, were computed by an iterative procedure since not all of the isomers are separately resolved. In fact, as shown in Fig. 5, only ocPDCH (the first peak) and ptPDCH (the last peak) are clearly resolved. The mt- and pc-PDCHs are nearly coincident as are the ot- and mc-PDCHs. In addition, the PECH is right between the two meta-PDCH peaks.

Referring back to the ambient air chromatograms of Fig. 3, it was now possible to identify the PDCH isomers in the last four peaks, which have been appropriately labelled. The ambient air concentration of each of the six PDCH isomers listed in Table 8 was computed as follows. The ocPDCH concentration was determined directly because it was separately resolved. Then the otPDCH was computed assuming the same ratio as in the manufactured ortho-PDCH (Table 9) which was then subtracted from the next-to-the-last peak to leave the ambient concentration of the mcPDCH. Note that the peak in Fig. 3 was only labelled as mcPDCH because the otPDCH was trivial (compare the ocPDCH peak).

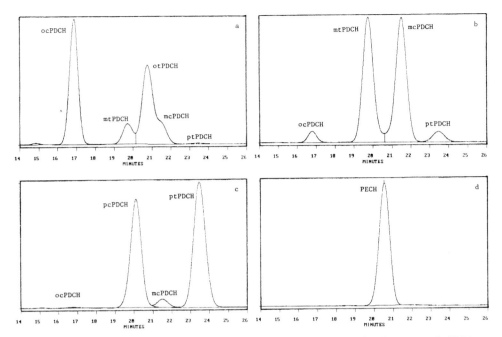

Fig. 5. Identification of the relative elution order of the isomers of PDCH as well as the location of PECH (see Table 8 for tracer names). (a) ortho-PDCH. (b) meta-PDCH. (c) para-PDCH. (d) PECH.

TABLE 9

Composition[a] of PDCH isomers, percent (mol)

Component	Ortho-PDCH	Meta-PDCH	Para-PDCH
PMCH	2.4	1.0	–
ocPDCH	46.4	3.1	0.1
mtPDCH	6.6	38.7	3.8
pcPDCH	0.3	4.1	36.7
otPDCH	38.0	2.6	0.1
mcPDCH	5.6	44.4	4.2
ptPDCH	0.4	6.1	55.1
Totals	99.5	100.0	100.0
Primary cis and trans:	84.4	83.1	91.8

[a]Analyzed by thermal conductivity (TC)GC which, unlike the ECD, gives the same response for each isomer.

Similarly, the ptPDCH was directly determined (it stands alone), and the computed pcPDCH was then subtracted from the combined mt- and pc-PDCH to yield the mtPDCH.

The resultant ambient air concentrations for each of the PDCH isomers are only good estimates because the relative response of the ECD to each of the isomers is not the same. The ECD was originally calibrated with the meta-PDCH, which is the predominant isomer in the ambient air at 15.9 of the 22.0 fL/L total (see the slope from Fig. 4d) or 72%; thus the meta-PDCH concentrations should be reasonably accurate. A preliminary evaluation of their ECD response showed that only the ptPDCH, the last peak, had a significantly lower response, indicating that its ambient concentration might be higher. Further laboratory studies will be conducted to accurately determine this important variable of ECD response.

From the ambient chromatograms in Fig. 3 and the known concentrations of each PFT (Table 8), it is possible to see that the limits of chromatographic detection range from 0.2 fL for PDCB and PMCP, and 1 fL for PMCH and ocPDCH, to about 1.5 fL for ptPDCH.

(ii) <u>New and future PFTs</u>. There are two basic processes that have been used commercially for the production of PFTs, here restricted to the family of per-fluoroalkylcycloalkanes because they have the maximum response to the ECD. Other perfluoroalkanes and other perfluorocarbons are two or more orders-of-magnitude poorer in detection capability (ref. 1). The one process already mentioned, cobalt trifluoride catalyzed fluorination, is available from ISC Chemicals Limited in England. The purity of their tracers has been from 85 to 99%, with a limited amount of the other existing and identified PFTs as impurities, generally less than 1%. This can be important in a tracer experiment if the PFT being released has a 1% impurity of another also being released. Of coarse, correction can be made based on the analyses of the impurities in the released PFTs, but that correction becomes more significant as the number of tracers used in any one experiment increases. Fortunately, ISC has been able to keep the purity of the PFTs they supply quite high for these applications. All of the PFTs in Table 8 with the exception of No. 1, PDCB, are supplied by ISC.

The other process for making PFTs is the dimerization of perfluoroalkenes at high pressures (up to 3000 atmospheres) and moderate temperatures (400°C). Originally patented by E.I. duPont in Wilmington, Delaware, more than 19 years ago, the technique was used at one time to make the PDCB. They abandoned the technology more than five years ago and other small companies can now produce a number of the dimerization products, generally perfluorodialkylcyclobutanes, but at costs up to ten times or more those of the PFTs from ISC. However, the PDCB is a potential continental scale tracer because it has the highest ECD response of any of the PFTs and has a low ambient concentration. Recently, the Flora Corporation in North Carolina has indicated an interest in supplying tracers made by this process.

Beginning in January 1987, a major tracer experiment, ANATEX (Across North America Tracer Experiment), will involve the release of two new PFTs never before intentionally released. This three-month-long experiment requires background concentrations to be less than 0.4 fL/L at a tracer cost of less than $200/kg if the total tracer cost is to be less than $500,000 for daily average PFT concentrations no higher than 10 times background. One tracer meeting that need was the ocPDCH which can be supplied by ISC.

A number of alternative compounds were suggested to ISC, one being per-fluorotrimethylcyclohexane (PTCH) and another, perfluoroindane (PI). A mixture of these two new tracers plus four of the earlier PFTs was analyzed as shown by the chromatogram in Fig. 6a, which showed the elution of the 1 PI (at 4.5 min) and seven isomers of PTCH at 6.3 to 11.2 min. The two major isomers of PTCH were arbitrarily named. Ultimately they will be identified to locate potential

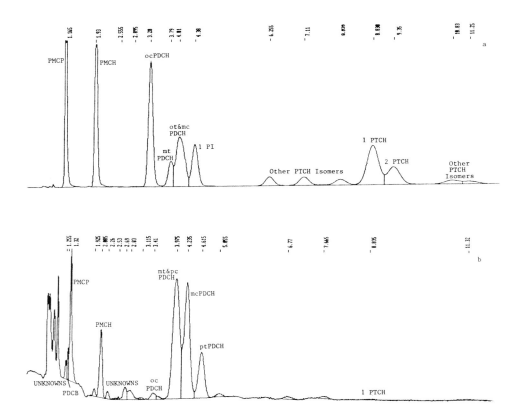

Fig. 6. Chromatograms of (a) about equimolar standard of PMCP, PMCH, o-PDCH, m-PDCH, PI and PTCH, and (b) about 50-L of ambient air, but with the column of Fig. 3 at 160°C.

future tracer types. The retention or elution time of all the PFTs at the 160°C column temperature are shown in Table 8.

A 50-L ambient air sample was analyzed under the same GC conditions. The normal PDCH isomer distribution was attained as shown in Fig. 6b. Note that the 1 PI in the standard had the same elution time as the ptPDCH in the air sample. This alone would preclude PI as a continental-scale tracer because the background ptPDCH would interfere. But it was also determined that the PI was sensitive to catalyst destruction depending upon its operating condition and cleanliness. Thus PI was temporarily abandoned as a tracer pending further research on alternative catalysts.

In Fig. 6b, it is difficult to see if any PTCH exists in the ambient air. It certainly would appear to be less than the ocPDCH level. Fig. 7 shows an expansion of that chromatogram in two regions. The early region shows that the PDCB is still separated from the PMCP. The late region shows that at the 1 PTCH elution time, the very small peak would correspond to an ambient air concentration of about 0.07 fL/L, which would make it a viable continental-scale tracer.

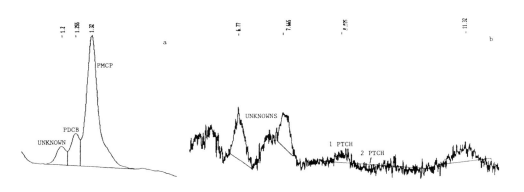

Fig. 7. Expansion of Fig. 6b (a) Early region showing resolution of PDCB and PMCP. (b) Late region indicating the detection of negligible quantity of PTCH in ambient air.

(iii) <u>Summary of PFT investigation</u>. It was shown that the family of PFTs can be uniquely detected on a single analysis of an air sample by a special gas chromatograph to be described in detail later. Table 10 lists the currently usable PFTs, their limits of detection, and the principal supplier. There is a good likelihood that additional research into fluorination of alkylcycloalkenes

230

TABLE 10
Currently usable PFTs[a]

No.	Tracer	Limit of detection, fL
1	PDCB	0.2
2	PMCP	0.2
3	PMCH	0.9
4	o-PDCH	1.0
5	m-PDCH	4[b]
6	p-PDCH	1.5
7	PTCH	3.5

[a]All PFTs supplied by ISC Chemicals Ltd.
(England) except PDCB, which is available
in small quantities elsewhere.
[b]Determined by difference.

and the dimerization of perfluoroalkylcyclobutanes will lead to a much larger
family of useful PFTs.

2.2 Air samplers

The PFTs can be collected as whole air samples using conventional means such
as bags, bottles, syringes, etc. (ref. 1). However, the PFTs can be adsorbed
onto the surface of carbon steel surfaces and, of course, they are soluble in
fluoroelastomers (e.g., Viton), silicone rubber, and fluoropolymers (e.g.,
Teflon) so these materials should be avoided. We have found that polyolefins,
polyurethane, and nylon are quite compatible, as are stainless steel and spe-
cially treated aluminum (ref. 1).

Like almost all hydrocarbons, PFTs readily adsorb onto charcoal-like materials. Unlike most hydrocarbons, however, the PFTs are thermally stable to quite
high temperatures in the absence of any reducing catalysts, thus permitting
thermal recovery for subsequent analysis. Based on these considerations, two
adsorbent samplers were designed and built.

(i) Programmable PFT sampler. This was developed by Dietz at Brookhaven and
commercially manufactured for NOAA by Gilian Instrument Corporation as the
Brookhaven Atmospheric Tracer Sampler (BATS). The entire unit shown in Fig. 8a
measures just 14 x 10 x 8 inches and weighs 7 kg. The lid contains 23 sampling
tubes, each containing 150 mg of 20-50 mesh type 347 Ambersorb (Rohm and Haas
Co.) which can retain all the PFTs in more than 30 L of air. Internal batteries provide power for up to one month of unattended operation of all the

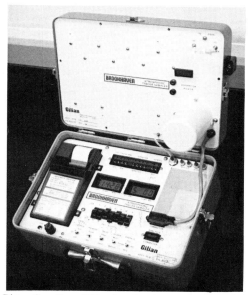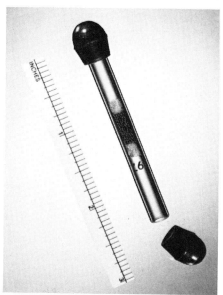

Fig. 8. Two PFT adsorbent samplers. (a) The programmable Brookhaven Atmospheric Tracer Sampler (BATS). (b) The passive capillary adsorption tube sampler (CATS).

automatic sampling and recording features. In the laboratory, sample recovery was accomplished by direct ohmic heating of the adsorption tube to 400°C, with the PFTs being purged from the BATS tube through an automated ECD-GC system, analyzing all 23 tubes in about 3 h (ref. 6).

The ability to collect variable volume air samples with good precision is indicated by the results shown in Fig. 4 for samples collected on a BATS. However, since their first commercial production in 1980, they have been used in no less than seven major atmospheric tracer experiments and numerous other field projects. The major weak points found in the instrument were the pump and the paper tape printer. In preparation for ANATEX, all of the BATS bases are having an EML-developed pump (ref. 7) called TAPS (tethered air pump system) installed and, in the BATS lids, a solid state memory device which can be read directly into a computer. These modified BATS units should have a substantially improved reliability.

A new prototype BATS base is under development to provide both computerized control of all the sampling functions as well as real-time determination of the pumping rate based on a patented pump and pressure and temperature measurements across a computer-selectable orifice.

(ii) <u>Passive PFT sampler</u>. Originally developed as a means to measure the indoor PFT concentration during the determination of air infiltration and air exchange rates in homes and buildings using miniature PFT permeation sources (refs. 8-9), the passive sampler has also been used in atmospheric tracer studies (ref. 10).

In its first configuration, one end of the sampler contained a 1-mm capillary tube and so was coined the Capillary Adsorption Tube Sampler (CATS). The present configuration of a CATS is shown in Fig. 8b. The passive sampler, which is made from 6 mm OD by 4 mm ID glass tubing exactly 2.5 inches (6.4 cm) long, contains 64 mg of Ambersorb 347. Sampling occurs by the process of Fickian diffusion when one cap is removed as shown. From the depth to the bed (2.76 cm), the cross-sectional area (0.126 cm^2), and the empirically derived diffusion coefficients of the PFTS in air, it was determined that the CATS sampled at a rate equivalent to about 200 mL of air per day for PMCH.

Numerous comparisons with the BATS and the CATS have shown that they determine the PFT concentration to within the same precision of about ±5%. A special rack was built to sequentially desorb 23 CATS for GC analysis in much the same way that the 23 tubes in the BATS lid were automatically stepped (ref. 9) for analysis.

2.3 Vertical atmosphering sampling cables (VASC)

Although aircraft platforms with real-time analyzers or with high-flow-rate sample collection systems can provide useful pictures of the plume aloft, in certain complex terrain studies, for example, near mountains and valleys and especially at nighttime, the well-defined directions of the valley flows lend themselves to be monitored with balloon-borne vertical sampling devices for the vertical definition of the transport and dispersion of pollutants.

Several types of ballon-borne systems have been designed and used in the ASCOT (Atmospheric Studies of Complex Terrain) program in which tracers, including SF$_6$, heavy methanes, and perfluorocarbons, were measured in air samples collected to a height of 500 m above the ground (refs. 11-12). Traditionally, these systems have used tethered balloons, from 5 to 100 m^3 in volume, to suspend whole air sampling packages such as plastic bags with pumps and automated syringes at several altitudes along the cable or to collect sequential whole air samples over time and space as the balloon is moved from the ground to its maximum altitude.

In this section, a brief description will be given of two novel balloon-borne VASCs - a multitube, ground-based cable (VASC-I) with multiple sampling pumps and a single-tube, multi-level adsorbent sampler cable (VASC-II) with a single pump on the ground. Each was carefully designed by considering all aspects of the flow of a compressible fluid (air) through long tubing,

computing the pressure drop by taking into account laminar and, where necessary, turbulent flow friction effects, the effect of gravity since the air was
being pulled down a vertical column, the impact of density changes due to
changes in altitude as an air parcel moves down the cable as well as effects of
expansion and contraction when the air flow is controlled by a capillary tube
or a sub-miniature orifice plate - a small hole in a nearly flat plate such as
a small hole drilled through the wall of the tubing cable. Laboratory validation tests conducted on the models used to compute flows at various pressures
through tubing (Fig. 9a) and orifice holes (Fig. 9b) showed good agreement
between measured points and computed curves. Also taken into account were the
characteristics of the vacuum pumps used to pull the air through the cables.
Matching the total flow rate and the pressure drop of the cable to the pump
capacity required multiple, step-wise computations from the top of the cable to
the pump at the bottom.

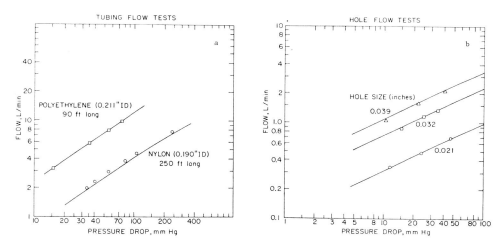

Fig. 9. Flow versus pressure drop tests for the validation of the model used
in the design of the vertical atmospheric sampling cables (VASCs). (a) Flow
through long plastic tubes. (b) Orifice plate model for flow through holes
drilled in a plastic tube wall.

The process consisted of assuming an orifice dimension and orifice flow rate
at the uppermost sampling location. Using the barometric pressure at that
altitude, the required pressure inside the cable necessary to cause that flow
was computed. Then, using that flow rate and the dimensions of the vertical
tubing cable, the pressure inside the cable at the next sampling location was
computed. From that pressure and the ambient pressure at that new location,
the dimension of the next orifice was computed assuming an orifice flow rate
equal to that of the first. This process was repeated, adding the flow from
subsequent sampling locations to the total cable flow when calculating the

234

cable tubing pressure drop from position to position. At the end of the cable
on the ground, the final total flow rate and inside tubing pressure had to meet
the design flow rate versus pressure characteristics of the pump. The iterative solution was repeated until those characteristics were met.

(i) VASC-I. Four separate 1/4-inch OD polyethylene sampling cables were
bundled together in a braided Kevlar sheath (Cortland Line Company) and suspended from a 100-m^3 balloon tethered for an altitude of 1600 ft above the
ground (ref. 12). The cable was designed at Brookhaven to sample four separate
400-ft layers from 0-400 ft up to 1200-1600 ft as shown in Fig. 10. Each cable

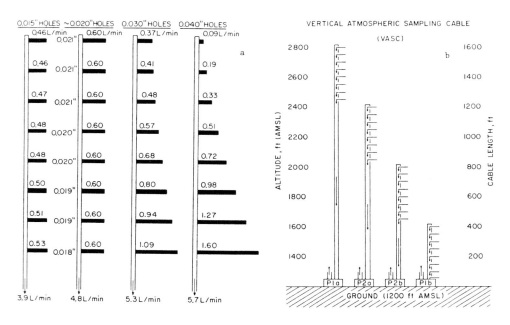

Fig. 10. Design of the vertical atmospheric sampling cable I (VASC-I). (a)
Calculated effect of hole diameter on flow through 8 holes spaced 50-feet
(15.2-m) apart for the four hole diameters shown. (b) Conceptual schematic of
VASC-I.

had eight intake holes at 50-ft intervals over the designated sampling span.
Fig. 10a showed that it was necessary to select an orifice diameter of about
0.020-inches (0.50-mm), varying slightly from top to bottom to have a uniform
flow rate into each hole of 0.60 L/min for a total flow rate of 4.8 L/min at
the pump. If the hole size had been 0.040-inches (1.0-mm), it was shown that
the sampling would have been biased to the lowest portion of the sampling
zone. Pumps on the ground continuously pull air through each of the four
cables to air sampling devices on the ground, certainly a significant
attribute.

(ii) <u>VASC-II</u>. A major disadvantage of VASC-I was its more than 18-kg weight requiring the large 100-m³ balloon and gasoline-powered winch. For sampling in complex terrain where access to sampling sites for the tethering of vertical cables is limited, a much lighter cable was desired. In addition, because flow in complex terrain is usually more variable near the ground, it was desired to have a greater number of sampling locations in the lower region of the cable.

The final design consisted of a 500-m-long polypropylene tubing, 0.062-in. (1.6-mm) ID by 0.010 in. (0.25-mm) wall, with 15 orifices located 50-m apart from 500 m down to 200 m and then decreasing to only 10 m apart near the ground. The orifice lengths (Fig. 11a) were individually sized using the flow model to allow the air to be sampled at closer intervals near the ground, but at the same flow rate, about 4.5 ml/min, through each sampling tube at every altitude. Each orifice was designed to be equipped with a CATS passive sampler of the type described earlier, but used in an active mode. Flow through each sampling tube was accomplished with a single small pumping station on the ground, pulling a total of 67.5 ml/min at about 0.5 atm vacuum. The whole cable, with 15 samplers, weighed just 0.6 kg, which, together with an airsonde for altitude and temperature requirements, was readily supported by a 5-m³ balloon (Fig. 11b).

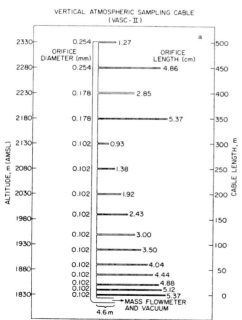
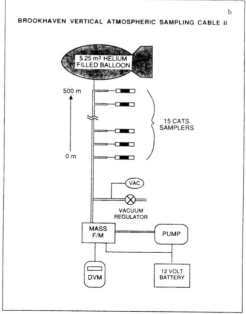

Fig. 11. Design of the vertical atmospheric sampling cable II (VASC-II). (a) Schematic of computed orifice dimensions necessary to achieve the same flow through each sampler. (b) Balloon and pumping system.

Each orifice was cut to the proper length (not a trivial task to prevent closing of the holes) and its effective diameter was computed using the flow model. Each of the measured tube diameters from individual flow measurements was larger than the manufacturer's claimed value (Table 11). After installation on the cable, the diameters were again determined by measuring the individual flow while pumping on the entire cable. The agreement on- and off-cable was excellent.

TABLE 11

VASC-II Calibration of orifice diameters

Sampler height	-----Orifice diameter (mil)[a]----		
	Claimed	Off cable	On cable
500	10	11.66	11.52
450	10	11.33	11.39
400	7	7.84	7.88
350	7	7.96	8.06
300	4	5.09	5.17
250	4	5.13	5.17
200	4	5.22	5.26
160	4	5.29	5.33
120	4	5.22	5.31
90	4	5.43	5.49
60	4	5.37	5.39
40	4	5.11	5.20
20	4	5.31	5.37
10	4	5.30	5.29
0	4	5.42	5.46

[a]A mil is 0.001-in or 0.025-mm.

The advantages of VASC-II were its low weight, multiple (15) sampling positions, and the fact that the sample did not flow through the cable. However, a disadvantage compared to VASC-I was that the sampling was not truly continuous. An integrated 0.5-L sample could be obtained in just 2 h, sufficient to measure down to the ambient background of PMCP and PMCH. But it requires about 1 h to replace the CATS before sampling can begin again.

2.4 Analysis tools

For application to long-range tracing, the principal PFT analyzers are the laboratory gas chromatograph system for the analysis of samples collected on the programmable and passive samplers and a real-time dual-trap analyzer for in-the-field collection and analysis of PFTs.

Another analyzer which continuously responds to electron capturing tracers by combusting the oxygen with hydrogen over a catalyst bed, will respond to PFTs. But its detection capability is limited to about 1 to 30 pL/L (parts-per-trillion) and is generally not economically useful (considering tracer costs) beyond 100 km. Details of the operation of this continuous analyzer have been presented earlier (refs. 13-14) and will not be discussed here.

(i) The laboratory gas chromatograph system. The system is comprised of the gas chromatograph (GC), the data handling devices, gas standards, and the PFT adsorbent samplers, the latter of which have already been described. The use of the system requires a temperature-controlled room and an uninterruptable power supply capable of supporting the data handling system during brief power failures. The GC has internal battery backup for its microprocessor.

The operation of the GC can first be given in a simple overview. Whether from programmable or passive samplers, the sample is automatically thermally desorbed and passed through a Pd catalyst bed, permeation dryer, and a pre-cut column before being re-concentrated on an in situ trap. The trap prevents the collection of unwanted low molecular weight constituents, the pre-cut column prevents the passage of unwanted high molecular weight constituents, and the dryer removes moisture from the ambient samples. After thermally desorbing the trap, the PFTs are separated in the main column after passing through another Pd catalyst bed and detected in the ECD.

The current configuration of the GC system (Fig. 12) was used to produce the chromatograms shown in Fig. 6. Two processes occur during a single cycle, analysis of a previously collected sample and the loading of a new sample onto the trap. At the start of the cycle, the FD valve goes on as well as the Florasil trap valve (FS). Note that all the valves are shown in their "off" position; "on" means the FS valve rotor turns 90 degrees and the others, 60 degrees. Thus, when heat is applied to the FS trap, the adsorbed PFTs are flushed out through catalyst bed "A," catalyst bed "B," the dryer, the main column, and the detector (ECD). As shown by the chromatogram in Fig. 6, the entire process for the last PTCH isomer to elute is under 12 min; the cycle time was set for 12 min.

However, during this time, another sample tube is desorbed, processed, and collected on the trap. For the first 3 min, the sample tube is purged of oxygen by the carrier gas (5% H_2 in N_2). Then both the PC and SV are turned on and heat is applied to the sample tube to sweep the PFTs into the pre-cut column, a 22-in. (56 cm) by 0.113-in. (2.9 mm) thin-walled stainless steel column packed with Unibeads 2S (Alltech) at 85°C. The unknown, early-eluting interfering compounds will flow out of purge vent #2. Just before the first PFT elutes from the pre-cut column, the FD valve is turned off and the FS trap is opened. This allows the PFTs to be collected in the Florasil trap as they leave the pre-cut column, which is ohmically heated to a higher temperature. When the last PFT component has entered the trap, all the valves go to their

238

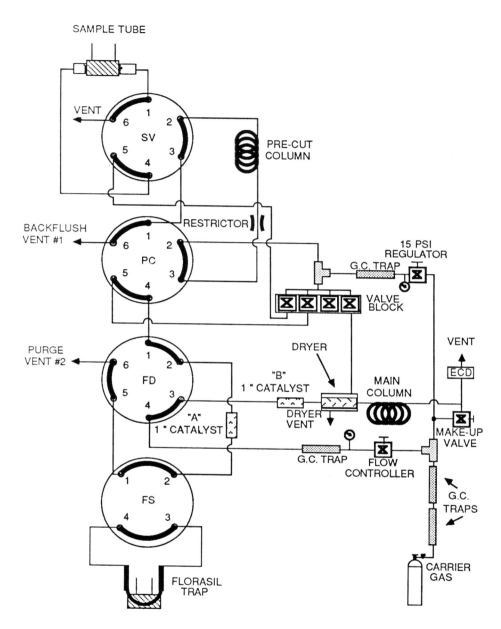

Fig. 12. Schematic of the laboratory GC plumbing. (SV) Sample valve. (PC)
Pre-cut column backflush valve. (FD) Flow direction value to isolate the
chromatography occurring in the main column from the loading of the next
sample. (FS) Florasil trap valve. All valves are shown in their "off"
position.

off position (at 9.6 min), which allows the pre-cut column to be backflushed at the higher temperature for more than 2 min to eliminate any heavy components.

The pre-cut column system prevents components lighter than the first PFT from seeing the catalyst bed "A" or from entering the trap. Components heavier than the last PFT selected for analysis are also precluded. By tailoring the pre-cut column temperatures, the PFT "window" can be increased or decreased at either the beginning or the end.

As mentioned earlier, the catalyst is important in removing interfering compounds. With this system, the PFT sample passes once through catalyst bed "A" on its way into the trap, once again upon recovery from the trap, and once through catalyst bed "B." This assures a good cleanup of the sample.

The data handling system is a basic Nelson Analytical 3000 Chromatography system with an IBM PC/AT, an ink-jet printer, the Series 860 A/D converters, and the Nelson 2600 Chromatography Software, set up in the laboratory as shown in Fig. 13. Other appropriate software and hardware complete the system,

Fig. 13. Gas chromatograph laboratory system. The IBM PC/AT is on the left followed by the printer, the Nelson A/D converter, the GC with recorder on top, the CATS desorption rack, and a BATS.

which currently handles three GCs for the analysis of PFTs. Once a data collection method has been stored in the A/D box, the computer can be used for other operations, such as checking previous data for proper peak integration, while the A/D box stores the current analysis output. The system data storage is sized such that chromatograms for up to 25 to 35 analyses can be stored on one floppy disk, thus accommodating an entire BATS lid or CATS rack.

Gas standards for calibrating the GC have been prepared in-house as well as commercially (Airco Industrial Gases, Riverton, New Jersey) in their Spectro Seal aluminum cylinders in the working range of 1 pp 10^8, 1 pp 10^{10}, and 1 pp 10^{12}. Brookhaven prepared primary standards in the range of 100 to 1000 pp 10^6 in He using pressure-volume techniques, which were corroborated by analyzing on a thermal conductivity detector (TCD) GC. The working range cylinders were then prepared by pressure-dilution in steps of 100-fold using ultrapure air (BNL standards) or nitrogen (Airco standards). Airco has prepared the cylinder standards on a very large manifold accommodating up to 24 cylinders. Thus they can make one large batch of a PFT standard to be distributed in cylinders to many users of the PFT technology such that every one has the same working standards.

BNL has prepared working standards containing different mixtures of PFTs-- one containing PDCB, PMCH, and m-PDCH and another containing PMCH and m-PDCH. Airco had previously prepared one set of standards containing PMCP, PMCH, and m-PDCH. They are currently preparing a large batch of another set containing PMCP, PMCH, o-PDCH, p-PDCH, and PTCH.

The working standards are corroborated by comparing them to dynamically prepared PFT mixtures by passing nitrogen at a measured flowrate over temperature-controlled, gravimetrically calibrated PFT sources which were originally developed to conveniently tag homes when performing air infiltration measurements (ref. 12).

Sampling of either of the standards, working or dynamic, is accomplished by passing a set flow rate of the PFT standard through a BATS tube or CATS for a known period of time and then analyzing on the GC system. Using preset flow restrictors to deliver 50 mL/min of the gas standard at tank pressure, a one-minute sample of the 1 pp 10^{12} standard would contain just 50 fL of each tracer. By increasing the standard loading time automatically, using a BATS base to increment the time, and by switching to the higher working standards once a 30- to 50-min sample of two decades lower had been loaded, it is readily easy to prepare sample tubes containing PFTs from 50 to 10^7 fL (the latter is 20 min of 1 pp 10^8 at 50 mL/min). Since the standards can readily be loaded onto CATS tubes, boxes of such samplers are routinely stored and used in the laboratory for calibrations about two to three times daily.

A complete calibration curve can be run using the prepared standards, but in two modes as shown in Fig. 14. Normally, as the quantity of tracer analyzed

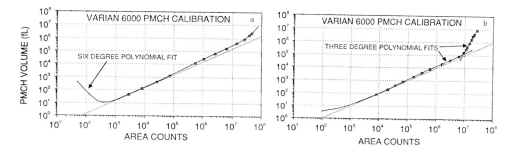

Fig. 14. PMCH calibration data obtained and correlated in two ways. (a) The electrometer gain was decreased 10-fold above 10^5 fL. (b) The gain was not reduced and the overload region was separately fitted.

increases, the electrometer gain is reduced an order-of-magnitude when its output nears 1 volt, the Nelson system voltage input limit. By multiplying the range 10 areas by 10, the tracer volume versus peak area data continue on a smooth curve above the range 1 overload point which, for PMCH, occurs at about 10^5 fL. It is apparent from Fig. 14a that the data in the region up to about 10^3 fL (1 pL) falls on the unity slope line. But from 10^3 to 10^4 fL, the curve displaces to a new unity slope line before starting to curve again at the overload of the PFT-electron reaction in the ECD which occurs at about 5×10^5 fL. This data is best fitted by a 6-order polynomial with a usual tracer quantity prediction of about ±4% or better. Below 50 fL (the limit of the usual low end of the calibration), the polynomial deviated markedly from the unity line which is what is used in that region. Five femtoliter standards (1 min of 1 pp 10^{12} at 5 mL/min) confirmed the linearity.

The alternative approach which is currently the favored method is shown in Fig. 14b. The electrometer range is not changed at the range 1 overload. Although the peak heights do not increase above 1 volt (the Nelson A/D cutoff), the areas increase, because the peaks are broadening, but at a higher quantity-per-unit area rate. A third-order polynomial was fitted to the data below the 1 volt overload (clearly evident at about 10^5 fL of PMCH), and another above that level. The precision of this approach appears to be as good as the former, but here there is no concern for guessing which electrometer range should be used when analyzing unknown samples.

Remembering that the limit-of-detection for PMCH was about 1 fL, the dynamic range would appear to be about seven orders-of-magnitude. However, the GC has a memory of about 0.01 percent of the previous sample when analyzing the next, which then decreases at a slower rate upon subsequent analyses of lower concentration samples.

(ii) <u>Dual-trap analyzer</u>. The peak PFT concentration typically encountered in long-range tracer experiments range from 1 to 5 pL/L at 100 km downwind to about 100 to 400 fL/L at 800 km downwind. These concentrations can easily be measured with BATS located both on the ground and in aircraft. The expense of an aircraft platform, however, mandates that the sampling occur where the plume is actually located. Thus, the need exists for a real-time PFT analyzer.

In the late 1970s, Lovelock, under a contract to NOAA, built a prototype instrument which was subsequently modified at Brookhaven. The unit consisted of two adsorbent traps, packed with the same material used in the BATS and CATS, and an <u>in situ</u> ECD chromatograph. While one trap was sampling at 1 L/min for 5 min, the other was heated to recover and analyze the collected PFTs. Since the traps reversed position every 5 min, no tracer was lost.

A new version of this real-time analyzer was built in 1983 for the fall CAPTEX experiment. Improvements allowed the separation of 3 PFTs in a 4-min chromatogram of a 4-min air sample collected at the rate of 1 L/min. The unit was able to see down to the ambient levels of PMCP and PMCH, indicative of the limit of detection of about 10 fL (ref. 15).

3 APPLICATIONS OF PFT TECHNOLOGY

The ultra sensitive detection of the PFTs through the use of some simple and inexpensive field tools as well as with sophisticated laboratory analyzers has lead to an extensive number of applications.

The principal supporter of the development of the technology has been the U.S. Department of Energy (DOE) because of its application in long-range atmospheric tracing (Table 12). Another DOE office has been funding the

TABLE 12

Supporters of PFT tracer technology

U.S. Department of Energy
- long-range and complex terrain atmospheric transport
- building infiltration and ventilation

Utilities
- oil leak detection from urban buried high-voltage cables
- air infiltration, weatherization and indoor air quality in homes and buildings
- leak detection in boiler and condenser tubes
- integrity of nuclear facilities, both safeguards and emissions

Oil and Gas Industries
- geophysical exploration
- leak detection in storage and pipe transport
- process flow measurements

Federal, State and Local Agencies
- indoor air quality and air infiltration
- atmospheric and oceanic transport and dispersion
- detection of tagged explosives

development of the passive PFT source and sampling device for the inexpensive determination of air leakage rates in homes and buildings.

In the last two years, the utility industry has expressed interest in using the technology to detect oil leaks from underground high voltage transmission cables. The industry costs for finding such leaks each year can run to $1 million for just one major city. Utilities supplying gas for home heating and cooking are concerned about the magnitude of pollutant emissions from various appliances and the increase in pollutant concentrations that might arise as customers attempt to conserve energy by reducing the leakage of heated air in the wintertime through weatherization efforts which reduce air infiltration and exfiltration. The PFT technology can quantify the energy savings of weatherization and can pinpoint and quantify the source of the various pollutants found in the home as well as provide a model for predicting the consequences of weatherization on the indoor pollutant levels. Concepts have been considered for detecting, quantifying and locating leaks in power plant condenser tubes and systems, but no active work is currently taking place.

Despite the apparent oil glut in the world economy today, the oil and gas industry is actively involved with the search for new sources of these natural resources and applying new technology in attempts to improve the yields from existing oil and gas fields. Understanding the geophysical characteristics of these petroleum reservoirs through the application of radioactive tracers has been occurring for more than 15 years, but recently the PFT technology has been applied because the PFT quantity required for the same signal-to-noise would cost at least two orders-of-magnitude less, saving nearly one hundred thousand dollars in just one experimental field measurement.

Other groups of federal, state, and local agencies have been supporting the use of PFTs for quantifying and locating the sources of radon contamination in homes and demonstrating the applicability of various mitigating strategies, for studying other indoor pollutants (it is always necessary to know the air leakage rate into the building), for determining the effectiveness of air handling equipment in commercial buildings, for atmospheric and oceanic transport studies on all scales, and for the detection of tagged explosives in clandestine bombs at airport environments. This latter research has been more-or-less ongoing since 1972. By attaching PFT vapors to blasting caps during their manufacture, the PFT technology has been demonstrated to have sufficient sensitivity to detect a tagged bomb in a suitcase on a conveyor belt, on a plane, or secreted into a building.

3.1 Atmospheric tracer experiments

Table 13 provides an outline of the atmospheric tracer field experiments in which PFT technology has been used.

Some details on the results will be presented during the lecture.

TABLE 13

Atmospheric tracer experiments using PFT technology

No.	Name	Date	Scale of transport	PFTs used	PFT sampling used
1	Multitracer atmos. exper.-Idaho (ref. 16)	April 1977	100 km	PDCB, PMCH, m-PDCH	Prototype (PT) adsorbent sampler, bags and bottles, PT Dual-trap
2	Long-range atmos. demo.-Oklahoma (ref.6)	July 1980	600 km	PMCH, m-PDCH	60 BATS, modified dual-trap (MDT), bags
3	ASCOT 80 and ASCOT 81 (ref.12, 17)	Sept. 1980 Sept. 1981	Complex terrain (15 km)	PMCH, m-PDCH	60 BATS, MDT, bags, VASC-I
4	Mini-CAPTEX 82	1982	10 km	PMCH	5 BATS, CATS, MDT
5	CAPTEX 82	Sept. 1982	800 km	PMCH	60 BATS, CATS
6	CAPTEX 83 (ref. 18)	Sept. 1983	1200 km	PMCH	80 BATS, New dual-trap (NDT)
7	ASCOT 84	Sept. 1984	Complex terrain (25 km)	PMCP, PMCH, m-PDCH	80 BATS, CATS, VASC-II
8	METREX (ref. 10)	Nov. 1983-Dec. 1984	Urban (15 km)	PMCP, PMCH, m-PDCH	3 BATS, 93 CATS sites
9	SCCCAMP	Sept. 1985	Coastal (200 km)	PMCP, PMCH, m-PDCH	30 BATS, CATS
10	ANATEX	Jan. 1987	3000 km	o-PDCH, PTCH	75 BATS, CATS, NDT

3.2 Building air infiltration and indoor air quality

Structures, from the smallest homes to the largest types of commercial buildings, are very rarely capable of being modeled as single, well-mixed zones, but rather are comprised of multiple zones each of which may be considered as well-mixed if care is given in the partitioning. Thus, tracer techniques for measuring air infiltration rates into each zone as well as the air exchange rates between zones must be able to provide multiple tracers, one for each physically distinct building zone, and the means to simply determine each of those tracers from a single air sample collected in each zone.

The passive sampler (CATS) provides the means to measure the indoor air concentration of PFTs. Miniature PFT permeation-type sources (ref. 9) can provide the tracer source. With up to seven PFTs now capable of being sampled and analyzed, each of up to seven zones in a home can be separately tagged.

By measuring the concentration of a pollutant in each zone of the house with a separate sampling device, for example, the passive Palmes tube is used for NO_2, it is possible to compute the source strength of the pollutant in each zone. For NO_2, if the source was found to be a positive value in the kitchen and about zero in all other zones, then the source might likely be the unvented gas range. If, however, a positive source term was computed in another zone as well, for example, a bedroom, then the NO_2 source might be from a leak in the flue duct from the home heating device. Such results have indeed been found and corrective action was initiated on the basis of this work. A summary of results in 30 homes will be shown.

The penetration of radon in homes is another important area where multiple PFT technology can identify the location of the source (usually the basement or crawl space) and the magnitude of the source. Studies on developing mitigating strategies based in part on the application of the PFT technology are underway.

3.3 Potential new applications

A two year research project has just been initiated with the Electric Power Research Institute (EPRI) in the U.S. to demonstrate the potential of detecting oil leaks in underground high voltage cables. By dissolving just 0.1% by weight of a PFT into the oil, which would increment the cost of the oil about 5 to 10%, it is projected that the PFT samplers will be able to pin-point and quantify the magnitude of oil leaks as low as 1 L/h, more than one order-of-magnitude better than prevailing methods, and be able to make that determination remotely whereas the other approaches require an attempt to physically measure the leak-induced low oil flow rate which requires removing the cable from service.

In another technique, the geophysical study of oil and gas movement two to three thousand meters under the North Sea are underway, but the samples have yet to be analyzed. The application of SF_6, another tracer, to gas movement 2500 m below the North Slope, Alaska, oil field was successful.

4 CONCLUSIONS

The PFT technology, consisting of the tracers, the samplers, and the analyzers, comprise a system which, because of its sub-femto-liter per liter detection capability for multiple tracers, has significantly advanced the ability to perform long-range and continental-scale atmospheric transport experiments. The data obtained from such studies will provide a valuable resource for modelers.

The break through in the passive PFT source and sampler has established the technique on a world-wide basis as the most economical and yet perhaps the most powerful multitracer technique for studying air flow patterns in multizoned homes and buildings.

The technology appears to be applicable to other transport and dispersion problems such as underground gas and oil movement as well as the numerous leak detection problems.

ACKNOWLEDGMENT

Many have contributed to the development of the PFT technology and are as well responsible for the success of the many applications which has lead to the establishment of the Tracer Technology Center at Brookhaven.

Appreciation is expressed to all my colleagues in the Center and to those in other government and private offices and laboratories who have played a key role in inspiring our work.

REFERENCES

1 W.F. Dabberdt and R.N. Dietz, Gaseous tracer technology and applications, in: D.H. Lenschow (Ed.), Probing the Atmospheric Boundary Layer, American Meteorological Society, Boston, Mass., 1986, pp. 103-128.
2 R.N. Dietz and G.I. Senum, Capabilities, needs, and applications of gaseous tracers, in: Atmospheric Tracer Workshop, Santa Fe, New Mexico, May 21-25, 1984, Los Alamos National Laboratory, LA-10301-C, December 1984, pp. 123-173, organized and compiled by S. Barr, W.E. Clements, and P.R. Guthals.
3 J.E. Lovelock, Improvements in the experimental methods for a long range tracer experiment, unpublished report, NOAA Air Resources Laboratories, Rockville, Maryland, 1974.
4 S.E. Schwartz, D.F. Leahy and S. Fink, Aircraft release of sulfur hexafluoride as an atmospheric tracer, J. Air Pollut. Cont. Assoc., 35(5) (1985) 555-557.
5 W. Engewald, J. Porschmann, T. Welsch and K.D. Sienbakova, The retention times of dimethylcyclohexanes determined by gas-adsorption and gas-liquid chromatography, Z, Chem. 17 (1977) 375-376.
6 G.J. Ferber, K. Telegadas, J.L. Heffter, C.R. Dickson, R.N. Dietz and P.W. Krey, Demonstration of a long-range atmospheric tracer system using per-fluorocarbons, April 1981, NOAA Technical Memorandum ERL ARL-101.
7 N. Latner, 'TAPS': tethered air pump system, EML-456, Environmental Measurements Laboratory, New York (1986) 11 pp.
8 R.N. Dietz and E.A. Cote, Air infiltration measurements in a home using a convenient perfluorocarbon tracer technique, Environ. Internat. 8 (1982) 419-33.
9 R.N. Dietz, R.W. Goodrich, E.A. Cote and R.F. Wieser. Detailed description and performance of a passive perfluorocarbon tracer system for building ventilation and air exchange measurements, in: Trechsel and P.L. Lagus (Eds.), Measured Air Leakage of Buildings, ASTM STP 904, American Society for Testing and Materials, Philadelphia, 1986, pp. 203-264.
10 R.R. Draxler, Metropolitan Tracer Experiment (METREX), October 1985, NOAA Technical Memorandum ERL ARL-140.
11 R. Woods, SNL vertical profiling of tracer concentrations, in: ASCOT Data from the 1980 Field Measurement Program in the Anderson Creek Valley, CA, UCID-18874-80, Lawrence Livermore National Laboratory, Livermore, CA, April 1983, pp. 1468-1475.

12 G.J. Ferber, K. Telegadas, C.R. Dickson, P.W. Krey, R. Lagomarsino and
    R.N. Dietz, ASCOT 1980 perfluorocarbon tracer experiments, in Ibid., pp.
    1202-1316.
13 P.G. Simmonds, A.J. Lovelock and J.E. Lovelock, Continuous and ultrasen-
    sitive apparatus for the measurement of airborne tracer substances, J.
    Chromatogr., 126 (1976) 3-9.
14 R.N. Dietz and R.W. Goodrich, The continuously operating perfluorocarbon
    sniffer (COPS) for the detection of clandestine tagged explosives, BNL
    28114, Brookhaven National Laboratory, Upton, New York, 1980.
15 T.W. D'Ottavio, R.W. Goodrich and R.N. Dietz, Perfluorocarbon measurement
    using an automated dual-trap analyzer, Environ. Sci. Tech., 20 (1986)
    100-104.
16 G.J. Ferber, R.R. Draxler, C.R. Dickson, G.E. Start, P.W. Krey, R.
    Lagomarsino, R.N. Dietz, J. Keller, V. Andrews, G. Cowan, P. Guthals, M.
    Fowler, W. Sadlacek, S. Barr, W.E. Clements, R.W. Bench, N. Daly, N.
    Campbell, D. Randerson, Experimental Design and Data of the April 1977
    Multitracer Atmospheric Experiment at the Idaho National Engineering
    Laboratory, Los Alamos Scientific Laboratory Informal Report, 1979,
    LA-7795-MS, compiled by W.E. Clements.
17. G.J. Ferber, K. Telegadas, C.R. Dickson, R. Lagomarsino, P.W. Krey and
    R.N. Dietz, Perfluorocarbon tracer experiments, in: P.H. Gudiksen (Ed.),
    ASCOT data from the 1981 coding tower plume experiments in the Geysers geo-
    thermal area, Lawrence Livermore National Laboratory, UCID-19332, 1982,
    Vol. II, part 7, pp. 49-82.
18 G.J. Ferber, J.L. Heffter, R.R. Draxler, R.J. Lagomarsino, F.L. Thomas,
    R.N. Dietz, C.M. Benkovitz, Cross-Appalachian Tracer Experiment (CAPTEX
    '83) Final Report, January 1986, NOAA Technical Memorandum ERL ARL-142.
19 P.H. Gudiksen, M.H. Dickerson and T. Yamada, ASCOT FY-1984 progress report,
    Lawrence Livermore National Laboratory, UCID-18878-84, 1984.

Regional and Long-range Transport of Air Pollution,
Lectures of a course held at the Joint Research Centre, Ispra, Italy,
15–19 September 1986, S. Sandroni (Ed.), pp. 249–280
© Elsevier Science Publishers B.V., Amsterdam — Printed in The Netherlands

THE REGIONAL TRANSPORT OF TALL STACK PLUMES

Millán M. Millán

1. INTRODUCTION AND HISTORICAL REVIEW

The systematic use of the tall stack as an alternate method to control air pollution started in the early 60's. This development was accompanied by: (1) an increase in the rated power output in power plants from tens to hundreds or thousands of MW, or (2) a consolidation of the emissions of several processes into one stack and (3) the use of lower quality fuels. In either case this translated into a significant increase in the emissions from these sources.

It is important to notice that the equations used to calculate the height and the dispersion from these structures were (still are) based on the experimental results of diffusion studies using non buoyant tracers released only at a few tens of metres from the ground (ref. 1). Their use, therefore, implied questionable extrapolations of height dependent parameters (ref.2).

Doubts about the applicability of existing theories triggered the onset of a number of experimental programs intended to characterize the real behaviour of the plumes and the atmosphere at heights above ~200 m. Among these were:

(1) The TVA (Tennessee Valley Authority) project, initiated with the operation of the first tall stacks (refs. 3-5).

(2) The LAPPES (Large Power Plant Effluent Study) program undertaken by NAPCA with the cooperation of the State of Pennsylvania and the TVA (refs.6/7).

(3) The Sudbury Study, initiated with the operation of the INCO 382 m stack in 1972 (refs. 2,8,38).

On the European side, programs of similar nature were initiated in the UK by the Central Electricity Research Laboratories (refs. 9,10). There are two important aspects associated with the work in this area. The first one concerns the programs' objectives and the second the evolution of the programs themselves. From their very onset these projects were designed to fulfill one or the other of the following general objectives:

(1) Characterize the atmospheric structure affecting the dispersion of buoyant gases emitted above 150-200 m as a previous and necessary step to advance the theoretical knowhow and the development of suitable dispersion models.

(2) Determine dispersion parametres to adjust existing models.

The legislation in the USA of a series of models based on the Gaussian concept of plumes (refs.11,12) and their indiscriminate extension to other countries resulted in a significant slowdown of research of the first type and a
desmesurate growth of modelling as the "easy way out" to fulfill the law. It has
only been after finding multiple problems with the verification of these models,
that the scientific community is recommending the reinitiation of programs
based on experimental measurements (refs. 13-16).

The second aspect of these projects is associated with the complex four-way
interaction between:

(1) the planning of field measurements,
(2) the status of "accepted knowhow" or "current wisdom" (ref. 17),
(3) the advances in instrumental development and measurement techniques,
(4) the (slow) process of data reduction, analysis and assimilation of results.

In the case of dispersion from tall stacks it is quite apparent, from the
very beginning, that the experimental results quickly overtake the level of
the "accepted knowhow", especially if seen from the Gaussian plume perspective.
In this manner the analysis of results obtained from programs designed to study
diffusion out to 20 Km revealed the existence of important effects starting from
~18 Km onwards and led to the discovery of regional and long range transport,
out to hundreds of Km, when a new experimental campaign was organized to charac-
rize dispersion within 25 to 40 Km from the source, etc.

One of the first reported observations of a plume at 400 Km from its source
is published in 1977 (ref. 18) and is followed by systematic observations of
transport between 300 to ~500 Km in Canada (refs. 19,20) and out to 1000 Km in
Australia (ref.21). The point to stress here is that the first observation takes
place while the instrumental technique was being optimized to study dispersion
within ~50 Km from the source (ref. 22).

1.1 Instrumental evolution
Advances in instrumental techniques concurrent with the measurement campaigns have been the 3rd factor in this interaction. Measurement methods available in the late 60's and early 70's for plume studies principally consisted of:

(a) Tracers, with little sensitivity and marginal results from 20 Km on (ref.23)
(b) Bubblers for the measurement of SO2 at ground level (ref.7)
(c) The first generation of fast response sensors (SIGN/X) for airborne studies
    (ref. 7)
(d) Densitometers for the detection of particles (ref. 24)

The needs for these programs accelerated the development of new instruments
and techniques with some specifically designed to track stack emissions to very
large distances (refs. 25-27). Among the most exotic instruments were the remote

sensors (ref.28,29).

As an active system, i.e. with its own light source, the laser begins to be used as an optical radar, LIDAR, almost simultaneously in the LAPPES (USA) and CEGB (UK) programs in the mid 60's (refs. 30-32). Its main application at this time is to study plume rise and diffusion near the source. As a passive sensor, using natural light, the other system that emerges in these years is the COSPEC (ref.33). Its utilization in the LAPPES program takes place in 1968-69 (ref.34).

Both types of instruments are intensively used in almost all atmospheric dispersion studies initiated in North America and Europe after 1971. One significant aspect of these techniques is their high complexity and that the majority of their most important initial applications have been achieved with advanced prototypes (research grade units) operated by their designers, i.e. by highly skilled and specialized personnel. Nevertheless, while the LIDAR systems have remained at this stage to the present moment because of their elevated cost, the COSPEC, being cheaper, begins to be produced commercially in 1970. This implies that its use becomes more widespread and the results more irregular.

The complexity of these instruments, high cost of operation, the difficulty of integrating their data into the most accepted theories, the reduced performance of the commercial versions, and the obvious differences in the quality of the results obtained by different groups: all of these factors have resulted in a certain (comfortable and "self-serving") disinterest or skepticism on the part of modellers to use this data.

1.2 Present problems

Finally, and what can be considered one the most problematic aspects in this interaction, the modelling community does not seem to have progressed much farther than ~10 - 15Km from the source in the parametrization of the most influential atmospheric processes, nor have they yet assimilated some of the most frequently observed phenomena, such as:

(1) the effects of wind shear.
(2) the convective cycles of coupling and de-coupling with a rotation of the ground impact area, in the majority of cases without reaching a steady state situation (ref.35).
(3) regional and long-range transport with spatial and time lags in the concentration fields.

These will be reviewed in this Chapter.

2. STABILITY CLASSIFICATION AND GROUND EFFECTS

One of the first problems in applying Gaussian models to describe the dispersión from tall stacks is to classify the field data according to the most common stability criteria, i.e. P-G, Turner, TVA, ASME-BNL, etc.(refs. 2,36,37).

The possible existence of this problem can be deduced from Fig.1. This shows three temperature soundings in the vicinity of the INCO stack in Sudbury. The early sounding shows an intense surface inversion ~110 m deep. A nearly iso-thermal strata follows up to 350 m and is topped by a slightly-stable-to-neutral profile up to the end of the sounding at 2.5 Km. Meanwhile the plume rises to its level of equilibrium between 500 and 700 m.

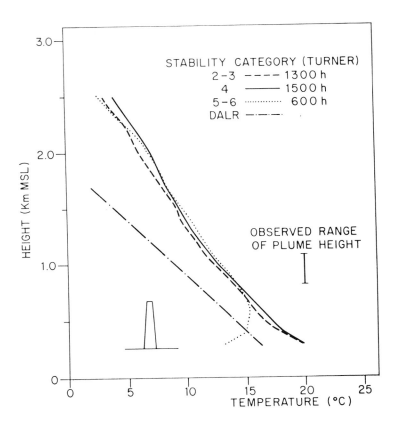

Fig. 1. Temperature soundings in the neighborhood of a tall stack (382 m), June 1973, in Sudbury. The three profiles show the variation of the stability during the day. The plume height was obtained by photography (ref. 38) and LIDAR.

The midday sounding shows a super-adiabatic profile up to ~200 m with the possible penetration of convective activity up to 450-500 m. Finally, the early evening sounding reveals the generalized heating of the air in the middle stra-tum (up to 400 m), while a super-adiabatic profile is maintained only up to 100-120 m, and the penetration of convective activity is reduced to ~ 350 m.

The "classical" variations of atmospheric stability which presuppose that a plume will pass through the phases of: (a) stable fan or ribbon during the night to (b) fumigation (transitory), with the change from stable atmosphere to unstable, and finally to (c) unstable "looping" during the hours of convective activity, are confined to a layer ~450 m deep and do not reach the plume. The problem, then, is not so much to determine what kind of stability must be applied but rather to determine whether, or not, the height and turbulent structure of the boundary layer reaches the plume emitted by the tall stack and, if they do, how long (or how far downwind) it takes to reach it.

A plume emitted at 50-60 m has a greater probability of being within a layer affected by "surface" stability changes, and its behaviour can be predicted more or less accurately by conventional procedures. However, this is not the case for the plume from a tall stack. This occurs because in certain periods, depending on the type of terrain and time of the year, the atmospheric turbulent structure which affects the plume is transported from other places upwind (see Fig. 15) and, not being of local origin, it is not "totally" predictable by procedures based on local observations obtained at the surface.

A key aspect of this problem is the fundamental importance which the annual variation in ground characteristics has in the diurnal cycle of surface heating and cooling. Fig. 2 schematically shows the range of variations in ground temperatures and of the atmospheric strata in contact with it.

Finally, the widespread assumption that the plume from a tall stack can be treated as any other plume after one period of convective activity must, therefore, be considered with extreme caution since the convective turbulence may not reach it at all.

3. OBSERVED PLUME BEHAVIOUR IN THE REAL ATMOSPHERE

From the point of view of the interaction between the plume and the ground, four situations are important:

(1) Dispersion in neutral conditions.
(2) Partial (and/or intermittent) coupling of the convective layer with the lower part of the plume.
(3) Full coupling (classical looping) over heated ground.
(4) Total decoupling from the ground. This situation is favored at night. However, in winter, or with very wet ground or snow cover, it can occur during the whole day, several days in a row.

3.1 Dispersion in neutral conditions

For neutral conditions to exist it is necessary that the air mass be in thermodynamical equilibrium with the ground. This situation occurs very rarely in the real atmosphere since a small (positive) temperature difference between the ground and the airmass is sufficient to trigger the convective activity. In

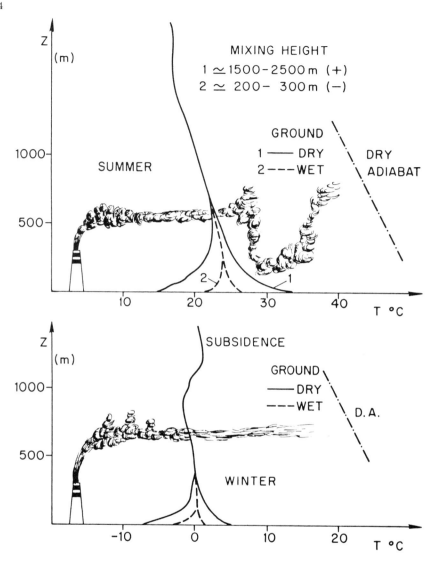

Fig. 2. Effects of ground wetness on the mixing height. Over dry ground, the range of temperature variations is maximum and consequently the depth reached by the convective activity. With wet soil, a large fraction of the incoming heat is spent evaporating water and the range of temperature variations is smaller (by as much as five times). In this manner, after a heavy downpour on a summer night over sandy soil, it may take one to two days before the convective layer can again reach the plume of a tall stack.

general, neutral conditions are assumed to be those in which mechanical turbulence can carry away the excess heat of the ground.

With increasing insolation, higher and higher wind speeds are required to carry out this process. In this manner winds of 5 to 7 m/s are required to maintain neutral conditions under overcast conditions, while speeds higher than

~12 m/s may be required to maintain quasi neutral conditions over dry ground under moderately strong insolation.

In a neutral atmosphere the dispersion of a plume from a tall stack is similar to that from a lower one under the same conditions. The fundamental difference, however, is associated with the greater depth of the plume from the tall stack. This implies that the upper and lower parts of the plume will be subject to different wind speeds and directions, i.e. to wind shear effects.

The tilt in the plume crossections caused by wind directional shear is one of the first characteristics reported by early researchers in this field (refs. 6,7). This effect was noticed in aircraft profiles of stable-to-neutral plumes and became particularly obvious with the application of remote sensors to plume studies (ref. 29), as shown in Figures 3 to 5. Less well documented but much more important are the wind shear effects under convective conditions.

3.2 Dispersion under convective conditions

(i) The convective turbulent field. To understand the plume behaviour under convective conditions, it is convenient to examine some of the characteristics of this type of turbulence. Basically, it consists of a field of updrafts and downdrafts with the following properties:

(1) The updrafts are overheated pockets of air which originate over "hot spots" such as: highway intersections, drier land areas, buildings, etc.
(2) The hot air pockets attain high ascensional velocities (of the order of metres per second).
(3) They have a reduced horizontal area or crossection (diameters of tens of metres, or less).
(4) If the hot areas are extensive enough and the excess heat large, permanent type "thermals" can form over them. In most cases, however, once the excess heat has been released, the updraft ceases over that point until a new updraft can be initiated, or a new one can originate over another hot spot. In this manner, updraft "feet" form and move in a random manner over the area.

On the other hand, the downdrafts or local subsidences required to maintain continuity of the mass flow, usually:

(5) Take place over much more extensive areas
(6) Are much more persistent
(7) Sink at much lower velocities (of the order of centimetres or of fractions of metres per second).
(8) They are more permanently fixed in space over "cooler" areas such as: bodies of water, green and/or grassy areas (as opposed to bare ground), etc.

Finally, the continuity required to maintain this kind of flow consists of: a radial convergence along the ground to feed the updraft, and a divergence at,

256

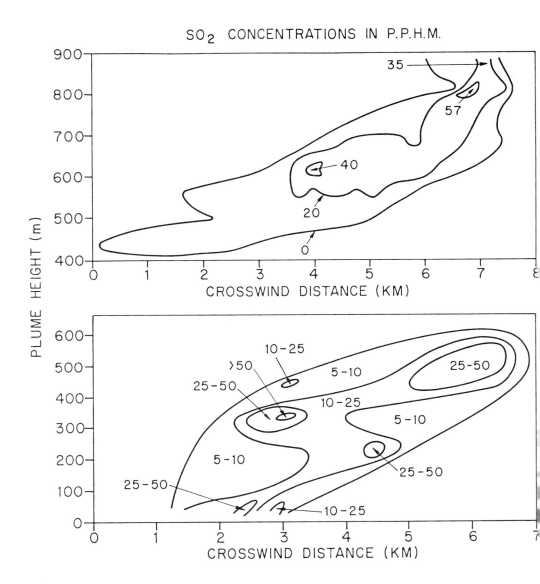

Fig. 3. Observed plume crossections under the effects of wind directional shear. (a) LAPPES project (ref.7), (b) NANTICOKE project (ref.35); in both cases the plume crossections were obtained by instrumented aircraft and are not truly instantaneous. The sections are presented as could be seen from the stack looking downwind.

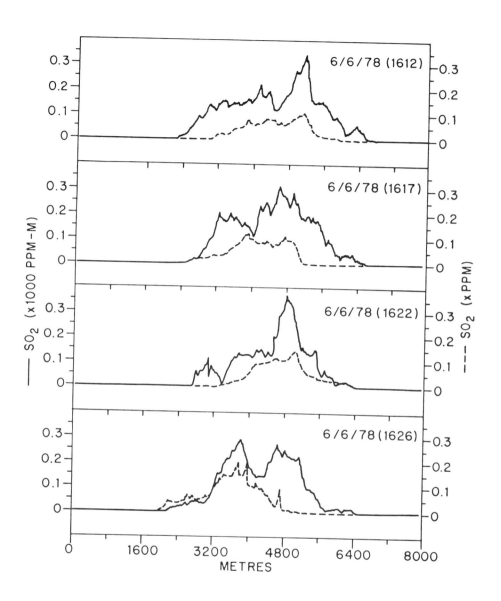

Fig. 4. Plume profiles at 16 Km from the source under nearly neutral conditions. These show the vertically integrated SO2 profile (COSPEC) and the SO2 ground concentrations obtained with a fast moving instrumented vehicle. The average time interval between profiles is 4 minutes and can thus be considered near instantaneous realizations. Under neutral conditions, both types of profiles are well conditioned (smooth) and reasonably well centered with respect to each other.

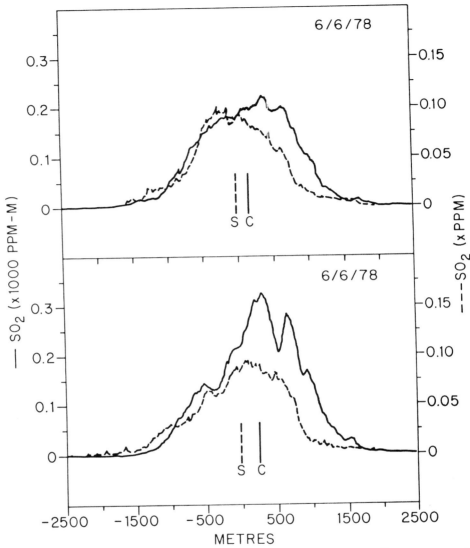

Fig. 5. Eulerian average of a tall stack plume at 16 km from the source under neutral conditions (NANTICOKE). This is the space average of seven profiles obtained in approximately 1/2 hour, and include those shown in the previous Fig. The vertical lines indicate the centres of gravity (COG) of the distributions represented, S for the ground SO2 average and C for the COSPEC average of the vertically integrated SO2 profiles (ref. 8). The displacement between these is indicative of wind directional shear effects.

approximately, the level reached by the thermal updraft. As soon as any wind exists, the convective structure just described can be tilted and/or shed along the wind direction.

Another important aspect of the convective activity is that it acts as a momentum transfer mechanism between the ground and the atmosphere by which the surface exerts a drag on the wind field. In this Chapter the term coupling - between plume and ground- is used with reference to whether the plume, or part of it, is in the layer where the exchange of momentum takes place.

(ii) <u>Effects on the concentration field.</u> With partial coupling, the observed ground concentration field is equivalent to a semi-continuous fumigation process in which the upper reaches of the convective activity interact with the bottom part of the plume and drag part of it to the ground. This situation is shown schematically and with field data in Figures 6 and 7, respectively.

This type of interaction can continue most of the midpart of the day and since the depth of the mixing layer experiences height fluctuations during this period, its reach of the plume can follow a similar variation. This is conveyed to the ground impact area with some delay. In general, the wind shear effects are more pronounced than in the neutral case. Important characteristics of this type of interaction are:that the skewness of the vertically integrated plume profiles are very pronounced and that the ground impingement area is located over the tail end of this distribution.

(iii) <u>Effects under strong convection.</u> Three other important aspects of plume behaviour in strong convective conditions are: (a) the lateral shearing of the vertical plume profile, (b) longitudinal tilting or differential stretching of the plume and (c) the sinking of the centreline. All these arise from the interaction between convective activity and the wind shear.

The first important effect of these interactions is the production of a time and space lag between the fraction of the emission that travels in the high part of the plume and that which is convected down to the ground. For example, with the "normal" veering of direction and increase of the wind speed with height, the parts of the plume dragged towards the ground are progressively delayed and displaced towards the left of the upper part emitted at about the same time, as shown schematically in Fig. 8.

Figures 9 and 10 show that strong convective activity results in much more marked and erratic lateral shear effects than those produced under neutral or partially coupled conditions. Finally, and as a consequence of the randomness of the convective process, the time necessary to obtain well conditioned averages increases to the point that it may not be possible to obtain them during the entire period of the convective activity (ref. 39).

An example of this situation can be observed by comparing the averages represented in Figures 5 and 7 obtained in neutral and semi-coupled conditions, with the averages shown in Figures 10 and 11, obtained under strong convective conditions. In Figures 5 and 7 approximately 1/2 hour of profiling (7 and 4 sections, respectively) is sufficient to obtain some smooth graphs. However,

260

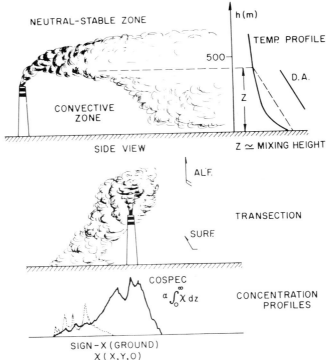

NEUTRAL-STABLE ZONE

TEMP. PROFILE

h(m)

500

D.A.

CONVECTIVE ZONE

Z

SIDE VIEW

$Z \simeq$ MIXING HEIGHT

A.L.F.

TRANSECTION

SURF.

COSPEC $\alpha \int_0^\infty X \, dz$

CONCENTRATION PROFILES

SIGN - X (GROUND)
X (X, Y, O)

Fig. 6. Semi-continuous fumigation process in a tall stack plume. The convective layer reaches only partially into the lower part of the plume. This part is brought down, with accused wind shear effects, and its impact on the ground appears well to the side of the vertically integrated profile. The photograph shows how this plume appears if it is visible..

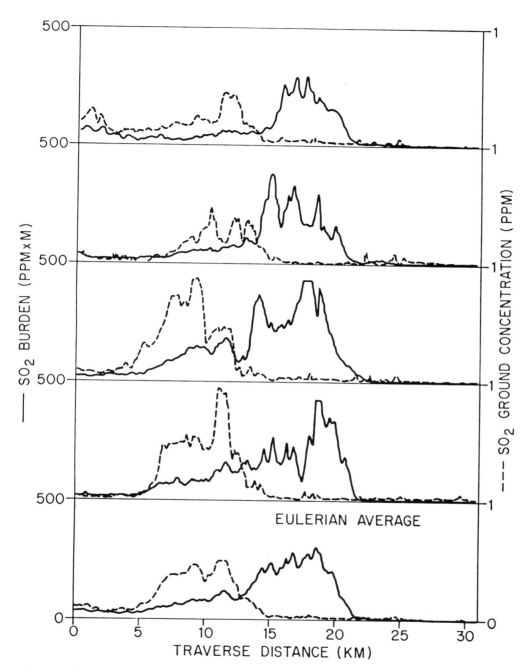

Fig. 7. Plume profiles of a tall stack plume at ~ 16 Km from its source under partially coupled conditions which generate a semi-continuous fumigation. These profiles are typical of the situations shown in Fig. 3(b) and Fig. 6.

262

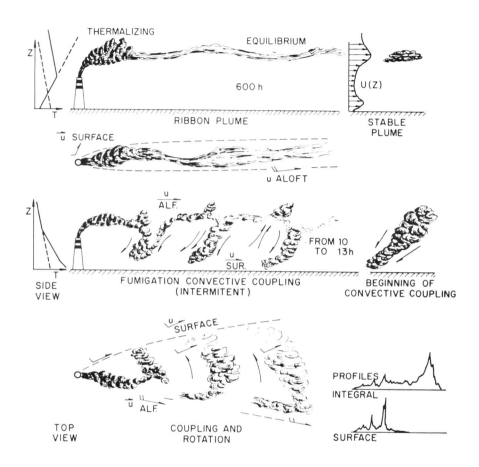

Fig. 8. Sketch of a coupling process. (a) Stable conditions previous to the onset of convective activity; the plume may be of the "ribbon" type (sec. 3.3). (b) When the convective activity reaches the plume the downdrafts bring down parts of it which are left behind and travel in a different direction than the the upper plume (to the left if the wind veers with height). The continuation of this process in a day with strong convective activity is shown in Fig. 13.

as Fig. 11 shows, this is not quite the case even after 28 profiles and ~4 hours of averaging under strong convective conditions.

Perhaps the most serious implication of these processes is the (almost) impossibility to adequately characterize them by means of a network of fixed sensors (ref.39).

Strong convective activity also makes the looping process of a tall stack plume be very assymetrical in its direction of propagation. As shown in Figures 12 and 13, plume turrets rapidly ascend to the limit of the convective activity while the contiguous sections, generally more extensive, sink at lower velocity and become delayed with respect to the upper parts of the plume.

As a result of these processes, the natural tendency of the sunken plume is to stay close to the ground and move along with the surface wind. This pollution cloud or "plume lump" can be locally perforated as it passes over areas with thermal updrafts, but the greater part of the plume remains close to the surface. As a result:

(1) at a certain distance from the stack, there may coexist concentrations emitted at different times and composed of some which have travelled slower along the ground, and some which may have been downdrafted just upwind of the observing point. In general, these lumps are separated angularly.

(2) On the average, the centreline or locus of the highest concentrations moves downward and to the left (if the wind veers with height) of the upper wind.

With a substantial number of measurements in the range from 10 to 80Km, this author has yet to find a convective situation where 1 to 2 hour Eulerian averages of the ground concentration are smooth or well centered with respect to the vertically integrated average profile as, for example, in Fig. 5 for the neutral case. This would indicate that the assumption of uniform mixing of a tall stack plume under convective conditions, at a certain distance from the stack, should also be seriously questioned. The experimental evidence tends to indicate that the dispersion of the plume from a tall stack under convective conditions is "lumpy" and remains so for as long as the convective conditions continue (4 to 8 hours, depending on the latitude, time of year and state of the ground).

Because the plume activity during the convective period also involves a substantial amount of rotation, this probably also indicates that the averaging period considered (the same 4 to 6 hours) does not define a steady state situation, on the basis of only one convective period, and that several periods (i.e. days) must be considered. This leads to the concept of scenarios which will be discussed in Section 4.

3.3 Dispersion under stable conditions and regional transport

On a sunny day the convective activity decays during the afternoon and ceases, quite suddenly, when the surface layer stabilization begins. The latter

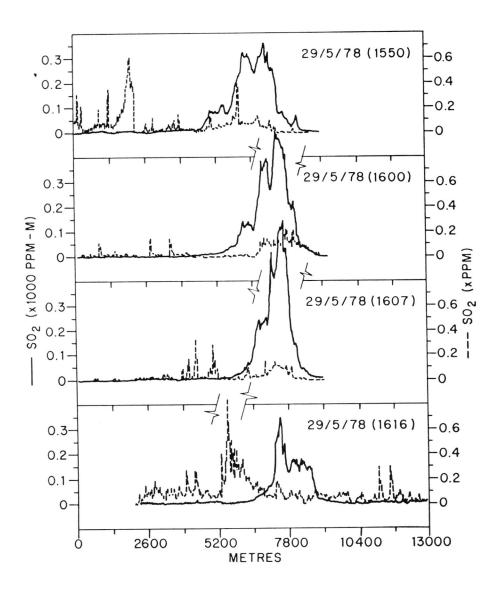

Fig. 9. Plume profiles under strong atmospheric instability and full coupling with the convective activity. The COSPEC profile is smoother as can be expected from its vertical integration of the SO2 concentration. The fast response SO2 ground measurements, however, show the erratic behaviour of the plume impact pattern under strong convection.

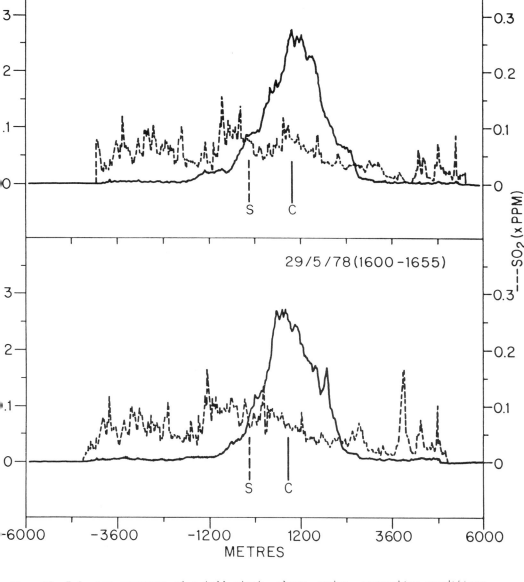

Fig. 10. Eulerian averages of a tall stack plume under convective conditions.
These include seven profiles obtained during ~ 1 hour. It is important to notice: (a) the irregularity of the ground average as compared to those obtained under different conditions (Figs. 5 and 7), (b) the centres of gravity of these
profiles are substantially displaced to the left of the vertically integrated a-
verage profiles (the wind veered with height). Distance to source ~ 13 Km.

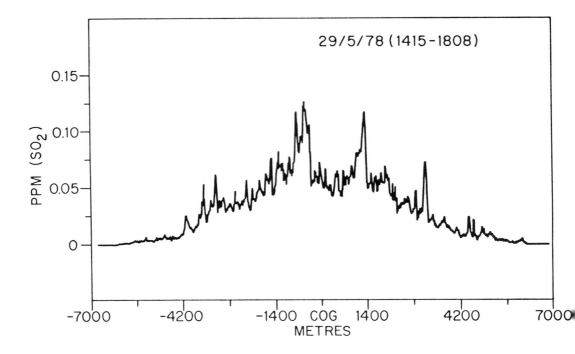

Fig. 11. Lagrangian average of the ground impact of a tall stack plume at ~13 Km from the source. This fixed-distance Lagrangian profile has been obtained by averaging, after re-centering, 28 plume profiles obtained during a period of ~4 hours. The recentering of the profiles effectively removes plume meandering and its rotation (~50 degrees) during this period (refs. 8,29). The interest of this profile is mostly academic in as far as it only proves that the central theorem of statistics holds, and the average of a large number of profiles converges towards a Gaussian type distribution. The point to note here is the irregularity of this average as compared to others obtained in more neutral conditions.

occurs as soon as the ground loses more heat than it receives from insolation; in very dry terrain this may start well before sunset. The stabilization process results in:(a) the formation of the ground based inversion,(b) the decoupling of the surface wind from the upper flow and (c) the onset of local drainage winds originated by the cooler (surface) air as it sinks towards lower areas. With respect to dispersion from tall stacks, the following processes take place:

(1) The upper wind starts to rotate very rapidly to align itself with the direction forced by the geostrofic balance between the pressure gradient and the Coriolis force.

(2) Associated with the decoupling of the upper wind from the ground, there is

an inertial overshoot of the upper flow. This gives rise to the formation of a nocturnal "jet" above the surface inversion. During anticyclonic conditions the high pressure center strengthens during the night (during the day the convective activity weakens it) and subsidence is encouraged while the ground-based inversion deepens. These three effects combine to reinforce the "nocturnal jet" which moves upwards and can end just below the subsidence inversion a few hundred meters above the ground.

(3) While these processes take place, a "new" plume forms, rises to its level of equilibrium, flattens out passing from "lofting" to "ribbon" and becomes realigned with the wind at its thermalized level. In sources with a high thermal output the "new" plume can easily reach the bottom of the subsidence inversion and be transported within the nocturnal jet. The onset of this process is schematically illustrated in the lower part of Fig.13.

(4) In this situation, Figures 14 and 15, the new plume is transported at higher speed during the night and can travel distances of several hundred Km from the source as shown in Fig. 16 (refs.18,19). Depending on whether the convective activity on the following day reaches it or not, this type of transport can last one night or longer.

(5) During the period when these phenomena take place, the "older" part of the plume, convected to the surface the previous afternoon, travels along the ground and becomes trapped within the surface-based stable airmass and participates in its motions.

(6) The following day, the processes of convective coupling between the ground and the upper wind will initially shear and later rotate the plume with respect to its direction during the night until a new equilibrium is established between the pressure gradient force, the Coriolis force and the convective drag of the ground. Finally, with the onset of stabilization the next evening, the decoupling process starts all over again. This cycle can recur every day and, depending on the local topography, the rotation of plume position during the day can be as large as 40 to 45 degrees or more.

(7) In most cases equilibrium conditions are only reached during the last part of the night, and the convective coupling part of this cycle is just a long transient during which the plume is never stabilized in a fixed (average) direction. Fig. 17 shows a coupling rotation situation of the type described under a lake breeze circulation. This is an extreme case, however, but similar situations occur on all sites and the transient can be more or less pronounced depending on the topographical setting.

Fig. 12. Photograph of a looping plume from a tall stack under strong convective conditions.

Fig. 13. (Opposite page) Schematic of plume transport under strong convective conditions followed by decoupling (this Fig. is a continuation of Fig. 8). With fully developed convective activity plume turrets, (A,B,C,D..) travel with the upper wind and overtake the parts of the plume convected down to the ground (A',B',C',D'..) which also travel in a different direction. The resulting concentration profiles are shown, to compare with Fig. 9. With the onset of stabilization and decoupling, a "new" plume forms aloft and realigns itself along with an upper wind which does not include the convective drag of the ground. Meanwhile, parts of the mixed plume remain close to the ground trapped within the stable layer. At certain distances from the stack, and just before sunset, it is possible to detect both plumes along a common traverse, i.e. as shown in section A-A'. After full stabilization, the upper plume reverts to a ribbon shape as shown in the top of Fig. 8.

(i) Resulting concentration field. Under strong convective conditions the upper and lower parts of the (near and midfield) plume travel in different directions, at different speeds and become progressively separated in space. This process reaches its ultimate conclusions if the diurnal cycle is considered as a whole and in its general perspective.

The transport of the "old or ground" part of the plume tends to be slower

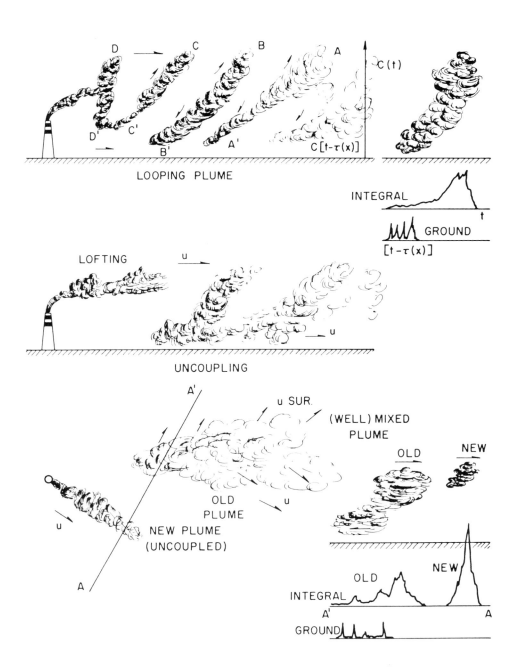

LOOPING PLUME

LOFTING

UNCOUPLING

(WELL) MIXED PLUME

OLD PLUME

NEW PLUME (UNCOUPLED)

Fig. 14. The Sudbury INCO plume travelling SW under stable conditions at 1900 h on September 1974. The estimated total length of visible plume ~80 Km. (Courtesy of Floyd Elder, CCIW).

Fig. 15. (Opposite page). Regional plume from a (350 m) tall stack early in the morning under very stable conditions. The ondulations in the plume are due to Kelvin-Helmholtz waves in the flow. These had been triggered by a hill range located some 25 Km upwind of the source.

than the upper plume and, depending on the local topography, it can become trapped in stagnated stable air pools in valleys and hollows. In this manner, the newer or "long range plume", travelling aloft, can overtake the "ground" plume, and its leading edge can reach the same distance as the ground plume emitted the previous afternoon and overlap (in distance) a sizable part of it. In complex terrain, the possibility exists that parts of the old plume will participate in local drainage flows and follow directions totally different from those of the plume aloft.

The processes described: involving convective coupling and plume rotation during the day, followed by the re-alignment of the "new" plume and regional transport at night,both at the surface and upper levels, have the result that at certain distances from the source, concentrations emitted at widely different times may coexist. These, however, will not necessarily be located in the same

direction. In this situation, one can no longer consider talking about "a plume" but rather about various plume "lumps" which follow their own evolution.

Depending on the local topography, the older and newer lumps of the plume may affect totally different areas in a more or less systematic fashion but on a discontinuous basis. For example, early in the morning a mountain ridge, at large distance from the stack, can be affected by aerodynamical effects, i.e. wake trapping or early fumigation of the new plume, while at the same time a populated area in the mouth of a valley many Km away can experience the fumigation of part of the old plume trapped the previous night on its local drainage flow. These effects can re-occur at about the same time of day for several consecutive days while the synoptic weather conditions persist. They will also recur on different days for similar meteorological conditions within the same season, i.e. similar insolation, vegetation cover and ground properties.

It is important to emphasize that unperturbed regional transport only lasts one decoupled period, and that the ground plume moves and participates in the diurnal cycles affecting the surface layer. The most conspicuous area as far as dry deposition is concerned is that where local topography favours the alignment of the old and new parts of the plume. Plume overlap distances will vary during the year. For instance, during the summer, with long periods of strong convective activity and a regional transport of short duration, overlap of the old and new plumes may occur in the range from ~80 to 200 Km from the source. In winter, with short, very limited or nonexistent periods of convective activity and a long time for regional transport, the overlap area may be further than 200(+) Km from the source and be very limited in extension or not exist at all.

4. MODELLING AND EXPERIMENTAL OUTLOOK

It is important to indicate that the usefulness of a dispersion model should be measured by its ability to represent the diffusive and transport processes in real life. The aim of Section 3 has been to illustrate some phenomena which must be considered in this endeavour. From this perspective, three basic approaches can be initially considered:

(1) The solution of the primitive equations with approppiate boundary conditions. This area is covered by other authors in this book.
(2) The parametric adjustment of existing models (refs. 15,41-43).
(2) The determination of local dispersion scenarios, their transition indicators from one to another and the development of pattern recognition techniques adapted to forecast their occurrence.

All of these approaches require extensive data bases.

Adjusting existing single stack models on the basis of experimental data was the approach taken by Weil in the Maryland Power Plant Sitting Program (refs.

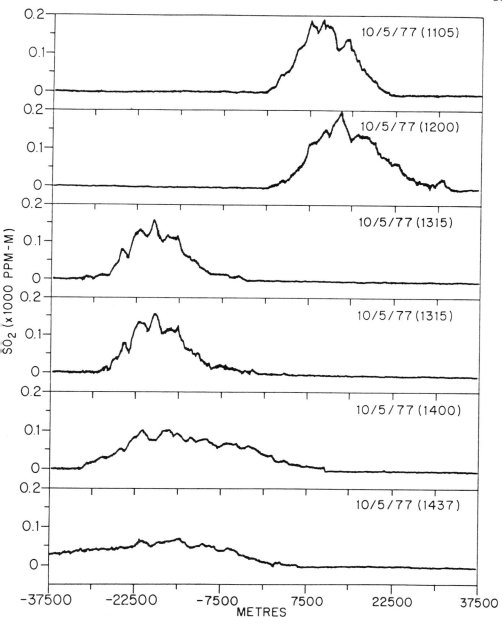

Fig. 16. Six plume profiles (COSPEC) of the INCO plume over Southern Ontario at
~300 Km from its source. Two instrumented vehicles were deployed to track the
plume. The middle profiles (at 1315 h) were obtained with the two vehicles following each other and show, essentially, the same plume. The profiles starting
at 1400h indicate that the convective activity had reached the plume and that
it had begun to fumigate and break up; by 1530 h the plume could no longer be
detected. Under persistent meteorological conditions, the same situation would
recur the following day.

37,41). It has been quite successful to describe dispersion within a few tens of Km from the source under (mostly) neutral conditions. The principal difficulties with this method are to account for the directional shear between the upper and ground plumes and that it tends to become somewhat unruly when dealing with convective conditions (refs. 15,16). Because of the lumpiness of dispersion under convective conditions in the real world, there is some doubt that convective scaling can be used for the averages of the concentration fields involving less than a few convective cycles (i.e. several days).

A scenario can be defined as a grouping of events which occur in the same sequence under similar boundary conditions. For example, the lake breeze cycle shown in Fig. 17 will be repeated under similar synoptic, ground and lake conditions. Because these have an annual cycle, similar dispersion scenarios will be favoured at certain times of the year (refs. 44,45). This concept can be considered as a smaller scale but more detailed approach to the "analog method" in weather forecasting (ref.46). It would have the advantages of identifying the final results of the dispersive processes, including the space and time lags, without the (inmediate) need to dwell on the mechanisms of the processes.

To put things in perspective, the Gaussian typing schemes can be considered as the simplest of scenarios in as far as they assume that for similar forcing or boundary conditions, classified as stability classes, the plume will behave (diffuse) in the same fashion. Among many others, one of the main problems with the Gaussian models is that they cannot deal with transient and "lumpy" stages which, in the dispersion of the plume from a tall stack, can be the norm rather than the exception. On the other hand, the classification of scenarios is highly dependent on extensive and well structured measurement campaigns followed by a long and careful process of data interpretation (refs.44,45).

4.1 Plume fates

In planning the experimental programs required to characterize the dispersion from a tall stack plume, several factors must be considered. These are related to the conditions presented in Section 3. Under neutral conditions the dispersion of a tall stack plume is similar to that of a lower stack except for:

(a) the more accused effects of wind directional shear, i.e. displacement of the ground impact plume vs. the total plume, and the shear enhanced lateral diffusion (refs. 8,40,47).

(b) the plume position at large distances from the source will experience wide and slow fluctuations as it is forced to follow the oscillations of the upper flow. The same process occurs for a de-coupled plume as Fig. 16 shows.

(c) because of its longer reach (before it can no longer be detected), its dispersion becomes more dependent on the changes of roughness of the terrain, whether these are mechanically induced, convectively origi-

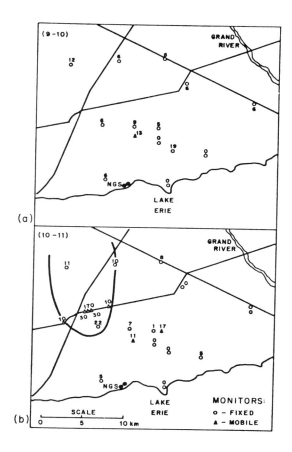

(a)

(b)

Fig. 17. Diurnal evolution of a coastal fumigation process. These results were obtained by combining the measurements of the fixed monitoring network (numbered points) and three mobile units (ref. 40). These were equipped with a COSPEC to locate and follow the overall (vertically integrated SO2) plume, and a fast response SIGN/X SO2 monitor to characterize its ground impact.
Graphs (a) and (b) show the early morning situation before the onset of the lake breeze and the initial fumigation after the turning of the plume.
Graphs (c) to (e), from 1130 to 1400 h. These show the intensive fumigation and the beginning of the plume rotation. Finally, graphs (f) to (h), from 1430 to 1700 h, show the generalized rotation of the ground impact area. The time and space lag between the near and mid field can be easily observed. By 1728 h, the plume had decoupled from the ground and was travelling along the shoreline. The overall rotation of the plume was ~ 120 degrees in about four hours.

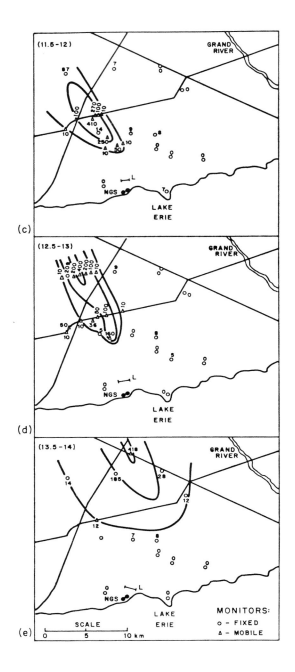

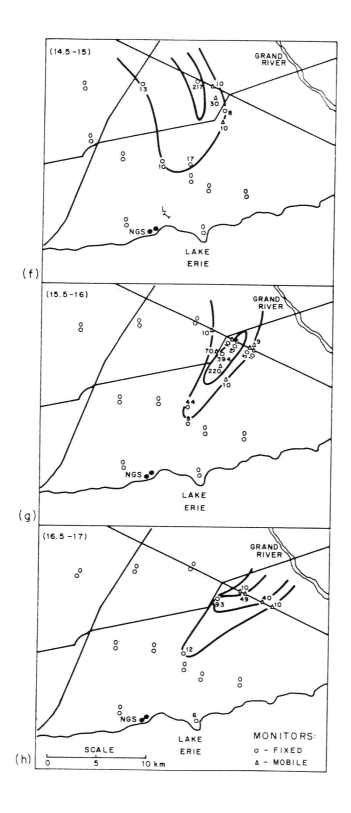

(f) (14.5–15)

(g) (15.5–16)

(h) (16.5–17)

GRAND RIVER

LAKE ERIE

NGS

MONITORS:
o – FIXED
△ – MOBILE

SCALE
0   5   10 km

nated (as it passes from one type of terrain to another), or both. Major changes in terrain features must be considered in planning the study and in the interpretation of the results.

With semicoupled conditions, the decoupled part of the plume can travel in a more cohesive fashion and be detected much further than in a beter mixed situation. Additional factors to be considered in an experimental program under these conditions are most likely to arise from the fluctuations in the depth of the lower, convective, layer. If it decreases, the plume decouples, narrows and rotates, rather fast, to align itself with the undisturbed flow. When the convective layer deepens, the plume is broken up; it usually becomes very sheared and rotates back or stops if the average flow is itself rotating. The intermittent rotation can become a wild experience for the measurement team in the far (80-100 Km) field as these effects are amplified along the flow. The plume may disappear for a while and reappear quite a few Km off to the side, while the ground concentrations follow an altogether different process. For measurements under these conditions, the traversing arches (ref.29) must be long, and the patience of the measuring team very great.

In the case of regional transport, the long plumes are likely to:

(a) interact with high topography or mountain ranges by becoming trapped within the wake of these obstacles. If mountain ridges are located across favoured paths of the stable plumes, important effects may be noted in the lee of the ridges (from where the stack cannot be seen !).

(b) become trapped in cloud formation mechanisms and participate in the rain-out process. This can also occur with orographic lift upon reaching high mountain barriers. In this case most of the noticeable effects will tend to occur in the windward slopes.

(c) be reached by convective activity if their transport has been towards drier and/or warmer areas. In this manner a tall stack plume will generally participate in rain-out processes upon travelling northward and be dry-precipitated if it travels southward.

These are orientative concepts, and in each case an evaluation must be made within the context of the local terrain and meteorology.

ACKNOWLEDGEMENTS

To Jose Ignacio Lazcano for his careful preparation of the drawings and to Dr.F. Mingot for his support.

REFERENCES

1   F.Pasquill and F.B.Smith, Atmospheric Diffusion. Ellis Horwood (John Wiley & Sons), Chichester, New York (third edition) (1983).

2   F.Fanaki and H.E.Turner, Plume dispersion from the Sudbury tall stack. 7th Int. Tech. Meetg. on Air Pollution Modeling and its Applications, Airlie, Virginia (1976).

3   F.W.Thomas, S.B.Carpenter and F.E.Gartrell, Stacks how high?. JAPCA 13, 198-204 (1963).

4   S.B.Carpenter, J.M.Leavitt, F.W.Thomas, J.A.Fizzola and M.E.Smith, Full scale study of plume rise at large coal-fired electric generating stations. JAPCA 18, 458-465 (1968).

5   S.B.Carpenter, L.W.Thomas J.M.Leavitt, W.C.Colbaugh and F.W.Thomas, Principal plume dispersion models: TVA power plants. JAPCA 21, 491-495 (1971).

6   F.Pooler Jr. and L.E.Niemeyer, Dispersion from Tall Stacks: an Evaluation. Proceedings 2nd International Clean Air Congress (H.M.Englund and W.T.Beery editors), Academic Press, New York (1971).

7   F.S.Shiermeier, Study of effluents from large power plants. Presented at the American Industrial Hygiene Assoc. Conf. May 24-28, Toronto, Canada (1971).

8   M.M.Millán, A.G.Gallant and H.E.Turner, The application of correlation spectroscopy to the study of dispersion from tall stacks. Atmos. Environ. 10, 499-511 (1976).

9   D.J.Moore, SO2 concentrations measurement near Tilbury power station. Atmos. Environ. 1, 389-410 (1967).

10   P.M.Hamilton, Plume height measurements at Northfleet and Tilbury power stations. Atmos. Environ. 1, 379-387 (1967).

11   D.B.Turner. Workbook on Atmospheric Dispersion Estimates. Office of Air Programs Publication No. AP-26, US EPA, N.C.(USA) (1970).

12   D.B.Turner. Atmospheric Dispersion Modeling, a Critical Review. JAPCA 29, 501-519 (1979).

13   M.E.Smith, Review of the attributes and performance of 10 rural diffusion models. Bull. Am. Meteorol. Soc. 65, 554-558 (1984).

14   D.G.Fox, Uncertainty in air quality modeling. Bull. Am. Meteorol. Soc. 65, 218-221 (1984).

15   J.C.Weil, An updated Gaussian plume model for tall stacks. JAPCA 34, 812-827 (1984).

16   J.C.Weil, Updating applied diffusion models. J. Climate Appl. Meteor. 24, 1111-1130 (1985).

17   JAPCA, Atmospheric dispersion modeling: Discussion papers. 29,927-941(1979).

18   M.M.Millán and Y-S. Chung, Detection of a plume 400 km from the source. Atmos. Environ. 11, 937-944 (1977).

19   M.M.Millán, Recent advances in correlation spectroscopy for the remote sensing of SO2. Proc. 4th Conf. on Sensing of Environmental Pollutants. Ameri Chemical Soc. NY.(1978).

20   R.M.Hoff and A.G.Gallant, The use of an available SO2 tracer during the 1983 CAPTEX experiment. Atmos. Eviron. 19, 1573-1575 (1985).

21   J.N.Carras and D.J.Williams, The long-range dispersion of a plume from an isolated point source. Atmos. Environ. 15, 2205-2217 (1981).

22   M.M.Millán and R.M.Hoff, How to minimize the baseline drift in a COSPEC remote sensor. Atmos. Environ. 11, 857-860 (1977).

23   R.M.Brown, R.N. Dietz and E.A.Cote, The use of sulfur hexafluoride in atmospheric transport and diffusion studies. J. Geophys. Res. 80, 3393-3398 (1975

24   R.M.Brown/ L.A.Cohen and M.E.Smith, Diffusion measurements in the 10-100 Km range. J. Appl. Meteor. 11, 323-334 (1972).

25   J.S.Nader, Source Monitoring. in Air Pollution Vol.III Measuring Monitoring and Surveillance (A.C.Stern editor). Academic Press, NY. (1976).

26   W.E. Herget and W.D. Conner, Instrumental Sensing of Stationary Source Emissions. Env.Sci. and Tech. 11, 962-967 (1977).

27   R.N.Dietz, Lectures in this Text

28   P.M.Hamilton, R.H.Varey and M.M.Millán, Remote sensing of sulphur dioxide. Atmos. Environ. 12, 127-133 (1978).

29 P.Camagni and S.Sandroni (editors), Optical Remote Sensing of Air Pollution. Elsevier Science, Amsterdam (1984).

30 P.M.Hamilton, The application of a pulsed light range finder (Lidar) to the study of chimney plumes. Phil.Trans.R.Soc. Series 8, 265, 153-172(1969).

31 W.B.Johnson Jr., Lidar observations of the diffusion and rise of stack plu mes. J.Appl.Meteor. 6, 443-449 (1969).

32 W.B.Johnson Jr. and E.E.Uthe, Lidar study of the Keystone stack plume. Atmos. Environ. 5, 703-724 (1971).

33 G.Newcomb and M.M.Millán, Theory, applications and results of the long-line correlation spectrometer. IEEE Trans. Geosc. Electr. GE-8, 149-157 (1970).

34 Barringer Research Ltd., Optical measurements of Sulfur Dioxide and Nitrogen dioxide Air Pollution Using Barringer Correlation Spectrometers. Contractors Report PB 193-485. NTIS, Springfield, Virginia, USA (1969).

35 R.V.Portelli, B.R.Kerman, R.E. Mickle, N.B. Trivett, R.M. Hoff, M.M. Millán, P.Fellin, K.S.Anlauf, H.A.Wiebe, P.K.Misra, R.Bell and O.Melo, The Nanticoke shoreline diffusion experiment, June 1978. Atmos. Environ. 16,413-466(1982).

36 J.C.Weil and A.F.Jepsen, Evaluation of the Gaussian plume model at the Dickerson power plant. Atmos. Environ. 11, 901-910 (1979).

37 J.C.Weil, Applicability of stability classification schemes and associated parameters to dispersion from tall stack plumes in Maryland. Atmos. Environ. 13, 819-831 (1979).

38 F.Fanaki, Experimental observation of a bifurcated buoyant plume. Atmos. Environ. 9, 479-495 (1975).

39 I.A.Singer, Personal communication with M.M.Millán during the 76th. APCA Meeting in Portland, OR.(1976).

40 R.M.Hoff, N.B.A.Trivett, M.M.Millán, P.Fellin, K.G.Anlauf and H.A.Wiebe, The Nanticoke shoreline diffusion experiment June 1978- III. Ground-based air quality measurements. Atmos. Environ. 16, 439-454 (1982).

41 A.F.Jepsen and J.C.Weil, Maryland Power Plant Air Monitoring Program: Preli-minary Results. Paper 73-157, 66th Annual Meeting APCA, (1973).

42 N.D.van Egmond and H.Kesseboom, Mesoscale air pollution dispersion models-I. Eulerian grid model. Atmos.Environ. 17, 257-265 (1983).

43 N.D.van Egmond and H.Kesseboom, Mesoscale air.... models-II. Lagrangian puff model and comparison with Eulerian grid model. Atmos. Environ. 17, 267-274 (1983).

44 M.M.Millán, E.Otamendi, L.A.Alonso and I.Ureta. Experimental characteriza-tion of atmospheric diffusion in complex terrain with land sea interactions. JAPCA (in press) (1987).

45 M.M.Millan, Meso-meteorological analisis of air pollution cycles in Spain. Presented at the COST 611 Meeting, Stresa, Italy, Sept. 23-25, 1986. To be published by Elsevier Science (1987).

46 R.G.Barry and R.J.Chorley, Atmosphere, Weather and Climate. Methuen, London, 4th. Ed.(1982).

47 G.T.Csanady, Turbulent Diffusion in the Environment. D.Reidel, Dordrecht-Holland/ Boston, USA, (1973).

Regional and Long-range Transport of Air Pollution,
Lectures of a course held at the Joint Research Centre, Ispra, Italy,
15–19 September 1986, S. Sandroni (Ed.), pp. 281–304
© Elsevier Science Publishers B.V., Amsterdam — Printed in The Netherlands

ATMOSPHERIC TRANSPORT OF AIR POLLUTANTS IN THE MESOSCALE OVER HILLY
TERRAIN: A REVIEW OF THE TULLA - EXPERIMENT

F. Fiedler

Abstract

   On the basis of the observations during the TULLA-Experiment (Transport und
Umwandlung von Luftschadstoffen im Lande Baden-Württemberg und aus Anrainerstaaten) the transport and distribution of SO_2 , NO, NO_2 und O_3 was studied
in a mesoscale area. During this experiment, which was supported by a large
number of scientific groups from Europe and the USA, emission rates, meteorological conditions and concentrations were measured at the ground and by
eight aircrafts in the lower troposphere during intensive periods of up to
36 hours length. These periods were embedded in the period of March 19 to
March 31 1985.

   As a measuring site the area of the state Baden-Württemberg, Federal
Republic of Germany, was chosen, an area of about 200 km x 200 km. The site
is a typical mountainous terrain with elevations from 110 to 1500 m above
mean sea level.

   Emission data from all sources with energy consumption > 10 MW are available from individual records on an hourly basis. Additionally, estimates of
area sources within a grid of 1 km x 1 km were calculated from energy use of
villages, size and types of houses and of traffic frequency counts.

   The meteorological conditions were determined by a temporary network including vertical soundings at several positions.

   Observations include easterly and southwesterly wind situations. The flow
field is strongly channelled especially in the Upper Rhine valley due to
topographical influences. In the region of topographical influence on the
lower atmosphere more than 50 percent of the masses of trace gases are transported. This leads also to a channeling of the mass transport for those
pollutants which are emitted at a few highly industrialized areas.

   Depending on the air flow conditions, the distributions of SO_2 , NO and
NO_2 are very inhomogeneous. Plumes from industry complexes can be detected
over long distances. The distribution of O_3 is instead determined by larger

scale processes and shows a rather homogeneous density pattern showing only reduced concentrations in those areas where NO is high.

Estimates of the mass balance for the whole mesoscale area show during southeasterly conditions consistent results with an export residuum, which depends on the strength of emissions.

1.   INTRODUCTION

Most studies on turbulent diffusion in the atmosphere have been concentrated on the single source case, in that area of the boundary layer which is accessible by ground-based measurement systems as high observation masts or other ground-based systems. The knowledge on the structure of the upper part of the atmospheric boundary layer and the spectrum of phenomena observable during different synoptic conditions is still rather crude. Although most of the work has concentrated in the past on the boundary layer structure under very idealized conditions especially over flat terrain, even for this case a wide gap of insufficient understanding of the processes exists in relation to un-stationarity as for example the passage of fronts, the influence of partly coverage of clouds, the rôle of convective systems such as cumulus clouds, thunderstorms, low clouds within the boundary layer or the influence of stratification and strongly layered thermal structures in night situations.

The distribution of material released into the lower layers of the atmosphere is strongly dependent on those processes, as can be seen, for example, from the transport distances material is advected during low-level jet situations with extremely low turbulence intensities.

Transport and turbulent diffusion processes become even more complicated over complex terrain. Diffusion may be altered by orders of magnitude, when the turbulence intensity, convection and secondary eddies are stimulated by hills and valleys or when the air becomes stagnant in hollows.

By increasing the distances of interests on the fate of air pollutants emitted from elevated sources and ground sources the flow and turbulence structure in the mesoscale must be regarded in greater detail. From previous studies (e.g. Georgii, 1982) it is well known, that the vertical distribution of pollutants with sources at the ground have a simular shape as water vapour. One main characteristic is that more than half of the masses are transported below 1500 m, a height which is comparable to many mountain ridges penetrating into the atmosphere.

In this respect it is of great importance to know the detailed structure of
the vertical profiles of concentrations under the varying meteorological
conditions since this determines in which direction the material is transported and where it is finally deposited. As will be demonstrated below, the
daily cycle of stability within the atmospheric boundary layer has a strong
influence on the transport direction of material released into the atmosphere
over complex terrain.

Still another aspect must be considered. Most of the emission sources are
not stationary. Especially, many large power plants operate in a daily
variing manner, which has an important influence on the concentrations and
deposition at the ground even at larger distances.

In order to learn to understand the behavior of measured concentrations of
trace constituents in the atmosphere all aspects have to be taken into account
at the same time in a balanced way. Ground concentrations measured in an area
may vary because of

a) a change of the emission rates of the numerous sources
b) a change in the transport and diffusion conditions in the atmosphere
c) due to the influence of chemical transformations e.g. as a function
   of solar radiation or air moisture
d) as a function of deposition conditions at the ground (e.g. dry or wet
   surface).

In a larger experiment called TULLA (Transport und Umwandlung von Luftschadstoffen im Lande Baden-Württemberg und aus Anrainerstaaten) which stands
for "Transport and transformation of air pollutants in the state Baden-
Württemberg and from surrounding states", these processes have been studied.

In this paper an overview on the field phase of this experiment will be
presented. Additionally, results are used to present some observed phenomena.

2.   SCIENTIFIC OBJECTIVES OF THE TULLA-EXPERIMENT

Discussions on the rôle of air pollutants in respect to observed damages
to forests and buildings and impacts to soil surfaces have strengthened the
interests on the paths air pollutants take after their release to the
atmosphere. This discussion was also stimulated by considerations on the

degree of improving the air quality by reducing emissions in a limited region only. Most of these problems are connected to the situation to measure concentrations of the leading trace gas constitutents which are very low compared to those in urban areas.

The TULLA-experiment was designed to determine the transport and the vertical and horizontal distribution of sulfur dioxide (SO_2), Nitrate Oxides (NO and NO_2) and Ozon (O_3) in the State of Baden-Württemberg in periods of 24 hours length.

These observations were planned

(1) to supply detailed information on the paths which trace substances take over complex terrain from source areas to the deposition areas or until they are transported out of the area

(2) to estimate the terms of the mass balance equation for SO_2 for the area of Baden-Württemberg

(3) to obtain an appropriate data set to validate mesoscale models for the atmospheric flow over complex terrain and for the diffusion processes.

The programme was confined to study the transport and the distribution of only a few chemical substances (SO_2 , NO_2 and O_3). However at most of the measuring stations concentrations of additional components have been recorded. The substances SO_2 , NO_2 and O_3 had been selected since a larger number of surface stations already exists in the area for these constituents.

It was of special interest to resolve the processes occuring during daily cycles, e.g. the change of hourly emissions, the meteorological conditions, the deposition and the distribution of concentrations.

This time scale is also selected because air takes about the same time to travel from the inflow side to the outflow side of the chosen area.

In addition an appreciable part of the released material is deposited within the area during the same time span.

3.   EXPERIMENT AREA AND MEASUREMENT SYSTEMS

3.1   EXPERIMENT AREA

The southwestern part of the Federal Republic of Germany is an area with
mountains and valleys typical for many parts of the world. Therefore the area
of the State of Baden-Württemberg was selected for the experiment (see Fig.1).

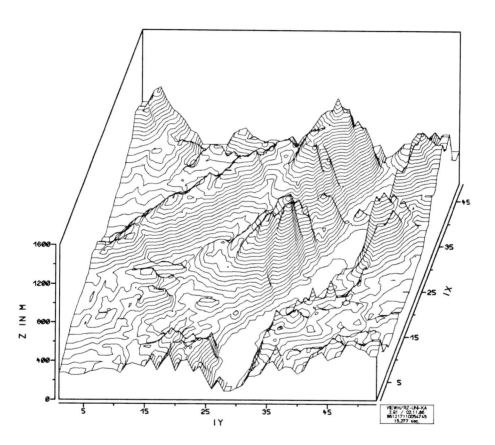

Fig. 1   Topography of the area for the TULLA-experiment

On its western side to France the Rhine valley forms a very marked structure with a great number of flow characteristics. Due to the Black Forest at the
eastern side, the Vosges at the western side, and North of it the Palatinate
forest and the Odenwald form a long north-south flow channel.

Other dominant topographical structures are the valley of the Danube and the Swabian mountains extending on the eastern side of the Black Forest in an north-easterly direction. The lowest levels of the area are at about 110 m, the highest levels are found in the Black Forest (Feldberg) with about 1400 m above mean sea level.

3.2   MEASUREMENT SYSTEMS

(a)   METEOROLOGICAL MEASUREMENTS

In Fig. 2 the locations of the various types of meteorological measurement stations are given (see Walk, 1986, Vogt and Fiedler, 1986).

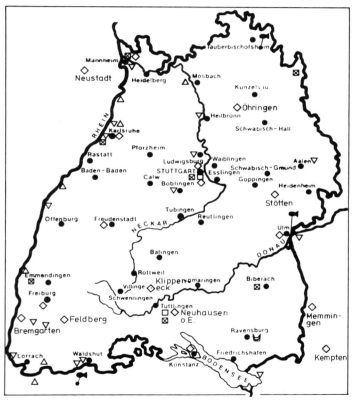

☐ Radiosondes   ◇ Synoptic stations
▽ Meteorological stations   △ Mast measurements
☒ Structure Microsondes   ☇ Tethered balloon
   Sodar

Fig. 2   Meteorological network for the TULLA-Experiment

In the meteorological program it was attempted to resolve the flow
characteristics in a more detailed form than it is measured from the routine
observing network of the weather service. Regular radiosonde stations are
located at Stuttgart and at Tuttlingen (Neuhausen ob Eck). An additional network was installed with structure sondes at Mannheim, Karlsruhe, Freiburg,
Stuttgart, Neuhausen, Bad Mergentheim and Biberach. These structure sondes
give a vertical resolution of 30 m for temperature, moisture, pressure, wind
speed and wind direction.

At the southern side, along the Swiss border, a motor glider equipped with
meteorological instruments and between Karlsruhe and Mannheim a smaller aircraft, sampled meteorological data.

In addition to the vertical soundings by sondes three tethered balloons
were operating and also one SODAR, 6 meteorological towers, some of it up to
200 m high and numerous surface station had been installed to receive detailed
information on the meteorological conditions. All station types are indicated
in Fig. 2 by different symbols. As an additional information base 6 hourly
synoptic objective analyses are available which contain the information of the
larger scale flow field approaching the experiment site.

(b)   CHEMICAL MEASUREMENTS

A network of ground based stations exists already in many urban areas for
measurements of SO_2 , NO, NO_2 and Ozon. Some additional stations are operated
in the vicinity of high stacks of power plants and in remote areas in connection with research projects of forest damage. In order to garantee a good
coverage of concentration measurements 10 additional stations for SO_2 and
particular pollutants had been installed (see Vogt and Fiedler, 1986).

On the western side of the area three correlation spectrometers were
opperating along the autobahn running south-north to measure the vertical
integral of SO_2 .

As the basic systems for determining the distributions of SO_2 , NO, NO_2
and O_3 in the lower atmosphere seven aircrafts were flying along predefined
flight tracks (see Fig. 3)

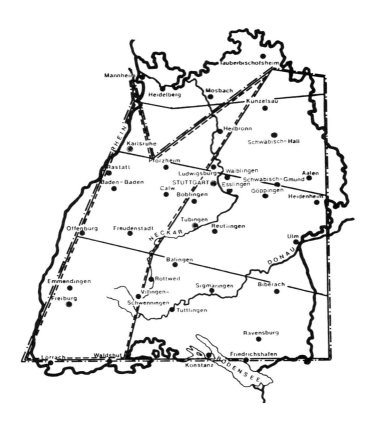

Fig. 3   Flight tracks for flight pattern B
          dashed line: blue formation
          dashed-dotted line: red formation
          Air flow from northwest or south-east

These flight tracks were chosen in such a way that the cross boundary
fluxes at the inflow and at the outflow side could be determined. Some
additional paths were selected to gain some information on concentrations
also inside of the area. The seven aircrafts flew in two groups four or
three seperated in the vertical. One aircraft was measuring concentrations
above the inversions if there was any, the others were fitted into the height
distance, between 150 m above ground and the inversion base.

An additional jet-aircraft cruised on a vertical butterfly pattern in order
to measure vertical profiles of some constituents (especially NO and CO) be -
tween the height range 150 m and 3000 m above ground. All aircraft were tracked
by a radar system which is normally used for following low flying aircraft.

(c)   EMISSION DATA

Of equal importance as meteorological information and the distribution of
chemical compounds is the knowledge of sources. Special affords have been made
to complete an inventory of SO_2 , NO and NO_2 which fits in its time and space
resolution to the processes going on in a limited area. Sources with energy
use of more than 10 MW have been treated individually by collecting the reports
on hourly emissions. Smaller point sources and area sources have been provided
by an objectiv procedure based on a huge amount of information data including
energy use, type of heating systems, type of houses and other factors. All
these emission data have been determined on a horizontal scale of 1 km x 1 km.
Details are given by Boysen et al. (1986).

4.   TIME OF THE EXPERIMENT AND INTENSIVE MEASURING PHASES

For the experiment a time span of forteen days (17.3.1985 - 29.3.1985) had
been chosen. The end of March was selected because at this time energy use for
heating is still high but at the same time, radiation energy from the sun is
already strong enough to produce strong daily variations of meteorological
diffusion conditions.

Within the forteen days period (see Fig. 4) intensive measuring phases have
been defined where all the meteorological sounding systems and the aircraft
operated. These intensive measuring phases were chosen only for those conditions where no precipitation and low clouds were expected.

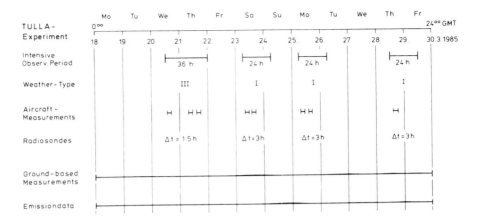

Fig. 4   Measuring period and intensive measuring phases

5.   SOME PHENOMENA OF EMISSIONS, TRANSPORT AND DISTRIBUTION OF AIR
     POLLUTANTS

5.1 EMISSIONS

In the following we will show only some results of the emissions for SO_2 ,
although emission data are also available with the same horizontal and time
resolution as for SO_2 . In Fig. 5 the integral of all emissions for the state
Baden-Württemberg during the time of the experiment are given (Boysen et al.,
1986). The total emissions are seperated into contributions by industry, power
plants, domestic heating and traffic. Dominant characteristics are rather
strong daily variations ranging between about 28 to 40 to h^{-1} . Beside the
daily variation one can also recognize a clear weekly variation with lower
emissions during the weekend. Since during a time of 24 hours the pollutants
are transported over longer distances, mixed completely in the atmospheric
boundary the rate of deposition will also strongly depend on the variable concentration field. It is clear from the nonlinear nature of transport processes
that deposition estimates based on mean values over a day or a month may give
a rather erroneous result. This factor will be increased by receptors also
having daily variations like the stomate of plants or the moisture on leaf
surfaces and on soil.

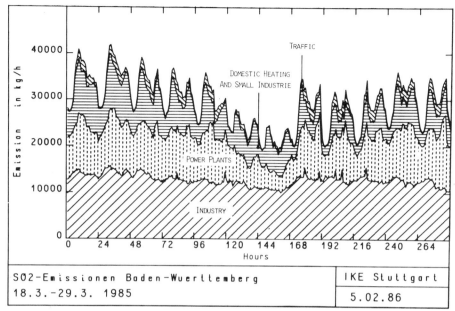

Fig. 5   Integral of hourly emissions of SO$_2$ for the TULLA-area,
          March 18 - 29, 1985 (Boysen et al. 1986)

Fig. 6 gives an example of the horizontal distribution of some larger
emission sources.

In this figure the emissions for one hour are shown. Although the relative
importance of the sources are changing during the days it is recognizable that
the major part of the emissions are released into the atmosphere only from a
few industry areas. Although emissions from domestic heating are also higher in
urban than in rural areas nevertheless they add more to the background concentration field together with the transboundary fluxes.

Similar results are available for the emissions of NO and NO$_2$ also for hourly time intervals and the 1 km grid. For the NO and NO$_2$ emissions traffic on
roads contributes a major part. Therefore the emission inventory reflects very
clearly the main traffic roads and the traffic density which has been the basis
for this inventory.

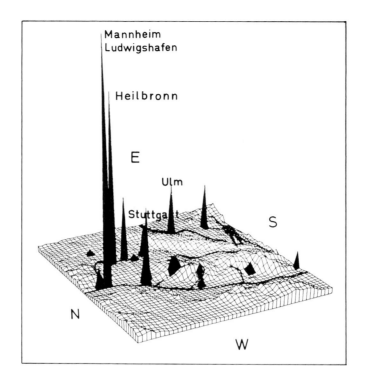

Fig. 6   Emission of single sources of SO_2 for the measuring
phase of March 25, 1985

5.2   CHANNELLING OF THE ATMOSPHERIC FLOW

It has been shown by observations as well as by the use of mesoscale models
that orography has a strong influence on the flow in the lower levels of the
atmosphere.

This phenomenon has been studied in greater details within the Upper Rhine
valley during the measuring compaign MESOKLIP (Fiedler and Prenosil, 1980).

In the area of the measurements the valley is 50 km wide, the surrounding
mountains are only 500 m high in the West and 150 m in the East.

The general picture of the flow conditions is characterized by a rather narrow wind direction distribution. All larger scale synoptic conditions with easterly directions are connected with a southerly flow in the Rhine valley, all easterly synoptic flow conditions force the flow within the Rhine valley coming from the north. A typical observation is shown in Fig. 7 where a rather homogeneous westerly flow is visible above the mountains at the western side. Within the Rhine valley appears a very remarkable turning of the wind along the valley. This situation was observed during morning hours (10:00 LMT) September 17, 1979 by a network of vertical radiosondes making hourly measurements with a horizontal resolution of about 7.5 km. As the time progresses during the day, through convectional influences a stronger coupling starts which brings up a rather surprising result. Instead of forcing the flow in the valley into the direction of the free atmosphere flow, the opposite is observed. This can be seen from Fig. 8 and 9 which give the situation for 12:00 and 14:00 LMT, which is the time of the maximum of this development. At this time the valley influence can be found even in heights which are about twice as high as the surrounding mountains of the valley. This phenomenon which is observable also in other valleys of similar size acts then as a barrier to the westerly flow and changes the frictional influences on the boundary flow drastically (Fiedler, 1983).

For the diffusion problem it is of main interest to know where the material is transported. Therefore, it will not be sufficient to rely the calculations of the concentration distributions on the low level wind field. Since intensive sources show heights of about 200 m diffusional effects will mix the material into the flow aloft which may show almost the opposite direction to the flow in the valley. Only taking synoptic data into account for determining the dispersion of pollutants will also not be sufficient when a resolution scale smaller than 100 km is approached.

5.3 DISTRIBUTIONS OF SO_2, NO AND NO_2 AT SOUTH-WESTERLY FLOW CONDITIONS (MARCH 23, 1985)

A main purpose of the TULLA-experiment was to give an insight into the present situation how the material released from a complex configuration of sources is distributed over an area of 200 km x 200 km. Although the data were taken also to serve other aims as estimating a mass balance of SO_2 and for validation of numerical simulation models some results will be shown here only of the distributions observed by the aircrafts.

294

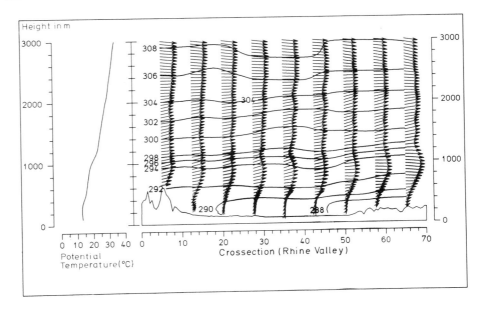

Fig. 7 Vertical profiles of the horizontal wind velocity across the
Upper Rhine valley at 10:00 LMT. Arrows from left to right mark
a west-wind and an upward directed arrow marks a south-wind.
Thin lines are the Montgomery potential.

Data from two cases will be taken. The first one is the mid-morning situation of March 21, 1985 and the second of March 23, 1985.

The case study of March 23, 1985 will discussed first. It resembles a
typical south-westerly synoptic flow condition.

In Fig. 10 - 15 the concentrations of SO_2 , NO and NO_2 are shown from two
aircrafts (named TULLA 6 and TULLA 7) flying within the blue formation.

The aircraft TULLA 6 flew at about 300 m above ground and TULLA 7 as the
lowest aircraft stayed at a height of 200 m above ground. Heights had to be
adjusted to the terrain height according to the operational plan.

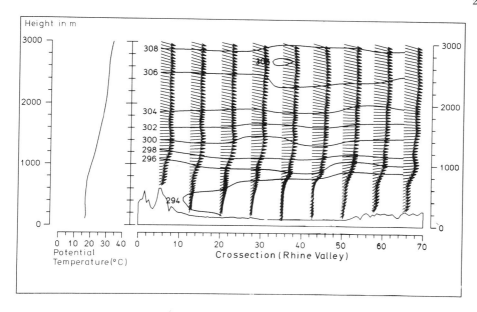

Fig. 8   Same as Fig.7 except for 12:00 LMT

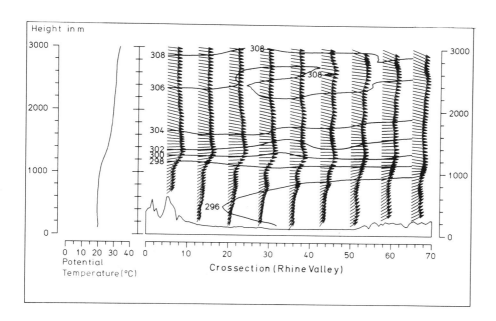

Fig. 9   Same as Fig.7 except for 14:00 LMT

The aircarfts started at Karlsruhe heading   south and landed from north
into Karlsruhe. In most of the area a rather low concentration of the species
had been found around 10 μg m^{-3} . Only in those areas where strong elevated
point sources are located a drastical increase of the concentrations have been
found. A consistant increase is detected for all components. Also from the
level of concentrations at the starting point it is detectable that during the
two hours of flying an influence of the daily cycle is already present leading
to an increase in the concentration at the higher levels for SO$_2$ and also for
Ozon at the lower levels (see Fig. 17). The heavily loaded air can be detected
until the air crosses the northerly boundary. A similar behaviour is also found
during the other measuring flights and by the other aircrafts. Over an area of
the size chosen for this experiment the distributions still reflect very clearly the source locations resulting in a strongly inhomogeneous level of concentrations.

5.4   OZON DISTRIBUTION AT SOUTH-WESTERLY FLOW CONDITIONS (March 23, 1985)

In Fig. 16 to 18 the measured ozon concentrations are presented as found by
the two aircrafts TULLA 6 and TULLA 7 from the blue aircraft formation and
additionally from TULLA 1 flying along the route of the red formation. From all
observations in the area a surprisingly high concentration of ozon was found
ranging up to 80 μg m^{-3} which is higher than is normally expected during this
period of the year. In comparison to the distribution of SO$_2$ and NO$_x$, Ozon is
rather homogeneously mixed over the area and no strong anthropogeneous variations
can be found. This behaviour is a consequence of the larger scale processes
which are responsible for the observed level of O$_3$ -concentrations. The only
exception is given in those areas which complement the concentration of NO
showing a remarkable destruction of O$_3$ within the plumes of industrial areas.
The daily cycle produces, however, variations of the same order of magnitude
as can be found between the larger scale concentration level and the concentrations within the city plumes.

5.5   DISTRIBUTION OF SO$_2$, NO AND NO$_2$ AT EASTERLY FLOW CONDITIONS
       (March 21, 1985)

During the intensive measuring phase of March 21, 1985 easterly flow conditions were prevailing.

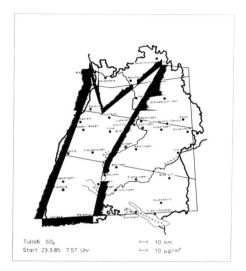

Fig.10 Distribution of SO$_2$ along
the flight track at 300 m above
ground

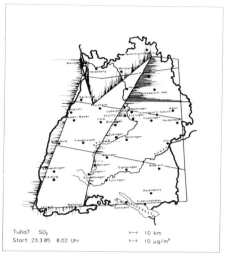

Fig.11 Distribution of SO$_2$ along
the flight track at 200 m above
ground

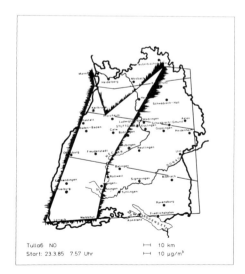

Fig. 12 Same as Fig. 10 except for
NO

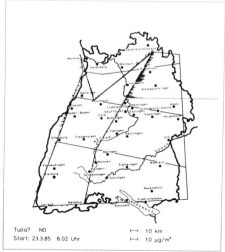

Fig. 13 Same as Fig. 11 except for
NO

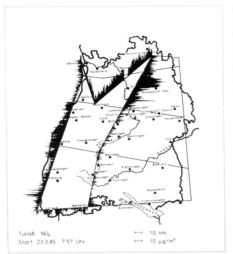

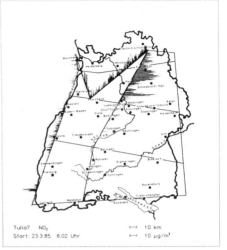

Fig. 14   Same as Fig.10 except for NO$_2$

Fig. 15   Same as Fig. 11 except for NO$_2$

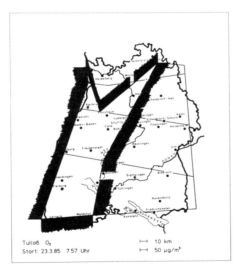

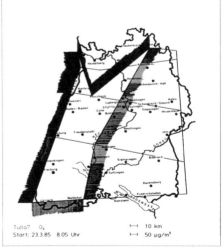

Fig. 16   Same as Fig. 10 except for O$_3$

Fig. 17   Same as Fig. 11 except for O$_3$

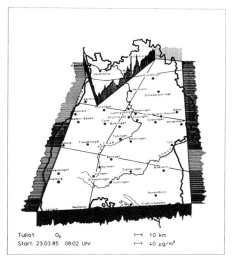

Fig. 18 Same as Fig. 10 except for O$_3$ and flight track of red formation at 550 m above ground

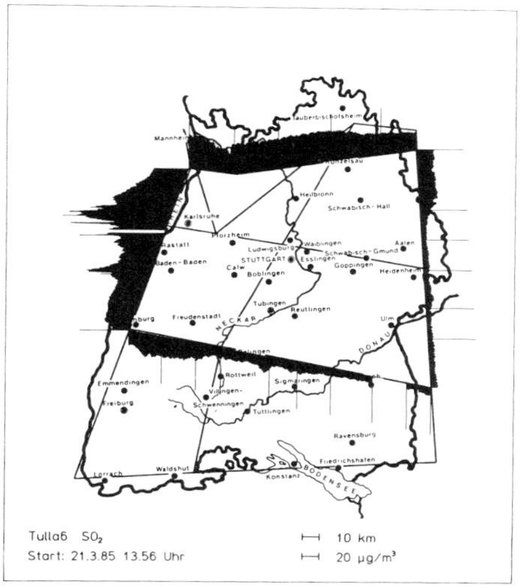

Fig. 19 Distribution of SO$_2$ along flight track of blue formation at above ground 400 m

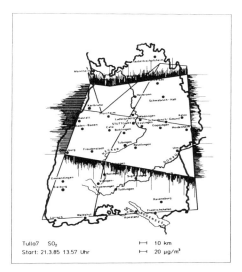

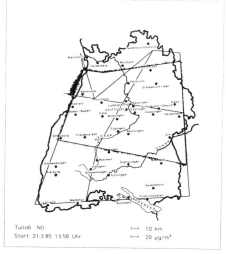

Fig. 20 Distribution of SO$_2$ along flight track of blue formation at 200 m above ground

Fig. 21 Same as Fig. 19 except for NO

300

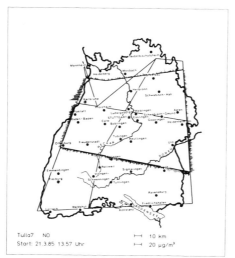

Fig. 22 Same as Fig. 20 except for NO

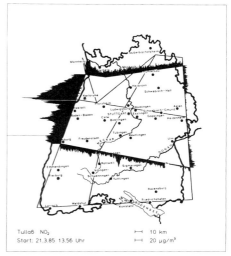

Fig. 23 Some as Fig. 19 except for NO$_2$

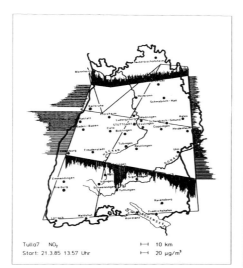

Fig. 24 Same as Fig. 20 except for NO$_2$

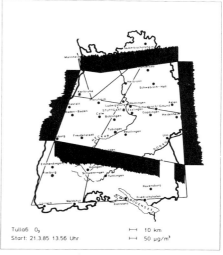

Fig. 25 same as Fig. 19 except for O$_3$

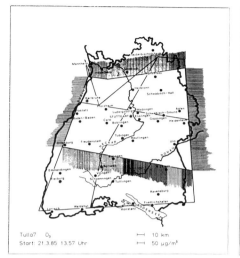 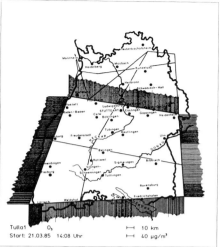

Fig. 26 Same as Fig. 20 except
        for O$_3$

Fig. 27 Same as Fig. 19 except
        for O$_3$ and flight track
        of red formation at 800 m
        above ground

At the inflow side (flight pattern C, south-east) only low concentrations
were observed (Fig. 19 to 24) at levels higher than 300 m above ground.
Rather high values were found on the north-western outflow side reflecting the
situation of the close location of strong sources in the northern part of the
area. Again, as it was found on other days, SO$_2$ and NO$_x$ concentrations are
very strongly dependent on the flow direction in relation to the sources of
industry and power plants. An area with side lengths of 200 km is to small to
expect a well mixed distribution of the constituents horizontally and vertically.

5.6   DISTRIBUTION OF O$_3$ AT EASTERLY FLOW CONDITIONS (MARCH 21, 1985)

The Ozon distribution during the intensive measuring period during afternoon of March 21, 1985 is given in Figs. 25 to 27. Also on this day the mean
value of the concentration was about 80 µg m which was evenly found along the
whole flight paths. Only in those regions again, where the level of NO$_x$ -

concentrations was high a considerable destruction of O_3 was found, especially at the western border. The TULLA 1-aircraft was flying at the highest level (800 m above ground) where the NO_x- concentration was rather low. Therefore the measurements of the aircraft TULLA 1 represents the larger scale concentration level outside of the boundary layer uneffected by local sources of NO within the experimental area.

6. ESTIMATE OF THE MASS BALANCE OF SO_2

Although a thorough determination of the mass balance for a mesoscale area needs to include elaborated methods for the different terms in the mass balance equation a rough and preliminary estimate will be given here.

At March 23 and March 25, 1985 similar synoptic conditions were observed. However there was a major difference in the amount of emissions due to the weekend day on March 23 (Saturday).

Equation (1) represents the vertically integrated mass balance of the experimental area, which is determined by the fluxes at the inflow and outflow sides, the emissions Q, the deposition and chemical transformations S:

$$\frac{\Delta M_{SO_2}}{\Delta t} = \nabla \cdot \mathbb{F}_{SO_2} + Q_{SO_2} + S_{SO_2}$$

Deposition taken as a residuum between the netflux and the source are in close agreement with those values which are determined via the measured ground concentration field and estimates of the deposition velocity in the area. At March 23, 1985 for the emissions of SO_2 a value of 24 to h^{-1} had been determined. Under the assumption of stationary conditions this is splitted up into a net-outflow of 10.3 to h^{-1} and a deposition of 13.7 to h^{-1}. The latter term include also the possible chemical transformations of SO_2.

In comparison to the Saturday conditions, on the following Monday (March 25, 1985) the emission rate increased to 30 to h^{-1} which divided up into a net-outflow of 13.8 to h^{-1} and a deposition of 16.2 to h^{-1}.

7. CONCLUSIONS

During the experiment TULLA a data set of the transport and diffusion conditions in a mesoscale area together with the vertical and horizontal distribution the constituents SO_2 , NO_2 and O_3 has been collected. Beside these main components several other species have been measured by the different participating groups. The data set gives now a valuable insight into the diffusion situation over a wider area, where the deformation of the wind field and the turbulence intensity especially due to topography plays an important rôle.

Although the experiment was not conducted primarily as an atmospheric chemistry project it gives some consistent information on the time and spatial variation of the measured constituents. The data set will however in the main part be the base for validating the results of numerical mesoscale models which at the end can only be the main tool for giving an integral information on the very complex physical and chemical processes involved in the field of air pollution.

References

Boysen, B., R. Friedrich, Th., Müller, N., Scheirle, A. Voss (1986
        Feinmaschiges Kataster der SO_2 und NO_x -Emissionen in Baden
        Württemberg für die Zeit der TULLA-Meßkampagne.
        In: 2. Statuskolloquium des PEF 4.-7.März 1986
        im Kernforschungszentrum Karlsruhe, KfK-PEF 4,
        Band 2 (available from Kernforschungszentrum)

Fiedler, F. und Prenosil, T. (1980): Das MESOKLIP-Experiment. Mesoskaliges
        Klimaprogramm im Oberrheintal.
        Wiss. Ber. Met. Inst. Univ. Karlsruhe Nr. 1

Fiedler, F. (1983): Einige Charakteristika der Strömung im Oberrheingraben.
        Wiss. Ber. Met. Inst. Univ. Karlsruhe, Nr. 4, S. 113-123

304

Georgii, H.W. (1982): The atmospheric sulfur-budget.
            In: Chemistry of the unpolluted and polluted troposphere
            (Edts.: H.W. Georgii, W. Jaeschke).
            D. Reidel, Dortrecht

Vogt, S. und Fiedler, F. (1986): Measuring activities of the TULLA-Experimant
            (Field Phase Report) KfK-Report (in preparation)

Walk, O. (1986): Meteorologische Sondermessungen
            In: 2. Statuskolloquium des PEF 4.-7. März 1986 im
            Kernforschungszentrum Karlsruhe, KfK-PEF 4, Band 2
            (available from Kernforschungszentrum Karlsruhe,
            Postfach 3640, 7500 Karlsruhe 1).

Regional and Long-range Transport of Air Pollution,
Lectures of a course held at the Joint Research Centre, Ispra, Italy,
15–19 September 1986, S. Sandroni (Ed.), pp. 305–336
© Elsevier Science Publishers B.V., Amsterdam — Printed in The Netherlands

PHYSICAL PROCESSES AND MODELLING APPROACHES

P. BESSEMOULIN

1. TIME SCALES RELEVANT TO THE TRANSPORT OF AIR POLLUTANTS SUBJECT TO CHEMICAL TRANSFORMATION AND DEPOSITION

Before tackling the problem of modelling meso-scale and large scale air pollution transport, it is necessary to examine :

- what time and space scales are involved, to determine the area which must be studied and the time during which pollutants must be considered ;

- what are the important factors which need to be looked at in the development of models of air pollution transport at these scales, by looking at the physics of the phenomena ;

- what meteorological information is available, and to what extent this information is consistent with the problem under study.

To derive an upper limit of pertinent time scales, we first consider a puff of pollutants, SO_2 for convenience. The depletion processes considered are transformation of SO_2 into sulphate, and dry deposition of both species. So here we ignore wet deposition.

Let m be the total initial mass of SO_2 in the puff. If we consider a first-order reaction as a good approximation, the evolution of the mass m is governed by the following equation :

$$\frac{dm}{dt} = - km \qquad (1.1)$$

where

$$k = k_t + \frac{V_d}{H} \qquad (1.2)$$

k_t is the SO_2 to SO_4 transformation rate (s^{-1})
V_d is the SO_2 dry deposition velocity (ms^{-1})
H is the depth of the well-mixed layer (mixing height)
The integration is straightforward and gives :

$$m = m_o \exp - kt \qquad (1.3)$$

where m_o is the initial mass of the puff.

In the same way, one can write an equation for the SO_4 mass content M in the puff :

$$\frac{dM}{dt} = - \frac{V_g}{H} M + k_t . m \qquad (1.4)$$

where V_g is the SO_4 dry deposition velocity.

Integration of this equation gives :

$$M = m_0 . k_t \left(k_t + \frac{V_d - V_g}{H} \right)^{-1} exp - \frac{V_g}{H} t . \left(1 - exp - \left(\frac{V_d - V_g}{H} + k_t \right) t \right) \qquad (1.5)$$

The behaviours of m and M are illustrated on fig. 1.a, according to typical values of the parameters V_d, V_g, k_t, H, whose ranges are the following in Laven u et al.' study :

Parameter	$SO_2 \rightarrow SO_4$ transf. rate (s^{-1})	SO_2 dry dep. vel.$(m.s^{-1})$	SO_4 dry dep. vel.$(m.s^{-1})$	mixing height (m)
Formulation	k_t	V_d	V_g	H
Value	7 to 280.10^{-6}	8.10^{-3}	$1.95.10^{-3}$	250 to 1000

Up to now, we have not considered diffusion at all. It is often assumed that the horizontal extent of a puff behaves according to a fickian diffusion law, so that the radius of the puff at time t is given by

$$r^2 = r_0^2 + Kt \qquad (1.6)$$

where r_0 is the initial puff dimension, consistent with the grid size on which the emission inventory is available. If the structure of the puff is assumed to be gaussian, and uniform in the vertical direction, the concentration is described by the following equation :

$$C = \frac{\varphi}{2\pi \sigma_H^2 H} exp - \frac{R^2}{2\sigma_h^2} \qquad (1.7)$$

where σ_h is the horizontal dispersion coefficient $(R \sim 3\sigma_h)$

R is the distance of the considered point from the center of the puff

Q is the pollutant release

A typical result is depicted on fig. 1.b : for travel times greater than 120h (5 days), SO_2 and SO_4 concentrations have become quite negligible (SO_2 more negligible than SO_4).

It must be stressed that the assumption of fickian diffusion may lead to underestimate of the horizontal spread since for small and moderate travel times, r behaves like t $(\sigma_h \sim 0.5 \, t)$ instead of $t^{1/2}$.

Considering possible advections, the length scales associated with the time scale quoted above range between some hundreds of kilometers to some thousands of kilometers.

2. VERTICAL CONCENTRATION PROFILES AND GROUND LEVEL CONCENTRATIONS

Again, let us consider simplifying assumptions, i.e. diffusion from a point source can be approximated by a gaussian plume :

$$C = \frac{\varphi}{2\pi \bar{u} \sigma_y . \sigma_z} . exp - \frac{y^2}{2\sigma_y^2} . \left\{ exp - \frac{(z - He)^2}{2\sigma_z^2} + exp - \frac{(z + He)^2}{2\sigma_z^2} \right\} \qquad (2.1)$$

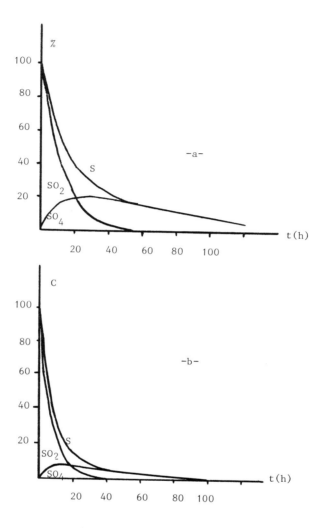

Fig 1 : a) Time evolution of the SO_2, SO_4, S mass content of the puff for typical values of the parameters: H=500 m, k_t=7.10^{-6}

b) Time evolution of the SO_2, SO_4, S concentration at the centre of the puff (same parameters as above)

where H_e is the effective height of the source.

It can be shown (Turner, 1970) that a good working approximation of the real plume when diffusion is limited in the vertical direction by an elevated inversion layer is to consider that after a distance $x = x_L$, where $\sigma_z(x_L) = 0.47 \, H$, the plume has become uniformly distributed between the ground and the inversion height H.

On the other hand, one can look at the ground level concentrations for various emission heights, but the same emission Q.

It can easily be shown (fig.2) that after some travel distance -depending on the vertical stability- ground level concentrations are no longer dependent on the emission height, and so are equal to concentrations due to ground level releases. For the neutral case presented on fig.2, this distance is about 100 km.

This is extremely fortunate from a modelling point of view, since for long range transport simulation it will not be necessary to take the emission heights into account, as long as diffusion is restricted to the atmospheric boundary layer.

These simple conclusions also allow us to make a distinction between meso-scale and large scale transport of air pollutants.

Meso-scale (or regional scale) deals with transport from some tens to some hundreds of kilometers (20-400 km). Corresponding transfer times lie between 1 hour to 1 day. At this scale, atmospheric pollutants due to a point source may have a non homogeneous vertical profile.

Transport over long distance concerns distances greater than some hundreds of kilometers. Corresponding transfer times are typically between one and five days because of the life time of the species considered here.

At this scale, one can consider that vertical pollutant profiles are uniform and that concentrations are not dependent on the emission height. Consequently, all sources are usually considered as emitting in such a way that pollutants can be considred as well-mixed in the original mesh.

Distances and characteristic times being different in the two cases, the modelling techniques will be quite different.

In particular, as will be seen later, the meteorological data available have an acceptable spatial coverage if one is interested in long range transport. On the other hand, that density (distance between stations of about 70 to 100 km) is notably too small in general for regional problems. In this case, it will be necessary prior to modelling diffusion, to set up a meso-scale meteorological model to compensate for the lack of information.

3.   THE METEOROLOGICAL INFORMATION AVAILABLE

Meteorological observations are performed in the framework of the World Weather Watch (WWW) which was approved by the 4th World Meteorological Congress

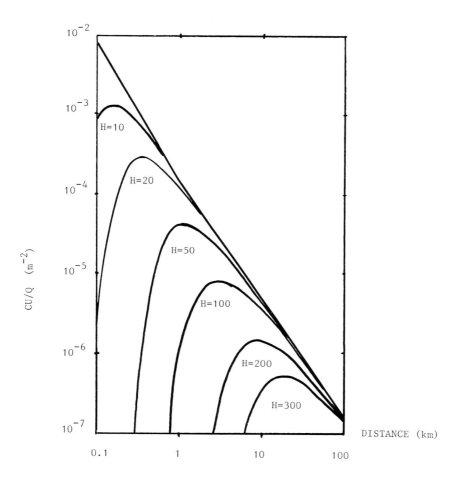

Fig 2 : CU/Q with distance for various emission heights.
D Stability (from Turner, 1970)

of the World Meteorological Organization in 1963. The WWW is divided into three main elements :

- the World Observing System (WOS), which includes all the systems which allow meteorological observations to be made, either over continents or over seas.

- the World Data Processing System (WDPS), which includes some very well equipped meteorological centers and the organization necessary to process data for immediate use, and for archiving it, and access to data.

- the World Telecommunication System (WTS) which is composed of all the systems necessary for the rapid collection and dissemination of raw and processed data.

The surface sub-system is composed of the synoptic network (including some automatic weather stations), and oceanic meteorological stations (on board stationary and selected ships, buoys...).

For altitude measurements, radio sounding stations, aircraft and satellite data are used.

According to the WMO rules,

- ground level meteorological measurements are performed and sent to the WTS every 3 hours, starting from midnight. The distance from synoptic stations in Western Europe lies between 70 and 100 km.

- radiosonde observations are taken twice a day (0 and 12 h UT) and for some stations altitude winds are also available at 06.00 and 18.00. The distance between stations is 300 to 500 km. All the data are collected at a national level and injected in the WTS. This latter is organized at three different levels :

- worldwide level : World Meteorological Centers (WMC) are located at Melbourne for the Southern Hemisphere, Moscow and Washington.

- regional level (in terms of WMO regions) : in region VI (Europe), regional centers (RMC) are Bracknell, Moscow, Norrköping, Offenbach and Rome.

- national level : national meteorological centers (NMC). The full objectives of the different centers are described in the WMO publication n°305 : Guide to the World Data Processing System.

The analyses and forecasts at large and/or global scale are the responsability of WMC and RMC's while NMC's are usually more involved in regional and local scales, and medium and short range forecasts.

The general list of products developed by WMC and RMC's is given in the WMO Publication n° 485 : Guideline of the World Data Processing System.

Of the meteorological analysed fields, the following are most relevant to LRT of pollutants :

- surface, 850, 700, 500 hPa analyses

- maps of precipitations

Reference hours are 00, 06, 12, 18 GMT. Similar information is also available for forecasts.

For an operational use of this data, it is interesting to know when the necessary information is available. The following table gives the expected time-lag between the time of the observations and the time at which they are received in different kinds of Centers.

Data received \ Receiving Center			WMC	RMC	NMC
Time at which observed data is received (1)	World Network	surface + altitude	H + 4 (7)	H + 4 (8)	H +4 (8)
	Regional Network	surface		H + 2 (3)	H + 2 (3)
		altitude		H + 3 (4)	H + 3 (4)
Minimum delay for recording observed data to be transmitted later		surface	H + 12	H + 12	
		altitude	H + 24	H + 24	

H is the time at which observations are performed.

The first number after H indicates the time necessary to collect data, to process it and to transmit elaborate material (for centers using modern equipment). The number in brackets is applicable when the WDPS is working under severe conditions.

The following table gives indications about the availability of "products" considered to be of a high priority for routine meteorological work.

Data received \ Receiving Center				WMC	RMC	NMC
Time at which analyses and forecasts from other Centers of the WDTS are received	WMC products (numerical data)	Analyses	surface + altitude	H + 5 (9)	H + 5(9)	H + 5(10)
		Forecasts	"	H + 8(11)	H + 8(11)	H + 8(12)
	WMC products (graphics)	Analyses	"	H + 6(10)	H + 6(10)	H + 6(11)
		Forecasts	"	H + 9(12)	H + 9(12)	H + 9(13)
	RMC products (numerical data)	Analyses	"		H + 4(5)	H + 4(6)
		Forecasts	"		H + 6(7)	H + 6(8)
	RMC products (graphics)	Analyses	"		H + 5(6)	H + 5(7)
		Forecasts	"		H + 7(8)	H + 7(9)

(1) The time at which observed data is received means the time at which a sufficient quantity of data necessary for analysis has been received

The European Center for Medium Range Weather Forecasts also has a role which fits well both with the objectives of a WMC and of a RMC.

4. DIFFERENT WAYS OF ESTIMATING TRANSPORT TERMS

One of the key parameters to be estimated is the advection. Before looking at the different ways of estimating the transport terms, it is interesting to examine the accuracy in the determination of the trajectories needed to ensure a correct source-receptor relation to be determined.

Fig. 3 shows the lateral position error that results from a given initial wind direction (dd) error, assuming a straight trajectory and a constant error with time.

For an initial error of 10°, a lateral displacement of 127 km (50 km) occurs beyond 750 km (285 km) of travel.

The two values 50 and 127 km are typical grid sizes used in operational LRT models. The consequence is that trajectories must be estimated very precisely (remember that ground-level wind directionsare usually reported according to a 36 directions wind-rose, i.e. every 10° !).

Let us start from the Navier-Stokes equations ; neglecting friction and turbulent terms ;

$$\frac{dV}{dt} + 2\,\Omega \wedge V = -\nabla \phi - \frac{1}{\rho}\,\nabla p \tag{4.1}$$

where Ω is the Coriolis rotation vector

$\quad$ V is the wind vector

$\quad$ ϕ = gz is the geopotential

$\quad$ P is pressure

Eq. 4.1 can be written slightly differently by developing $\frac{d}{dt}$:

$$\frac{dV}{dt} = \frac{\partial V}{\partial t} + V \cdot \nabla V \tag{4.2}$$

Using the general relationship $a \wedge (b \wedge c) = b(c.a) - c\,(a.b)$ it is possible to make the kinetic energy and the vorticity appear :

$$V \wedge (\nabla \wedge V) = \nabla (V \cdot V) - V (\nabla \cdot V) = \nabla \left(\frac{V^2}{2}\right) - V (\nabla \cdot V)$$

Finally, eq. 4.1 can be written :

$$\frac{\partial V}{\partial t} + \overline{T_a} \wedge V = -\nabla \left(\frac{V^2}{2} + \phi\right) - \frac{1}{\rho}\,\nabla p \tag{4.3}$$

where $T_a = \nabla \wedge V + 2\,\Omega$ $\quad\quad$ is the absolute vorticity

Multiplying this equation by V gives :

$$\frac{d\left(V^2/2\right)}{dt} + V \cdot \nabla \phi = -\frac{1}{\rho}\,V \cdot \nabla p \tag{4.4}$$

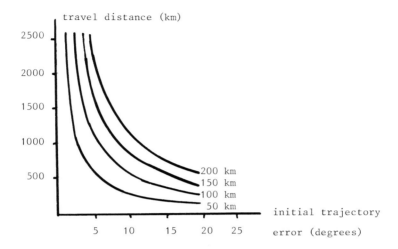

Fig 3 : Trajectory errors resulting from an initial direction error

(from Pack et al., 1978)

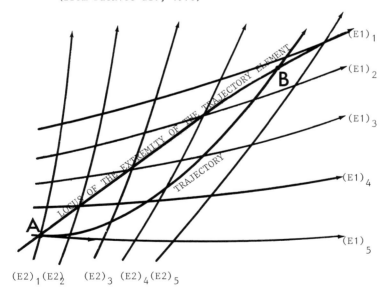

Fig 4 : Graphical construction of a trajectory.

(Ei)$_j$ are potential isoplethes (from Doury).

or

$$V. \frac{\partial V}{\partial t} + V. \nabla \left(\frac{V^2}{2} + \phi \right) + \frac{1}{\rho} V. \nabla p = 0 \qquad (4.5)$$

It is now possible to make different kinds of "generalized potential energy"
appear :

$$\frac{d}{dt} \left(\frac{V^2}{2} + \phi + \frac{P}{\rho} \right) = \frac{\partial}{\partial t} \left(\phi + \frac{P}{\rho} \right) + P. V. \nabla \left(\frac{1}{\rho} \right) \qquad (4.6)$$

$$\frac{d}{dt} \left(\frac{V^2}{2} + \frac{P}{\rho} \right) = \frac{\partial}{\partial t} \left(\frac{P}{\rho} \right) - V. \nabla \phi + P. V. \nabla \left(\frac{1}{\rho} \right) \qquad (4.7)$$

$$\frac{d}{dt} \left(\frac{V^2}{2} + \phi \right) = \frac{\partial \phi}{\partial t} - \frac{1}{\rho} V. \nabla p \qquad (4.8)$$

If one introduces enthalpy $H = C_p T$, it can be shown that one obtains a
fourth equation :

$$\frac{d}{dt} \left(\frac{V^2}{2} + \phi + H \right) = \frac{\partial}{\partial t} (\phi + H) + T. V. \nabla S \qquad (4.9)$$

where S is entropy.

The four preceding equations all have the same form :

$$\frac{d}{dt} (E_c + E_p) = \frac{\partial}{\partial t} E_p + a. V. \nabla b + c. V. \nabla d \qquad (4.10)$$

and it can be considered (Doury, 1968), that :

- the first term represents the variation of the sum of kinetic and potential
energy of a unit mass parcel of air along the trajectory ;
- the second represents the time variation of the potential energy along the
trajectory ;
- the last two terms are both proportional to the advection of b and d.

One striking property of equation 4.10 is that if a V. ∇b and cV. ∇d are
equal to zero simultaneously (in this case the streamline is perpendicular to
the gradients of b and d), the trajectory is simply defined between times t_1
and t_2 by :

$$\int_{t_1}^{t_2} \frac{\partial E_p}{\partial t} dt = \left(E_{p_2} - E_{p_1} \right) + \left(E_{c_2} - E_{c_1} \right) \qquad (4.11)$$

The meaning of this relation is that at the intersection of constant-b and
constant-d surfaces (or on constant-b or constant-d surfaces in cases 4.6, 4.8,
4.9), the integral of the variation of potential energy along a constant b and
d trajectory -with b or d as the vertical coordinate, respectively- remains
equal to the algebraic sum of the absolute variations of kinetic and potential
energies along this trajectory.

Eq.4.11 can be discretized between a starting point A at time t_1 and an
ending point B at time t_2 :

$$E_{B_2} - E_{A_1} = \frac{A_A E + A_B E}{2} - \frac{V_{B_2}^2 - V_{A_1}^2}{2}$$

$$= \frac{1}{2} \left(E_{A_2} - E_{A_1} + E_{B_2} - E_{B_1} \right) + \frac{V_{A_1}^2 - V_{B_2}^2}{2} \qquad (4.12)$$

Finally :

$$E_{A1} + E_{A2} = E_{B1} + E_{B2} + V_{B2}^2 - V_{A1}^2 \qquad (4.13)$$

which allows a graphical construction of the trajectography between two successive synoptic network data : neglecting first the correcting term $V_{B2}^2 - V_{A1}^2$, the extremity of the trajectory B is the diagonal line joining the intersections of the two equipotential networks (fig.4) $(E_1)_i$ and $(E_2)_i$.

The condition which allows the determination of the position of the point B is given by the availability of the wind speed. A geostrophic wind can be deduced from the potential E_p from :

$$V_g = \frac{\vec{k}}{f} \wedge \nabla_b E_p \qquad (4.14)$$

if b is taken as the vertical coordinate.

By use of an iterative process, the correcting term $V_{B2}^2 - V_{A1}^2$ can be taken into account.

This method has been extensively used in the past for "hand calculations".

As has already been seen, the potential E_p driving the streamfunction may take four different forms :

$(4.15) \quad E = \dfrac{P}{\rho_P} + \phi = RT + \phi \qquad$ (pressure divided by density + geopotential)

$(4.16) \quad E = \dfrac{P}{\rho_P} \qquad$ (pressure divided by density)

$(4.17) \quad E = \phi \qquad$ (geopotential)

$(4.18) \quad E = C_p T + \phi \qquad$ (Montgomery's potential : sum of enthalpy and geopotential)

From the <u>first expression</u>, trajectories occur on constant mass per unit volume surfaces ("isosteric" trajectories), as constant volume balloons behave for example. In this case the simplifying hypotheses $\nabla\left(\frac{1}{\rho}\right) = 0$ and $V.\nabla\left(\frac{1}{\rho}\right) = 0$ lead in taking the specific volume as the vertical coordinate. Because of simpler solutions, this has not been considered for LRT.

The <u>second expression</u> corresponds to trajectories on horizontal surfaces (cf $V. \nabla\phi = 0$) , when the fluid remains homogeneous ($V. \nabla\left(\frac{1}{\rho}\right) = 0$). Again, this formulation has seldom been used for air pollution studies.

The <u>third formulation</u> results from the single approximation $V. \nabla P = 0$ with pressure as the vertical coordinate.

So by definition, trajectories are isobaric. This procedure is very easy to use, because the current meteorological data is given at standard pressure levels. Its main deficiency is that it can only describe quasi-horizontal movements which fortunately are the most frequent.

Finally, the <u>fourth expression</u> implies that trajectories are considered on isentropic surfaces, i.e. on isopleths of potential temperature. The main advantage is that it is possible to take vertical movements into account in this way.

Although this procedure is certainly the most satisfying, it is not so widely used because it needs extra information in addition to data from the standard pressure levels, which is often not available.

Several comparisons of different methods of computation of trajectories have been conducted (ARTZ et al., 1985 ; KUOet al., 1985 ; TAKACS and STARHEIM, 1983;...).

There is a general agreement about the fact that the most realistic results are generally obtained with an isentropic model, mainly because in this case the vertical transport is taken into account correctly. Of these studies, KUO's appears most interesting. A mesoscale model simulation with a 80 km mesh size was used to test the accuracy of different trajectory models.

The results from the predictive model are assumed to represent the true atmosphere. The model data are then degraded to model observations with lower spatial and/or temporal resolutions. We repeat here the conclusions of this study.

"As the temporal and spatial resolutions of the data set are degraded, the accuracy of the trajectory model is quickly degraded. For the experiment that mimics the data base routinely available from the current observational network with imposed measurements errors, the mean absolute horizontal transport errors associated with trajectories originating from the PBL can be as large as 600 km after 3-day integration (fig.5).

This suggest that the current synoptic network, and current observational frequency are inadequate for calculation of LRT of episodic events. Comparisons of simulations with various spatial and temporal resolutions indicate that the major limitation of the current network is the 12 h observational frequency ... For three simplifying assumptions : isobaric, isosigma, and isentropic, the isentropic model is shown to be superior to the other two models".

A similar study has been conducted by MARTIN et al., 1986 ; they have performed a sensitivity study, with the assumption of a 10 % uncertainty on the three components of the wind vector, using the 3-D analysed winds from the ECMWF. This allow to draw envelope trajectories (fig.6).

An other purpose of the work was to define precisely the role and interest of taking into account the vertical velocity in trajectory models, i.e. to assess for which meteorological conditions the use of w is recommended. It was found that w has a noticeable effect when the atmospheric transport occurs in the free atmosphere (resp. boundary-layer) during an anticyclonic situation (resp. low pressure system).

On the other hand, trajectories were found to be rather close-at least included within the uncertainty trajectories- when the free troposphere transport occur during cyclonic situations or when transport was restricted to the

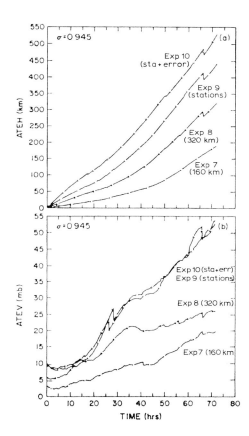

Fig 5 : Mean absolute error between different experiments

    a) horizontal deviation

    b) vertical deviation

-Exp 7 and 8: data are extracted from the numerical simulation at regular

 spatial intervals: 160,320 km with a 12h temporal resolution

- Exp 9:interpolates the 80 km data to the actual rawinsonde network

- Exp 10: same as 9 except that a 10%rms error is added to the u and v

  components and a 1 mb error is added to the surface pressure

  (From Kuo et al.,1985)

318

Fig 6 : Trajectories starting from Mount Etna (from Martin et al.)

---- error envelopes

—·—isobaric

——— 3D

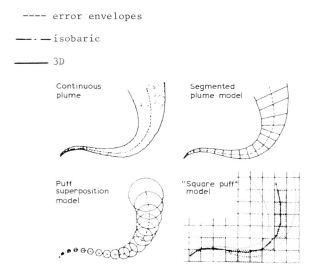

Fig 7 : Various techniques for simulating a continuous emission

(from Bass, 1980)

PBL during anticyclonic conditions.

5.   OPERATIONAL METHODS OF COMPUTATION OF WINDS FIELDS FROM ARCHIVED STANDARD
     LEVELS DATA

For air pollution purposes, computation of trajectories in France are based
on two types of information : observed meteorological parameters at standard
levels, or ECMWF analysed winds.

Several NMC's archive values of geopotential, temperature, relative humidity...at standard levels (surface, 850, 700, 500, 300... hPa) ; observations are
taken from radiosoundings performed at 00.00 and 12.00 GMT. These data are interpolated on the nodes of a regular grid. In France for example, the HEMIS
file has been constituted from the archived analyses on a square grid on a ste-
reographic projection, the length of which is 381 km at 60°N. The grid covers
the northern hemisphere to the North of 20°N.

From the gridded geopotentials, it is possible to determine a geostrophic
wind, using the following equations written in finite   difference form :

$$U_g = - \frac{g}{f} \frac{\partial Z}{\partial y}\Big|_p$$
$$V_g = \frac{g}{f} \frac{\partial Z}{\partial x}\Big|_p$$

(5.1)

the x axis being oriented to the East, and y to the North. Several procedures
for time and space interpolation can be used :
- one may consider that wind from a network at time t remains valid until the
next network at time $t + \Delta t$.
- or an interpolation between the two networks is performed using the parameter
α defined by

$$\alpha = \frac{(T+12)-T - (T+12)+t}{12}$$

(5.2)

and $u(t) = u(T) + \alpha [u(T+12) - U(T)]$
       $v(t) = v(T) + \alpha [v(T+12) - v(T)]$

if Δt is taken as 12 h.

As the geostrophic wind is only an approximation to the real wind, it is
better to correct it by taking into account the ageostrophic components, which
may be of importance especially when the curvature of the isobaric curves is
important (Lavenu et al., 1985).

Atmospheric trajectories representation of the mean transport in the ABL
are usually constructed by correcting both the modulus and the direction of the
geostrophic wind.

The most common corrections used are the following :

$$\vec{U} = \frac{\vec{U_{850}} + \vec{U_{surf}}}{2} \tag{5.3}$$

or $\quad |\vec{U}| = 0.75 |\vec{U_{850}}| \qquad \Theta = \Theta_{850} - 15° \tag{5.4}$

(Lavenu et al., for example)

To cope with the shortcomings of the geostrophic approximation, some meteorological centers perform 3D objective analysis.

As an example, the ECMWF operational analysis system is an intermittent data assimilation system consisting of a multivariate optimum interpolation analysis, a non-linear normal mode initialisation coupled with a high resolution spectral model which produces a firstguess forecast for the subsequent analysis. Data are assimilated with a frequency of 6 h.

Three dimensional wind fields are given on a grid with mesh size 1.875°.

The retrieval procedures of ECMWF archived fields are described in the ECMWF Bulletin entitled "Retrieval utility for ECMWF data bank for reports and fields on standard levels", by J.D. Chambers.

New retrieval directives are given in the MARS User Guide (revised 20.12.1985)

6. MAJOR TYPES OF LONG RANGE TRANSPORT MODELS

Several reviews have been published (Eliassen, Van Derhout and Van Dop, Johnson, Bessemoulin et al., ...)

Most of the models use a deterministic description of the processes involved the starting point being the classical diffusion equation :

$$\frac{dC}{dt} = \frac{\partial C}{\partial t} + u_i \frac{\partial C}{\partial x_i} = \frac{\partial}{\partial x_i}\left(K_i \frac{\partial C}{\partial x_i}\right) + \text{sources and sinks} \tag{6.1}$$

Traditionaly, models are classed into three main types :

- lagrangian trajectory models (LTM)
- eulerian grid models (EGM)
- statistical trajectory models which can be considered as a simplified LTM.

6.1 LAGRANGIAN TRAJECTORY MODELS

Diffusion, transformation and removal of pollutants are calculated in a moving frame along a calculated trajectory.

Again, two types of model exist, depending whether one is interested in the contribution of a source, or the pollution at a receptor point :

- receptor-oriented models use backward trajectories, i.e. trajectories finishing at the point under study ;
- source-oriented models deal with concentration/deposition computed forward in time from the release by the emissions sources.

A continuous plumecan be simulated by three techniques (fig.7) :

- assimilate the continuous emission by successive independant "puffs" travelling along trajectories (see Pack et al., 1978, for example for the NOAA model and Appendix 1) ;

- break up the trajectory into segments represented by a volume with a gaussian cross-section in both the horizontal and vertical directions. The along trajectory dimension of each volume is defined by the time increment of interest (ranging usually between 0.25 h to 1 h). The cross sections are defined accor- ding to known curves (Chan et al., 1979).

- keep track of the cell containing the puff at every time step and assign the contribution of the puff to this cell, ignoring horizontal spread due to diffusion (see Eliassen, 1980 for the structure of the EMEP model and Appendix 2 for a brief overview of this kind of model).

6.2   EULERIAN GRID MODELS

In this case a solution is fund by solving the Navier Stokes equations coupled with the diffusion equation mentionned above using numerical methods which are generally a finite-difference version of the equations. The solution is computed at the nodes of a regular grid in 3 dimensions. The general charac- teristics and major respective advantages/disadvantages of different models are depicted in Annex 3.

As the structure of these models will be described in detail elsewhere by other lecturers we shall focuss here on the problem of spurious numerical diffusion associated with the numerical integration of the advection equation.

In 2 dimensions, this equation reads :

$$\frac{\partial c}{\partial t} + u \frac{\partial c}{\partial x} + v \frac{\partial c}{\partial y} = 0 \qquad (6.2)$$

where : C is the concentration

U,V are the horizontal components of the wind speed. Many ($0 \sim (30)$ algo- rithms have been proposed, tested, and compared in the past to solve this equa- tion numerically (Long and Pepper, 1976 ; Chock, 1985 ; Schere, 1983 ; Sheih and Ludwig, 1984 ; Villouvier, 1982...).

Usually, the method used to test the numerical schemes consists of rotating a bell-shaped "hill" of concentration in a 2D circular-constant velocity field, or displacing it in a constant wind.

Before going further, consider the simplified transport equation

$$\frac{\partial c}{\partial t} + u \frac{\partial c}{\partial x} = 0 \qquad (6.3)$$

One of the classical methods used to solve this kind of equation is to use an upstream differencing method :

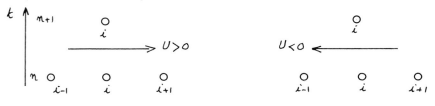

According to the direction of U, the difference forms are respectively :

$$\frac{C_i^{n+1} - C_i^n}{\Delta t} = - U_i \frac{C_i^n - C_{i-1}^n}{\Delta x} \qquad \frac{C_i^{n+1} - C_i^n}{\Delta t} = - U_i \frac{C_{i+1}^n - C_i^n}{\Delta x} \qquad (6.4)$$

Using a Taylor expansion, we obtain for U > 0 :

$$\left(\frac{\partial c}{\partial x}\right)_i = \frac{C_i - C_{i-1}}{\Delta x} + \frac{\Delta x}{2} \left(\frac{\partial^2 c}{\partial x^2}\right)_i + O\left(\Delta x^2\right) \qquad (6.5)$$

So, Eq. 6.4 is an approximation of :

$$\frac{\partial c}{\partial t} + U \frac{\partial c}{\partial x} = \frac{U \Delta x}{2} \frac{\partial^2 c}{\partial x^2} + O\left(\Delta x^2\right) \qquad (6.6)$$

The term $\frac{U.\Delta x}{2}$ is homogeneous to a viscosity ν .

Consider a physical example : U = 10 m.s^{-1}, Δ x = Δ z = 100 m so ν = 500 m^2/s. If we now consider turbulent diffusion along the vertical, we must consider an additional term $K_z \frac{\partial^2 c}{\partial z^2}$. In the ABL, K_z ranges between 0 and 100 m^{2}s^{-1}, which is only 20% of the coefficient due to the sole finite differencing contribution.

Although it is often claimed that these methods conserve mass fairly well, some rather sophisticated algorithms were found to conserve mass only within 24 % in the kind of experiments depicted above (Schere, 1983) !

With most of the existing algorithms, an error on the phase velocity is observed (the peak concentration is not advected at the right point).

On the other hand, the shape of the initial concentration distribution is often altered (fig. 8,9,10   from Villouvier).

Several parameters allow the estimation of the perturbation due to finite-differencing the equations :

- the position of the peak and its magnitude
- the conservation of mass, i.e. check of $r = \dfrac{\Sigma_{ij} C_{ij}(t)}{\Sigma_{ij} C_{ij}(0)}$

- the distribution of mass, i.e. check of $q = \dfrac{\Sigma_{ij} C_{ij}^2(t)}{\Sigma_{ij} C_{ij}^2(0)}$

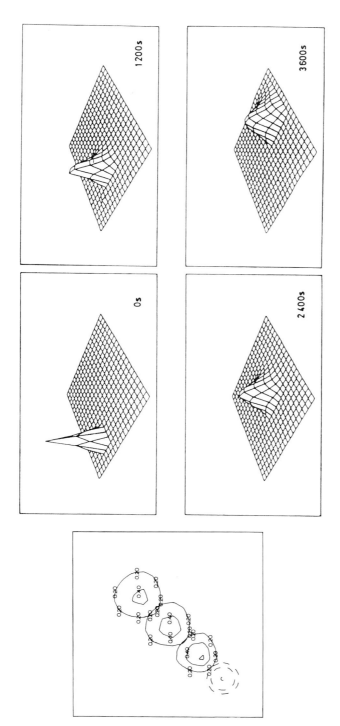

Fig 8 : MATSUNO scheme.

- left: Time evolution of isoplethes of concentration of a passive contaminant advected by a constant wind

- right: 3D representation of the time evolution of a passive contaminant advected by a constant wind

324

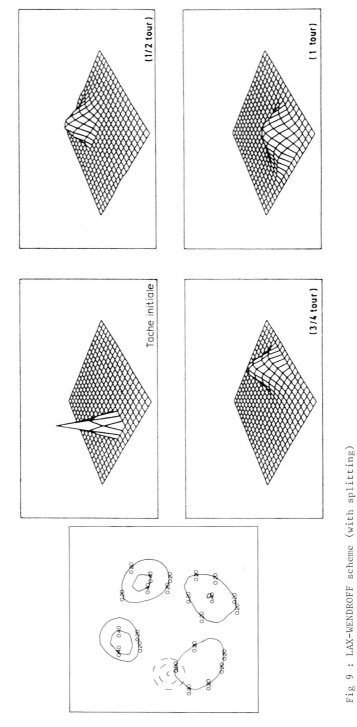

Fig 9 : LAX-WENDROFF scheme (with splitting)

- left: Time evolution of a passive contaminant advected by a rotating wind field

- right: 3d representation

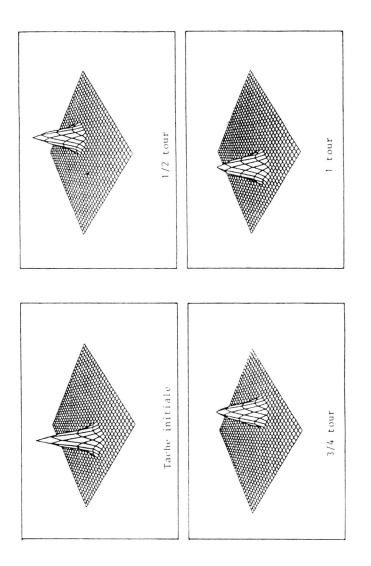

Tache initiale

1/2 tour

3/4 tour

1 tour

Fig 10 : EGAN & MAHONEY scheme

Caption same as Fig 9;

Some other parameters have also been used :
- the mean error : $\dfrac{\sum_{ij} C_{ij}(t) - C_{ij}^e(t)}{n}$

where C_{ij}^e is the exact value of concentration at point (i,j)

- the maximum error : $\max \left| C_{ij}(t) - C_{ij}^e(t) \right|$
- conservation of $\sum C^q(t), \sum \left(\dfrac{\partial c}{\partial x}\right)^t, \sum \left(\dfrac{\partial^2 c}{\partial x^2}\right)$ associated with flatness, local gradients, local curvatures of the C distribution.

Unfortunately, on many occasions, a numerical scheme is conservative for one quantity and <u>not</u> for another (fig. 11)

Apart from that, it must be said that different algorithms need different computation time and memory storage according to their degree of complexity. The following table gives some examples for several well known schemes (Villouvier).

NUMERICAL SCHEME	CPU/iteration	MEMORY STORAGE (a b.unit)
MATSUNO & TASU-MATSUNO	0.008	3.0
LAX WENDROFF	0.029	4.0
Cubic spline	0.040	2.3
Bicubic spline	0.075	5.3
EGAN & MAHONEY	0.120	12.1

Recently, Sheih and Ludwig, 1985 have evaluated the relative importance of numerical diffusion and physical diffusion for 27 algorithms in use in either urban, mesoscale or regional models.

Since many experiments testing the effects of numerical diffusion use the ratio of peak concentrations of the concentration distribution used at the final and initial times steps (resp.t and to), they have chosen to quantify the performances of each algorithm by computing the "attenuation factor" :

$$R_n = \frac{C_{peak}(t)}{C_{peak}(0)}$$

On the other hand, they have shown that the attenuation factor due to atmospheric diffusion can be expressed as $R_a = \dfrac{\sigma^2(t_0)}{\sigma^2(t)}$

where σ is the horizontal dispersion coefficient at time t.

Following Gifford, $\sigma(t)$ can be approximated by :

$$\sigma^2(t) = 2 k \cdot t + \left(\frac{v_0}{\beta}\right)^2 \left(1 - \exp -\beta t\right)^2 - \frac{k}{\beta}\left(3 - 4\exp -\beta t - \exp -2\beta t\right)$$

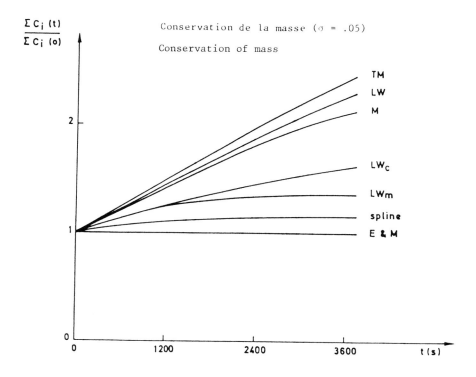

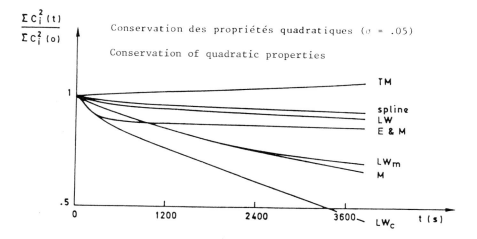

Fig 11 : Comparison of several advection schemes relative to the
conservation of mass and quadratic properties. Constant windcase

with $K = 5.10$ m^2.s ; $v_o = 0.15$ m.s and $\beta = 10^{-4}$ s^{-1}

So the ratio of peak concentration reduction due to numerical diffusion and atmospheric diffusion respectively can be estimated with the parameter $\dfrac{1 - R_n}{1 - R_a}$

It can be seen from Table 1 (from Sheih and Ludwig) that, depending on the magnitude of grid sizes used, differencing methods may produce quite large spurious diffusion.

The major findings of this most interesting study are the following :

- upstream differencing methods always produce larger dispersion than occurs naturally ;

- for grid sizes as large as those used in regional scale models, most of the numerical schemes produce larger numerical than atmospheric diffusion.

As an alternative to integrating the Advection-Diffusion equation numerically, and because of the increasing power of computers, another solution which avoids the problems mentioned above is to estimate diffusion using a large number of particles advected by the mean wind and its turbulent components. At this time, this technique has been used mainly for short-range transport (Baerentsen and Berkowicz, 1984 ; De Bass et al., 1986 ; Bessemoulin, 1986...) but meso-scale applications have already been presented (Blondin, 1982 ; Yamada, 1985). Nevertheless it must be said that this kind of modelling is not very well suited to handle complex chemistry.

Wind speed (ms⁻¹)

		5			20			
Reference	Numerical method	0.5	5	50	0.5	5	50	RANK
				$(1 - Rn)/(1 - Ra)$				
Molenkamp (1968)	Upstream N + 1	1.42	1.24	4.52	4.18	3.17	16.02	26
	Upstream N	1.23	1.08	3.93	3.63	2.76	13.93	25
	Leap Frog	0.29	0.25	0.92	0.85	0.64	3.25	19
	Lax Wendroff	0.51	0.45	1.64	1.51	1.15	5.81	22
	Arakawa Euler	0.29	0.25	0.92	0.85	0.64	3.25	18
	Arakawa Adams Bashforth	0.25	0.22	0.79	0.73	0.55	2.79	14
Anderson and Fattahi (1974)	Mac Cormack	0.44	0.45	0.82	0.70	0.63	2.09	12
	Rusanov	0.50	0.51	0.95	0.80	0.73	2.40	15
	Kotler Warming Lomax	0.51	0.53	0.97	0.82	0.74	2.45	16
McRae et al. (1982)	Fromm	0.47	0.49	0.81	0.68	0.64	1.87	10
	Crowley	0.28	0.29	0.48	0.41	0.38	1.12	8
	Finite élément (Chapeau)	0.15	0.15	0.25	0.21	0.20	0.58	6
	SHASTA	0.52	0.53	0.88	0.74	0.70	2.04	13
Long and Pepper (1976)	Upwind difference	0.97	1.00	1.55	1.30	1.25	3.41	21
	Fully implicit	0.64	0.66	1.03	0.87	0.83	2.27	17
	Crank Nicolson	0.41	0.42	0.65	0.55	0.52	1.43	9
	Second moment	0.03	0.03	0.05	0.04	0.04	0.11	3
	Cubic spline	0.09	0.10	0.15	0.13	0.12	0.33	5
	Galerkin Chapeau function	0.02	0.02	0.03	0.03	0.03	0.07	1
Lee and Meyers (1979)	Fully implicit.multi grid	0.49	0.50	0.82	0.69	0.65	1.92	11
	Crank Nicolson.multi grid	0.27	0.28	0.47	0.39	0.37	1.08	7
	Pseudospectral	0.02	0.02	0.04	0.03	0.03	0.08	2
Shannon (1979)	Gaussian	0.09	0.10	0.15	0.12	0.12	0.32	4
	Moment conservation	0.35	0.31	1.11	1.03	0.78	3.95	20
Pedersen and Prahm (1974)	Mass in cell	1.46	1.22	5.04	4.72	3.50	18.54	27
	Second moment	0.52	0.43	1.77	1.66	1.23	5.52	23
	Second moment with width correction	0.79	0.65	2.71	2.53	1.88	9.96	24

Table 1

APPENDIX 1

AN EXAMPLE OF PUFF MODEL : THE NOAA MODEL

A plume is represented by series of puffs simulating the effluent plume for a specific time interval (e.g. 6h if four trajectories are computed per day)
Each puff is diffused according to :

$$C = \frac{2Q}{(2\pi)^{3/2} \sigma_h^2 (2K_z \cdot t)^{1/2}} \, exp - \frac{R^2}{2\sigma_h^2} \qquad (A\ 1.1)$$

where :

C is the concentration (g/m^3)

Q is the puff emission (g)

$\sigma_h = 0.5\ t$ (m) is the horizontal dispersion coefficient

t is travel time (s)

K_z is the vertical diffusion coefficient $(m^2 \cdot s^{-1})$

R is the distance to the puff center

The initial dimension of the puff is adjusted according to the emission grid size used.

If it can be considered that the pollutant is uniformly mixed through the mixing layer, the above equation writes simply :

$$C = \frac{Q}{2\pi \sigma_h^2 H} \, exp - \frac{R^2}{2\sigma_h^2} \qquad (A\ 1.2)$$

Removal processes are treated in the following way.

Dry deposition along a trajectory is computed using the concept of dry deposition velocity : the mass M_d deposited is equal to :

$$M_d = C\ V_d\ \Delta t$$

where C is the local ground concentration, Δt is the time interval over which puff concentrations are computed (in the NOAA model, 3-h segment intervals are used for climatological applications).

The fraction of mass removed from the plume is thus the ratio of mass dry-deposited to the total mass present in a vertical cross-section of the atmosphere:

$$R_d = \frac{C \cdot V_d \cdot \Delta t}{C H} = \frac{V_d \cdot \Delta t}{H}$$

So, the air concentration at every time step is depleted by dry deposition, the "new" starting concentration in the puff at the end of the time interval considered being :

$$C' = C\ (1 - R_d)$$

in the case of dry deposition only.

Wet deposition processes are parameterized using an empirical scavenging ratio $E = \dfrac{C_R}{C_p}$

where C_R is the concentration at ground level <u>in the rain</u>

C_p is the average <u>air</u> concentration in the layer where precipitations occur.

Wet deposition along the trajectory is set to :

$$M_w = C_R \cdot P \cdot \Delta t = C_p \cdot E \cdot P \cdot \Delta t$$

P being the precipitation rate.

The fraction of mass removed is the ratio of mass wet-deposited to the total mass present in a vertical cross-section of the atmosphere :

$$R_w = \frac{C_p \cdot E \cdot P \cdot \Delta t}{C \cdot H}$$

which can be written as

$$R_w = \frac{C \cdot E \cdot P \cdot \Delta t \cdot H}{C \cdot H \cdot H_p} = \frac{E \cdot P \cdot \Delta t}{H_p}$$

if it is assumed that $H_p C_p = HC$, where H_p is the height of the precipitation layer.

Finally, puff concentrations depleted both by dry and wet deposition write

$$C" = C(1-R_d)(1-R_w)$$

APPENDIX 2

A SIMPLIFIED TRAJECTORY MODEL (OECD, EMEP)

The philosophy of the model is based on the mass balance for a given pollutant diffusing within the atmospheric boundary-layer :

$$\frac{d\varphi}{dt} = \left(\frac{\partial}{\partial t} + U\nabla.\right)\varphi = E - R \qquad (A\ 2.1)$$

where E are emissions and R are removal processes.

Horizontal diffusion is neglected because E is generally given in terms of area sources of large dimensions (100 km). This equation is integrated along trajectories arriving every 6 h at grid points in a 127 km grid covering Europe. Eq. A2.1 takes the following forms for SO_2 and sulfates :

$$\frac{d\varphi_{SO_2}}{dt} = -k_{SO_2}\,\varphi_{SO_2} + \frac{\varphi_{1\,SO_2}}{H} \qquad (A\ 2.2)$$

$$\frac{d\varphi_{SO_4}}{dt} = -k_{SO_4}\cdot\varphi_{SO_4} + \frac{3}{2}\left(k_t\cdot\varphi_{SO_2} + \frac{\varphi_{2\,SO_4}}{H}\right) \qquad (A\ 2.3)$$

where Q_{SO_2} and Q_{SO_4} are concentration of SO_2 and sulfates
Q_{1SO_2} is the effective SO_2 emission for long range
transport (αE is deposited in the emission mesh, with $\alpha \sim 0.15$)
Q_{2SO_4} is the SO_4 emission
$k_{SO_2} = \dfrac{V_d}{H} + k_t + k_{w\,SO_2}$
$k_{SO_4} = \dfrac{V_g}{H} + k_{w\,SO_4}$

V_d, V_g, k_t, H having the same meaning as in Eq. I.1 to I.4
k_w is the wet deposition rate.

Dry and wet sections of trajectories are determined according to analysed precipitation fields.

Eq. A 2.2 and A 2.3 are integrated to give concentrations estimates Q_{SO_2} (N Δt) and Q_{SO_4} (N Δt) for a transport time N Δt since the emission, along a trajectory crossing elementary source areas :

$$\varphi_{SO_2}(N\Delta t) = \varphi_{SO_2}(0)\,\exp{-k_{SO_2}N.\Delta t} + \sum_{i=0}^{N-1}\frac{(\varphi_{1\,i})_{SO_2}\,\Delta t}{H}\,\exp{-k_{SO_2}i.\Delta t}$$

$$\varphi_{SO_4}(N\Delta t) = \varphi_{SO_4}(0).exp - k_{SO_4}N.\Delta t$$

$$+ \frac{3}{2}\frac{k_t}{k_{SO_2}-k_{SO_4}} \sum_{i=0}^{N-1} \frac{(Q_{1i})_{SO_2}.\Delta t}{H} \times \left(exp - k_{SO_4}i\Delta t - exp - k_{SO_2}i\Delta t\right)$$

$$+ \frac{3}{2}\sum_{i=0}^{N-1}\frac{(Q_{2i})_{SO_4}\Delta t}{H} \, exp - k_{SO_4}.i\Delta t$$

These formulas simply express that concentrations at a given receptor point are the sum of the contributions of different emissions along the trajectory, exponentially weighted according to the transport time between the source and the receptor point.

334

APPENDIX 3

I. LAGRANGIAN TRAJECTORY MODELS

ADVANTAGES	DISADVANTAGES
- Not too expensive to run - No numerical diffusion - Capability to compute detailed source/receptor contributions : pollutant mass balance can be established - Adapted to treat point sources - Best suited for climatological applications (long term average concentrations/deposition)	- In general 1 level : no topography - Diffusion is highly parameterized as well as deposition - Non-linear chemistry is difficult to be treated - Errors can come from interpolating results available on receptors to an eulerian grid - Basic meteorological data only available every 6 or 12 h. Simple "interpolation" between two successive networks may be questionable.

II. EULERIAN GRID MODELS

ADVANTAGES	DISADVANTAGES
- 3-D physical treatments - 3-D fields produced on a regular grid - Non-linear chemistry can be incorporated - Best suited for case studies	- Expensive to run (large amount of computer time and storage) - Not well adapted to compute individual contributions from one source - Numerical diffusion associated with finite-differencing the A-D equation (especially for large mesh-sizes) - Input data is large and not always available - Boundary conditions may be complicated

REFERENCES

1   R. Artz, R.A. Pielke and J. Galloway, Comparison of the ARL/ATAD constant
    level and NCAR isentropic trajectory analyses for selected case studies,
    Atm. Environment, 19 (1985), 47-63.
2. J.H. Baerentsen and R. Berkowicz, Monte Carlo simulation of plume disper-
    sion in the convective boundary layer, Atm. Environment, 18 (1984),
    701-712.
3   A. Bass, Modelling long-range transport and diffusion, in Proceedings of
    the 2nd Joint Conference on Applications of Air Pollution Meteorology, AMS,
    (1980), 193-215.
4   P. Bessemoulin, C. Blondin, A. Despres, R. Rosset, J.P. Granier and A.E.
    Saab, A review of mesoscale and long range transport and dispersion model-
    ling of air pollutants. EDF, Bulletin de la Direction des Etudes et Recher-
    ches, série A, (1982), 45-57 (in french).
5   P. Bessemoulin, Monte Carlo methods for simulating diffusion in the atmos-
    pheric boundary-layer. Preprint of the 4th European Symposium "Physic o
    Chemical behaviour of atmospheric pollutants", Stresa 23-25/09/86, (1986),
    Commission of the European Communities.
6   C. Blondin, Preliminary tests of the introduction of lagrangian dispersion
    modelling in a 3D-atmospheric meso-scale numerical model in Proceedings of
    the 13th international technical meeting on Air Pollution Modeling and its
    application, Ile des Embiez, 14-17/09/1982, (1982), 8.1-8.14.
7   J.D. Chambers, Retrieval utility for ECMWF data bank for reports and fields
    on standard levels, ECMWF Bulletin, (1984).
8   M.W. Chan, S.J. Head and S. Machiraju, Development and validation of an air
    pollution model for complex terrain application, in Proceedings of the 10th
    international technical meeting on Air Pollution Modeling and its applica-
    tion, Rome, 23-26/10/1979, (1979).
9   D. Chock and A. Dunker, A comparison of numerical methods for solving the
    advection equation, in : Atm. Environment, 19, (1985), 571-586.
10 A.F. De Bass, H. Van Dop and F.T.M. Nieuwstadt, An application of the Lan-
    gevin equation for inhomogeneous conditions to dispersion in the convecti-
    ve boundary layer, in : Quart. J.R. Met. Soc., 112, (1986), 165-180.
11 A. Doury, Le problème des trajectoires atmosphériques à grande échelle trai-
    té par les équations d'énergie, in : La Météorologie, II.4, (1968), 471-491.
12 A. Eliassen, A review of long range transport modelling, in : J. Appl.
    Meteor, 19, (1980), 231-240.
13 F.A. Gifford, Horizontal diffusion in the atmosphere : a lagrangian dynami-
    cal theory, in : Atm. Environment, 16, (1982), 505-512.
14 W.B. Johnson, Interregional exchanges of air pollution : model types and
    applications, in : J. of the Air Poll. Control. Assoc., 33, (1983), 563-
    574.
15 Y.H. Kuo, M. Skumanich, P.L. Haagenson and J.S. Chang, The accuracy of tra-
    jectory models as revealed by the observing system simulation experiment,
    in : Mon. Wea. Rev., 113, (1985), 1852-1867.
16 D. Lavenu, S. Legouis, P. Bessemoulin, Sensitivity studies with a LRT model,
    in : Air Pollution Modelling and its Application IV, Plenum Press, (1985),
    425-434.
17 P.E. Long and D.W. Pepper, A comparison of six numerical schemes for calcu-
    lating the advection of atmospheric pollution, in : Proceedings of the
    third symposium on atmospheric turbulence, Diffusion and Air Quality,
    Raleigh, 19-11/10/1976, AMS, (1976), 181-187.
18 D. Martin, C. Mithieux and B. Strauss, On the use of the synoptic vertical
    component in a trajectory model. To appear in Atm. Environment, (1986).
19 Mars, User Guide, ECMWF Publication, Draft, (1985).
20 D.H. Pack, G.J. Ferber, J.L. Heffter, K. Telegadas, J.K. Angell, W.H.
    Hoecker and L. Machta, Meteorology of long-range transport in : Atm.
    Environment, 12, (1978), 425-444.

336

21 K.L. Schere, An evaluation of several numerical advection schemes, in :
    Atm. Environment, 17, (1983), 1897-1907.
22 C.M. Sheih and F.L. Ludwig, A comparison of numerical pseudo diffusion and
    atmospheric diffusion, in : Atm. Environment, 19, (1985), 1065-1068.
23 J.F. Takacs and F.J. Starheim, Intercomparison of meteorological trajecto-
    ry techniques available for use in long range transport modelling, in :
    Meteorology of Acid Deposition, (1983).
24 D.B. Turner, Workbook of atmospheric diffusion estimates, in : Environmental
    Health Series, USEPA, (1970).
25 K.D. Van Der Hout and H. Van Dop, State of the art of interregional model-
    ling, in : Nato/CCMS, Panel 2, (1981).
26 V. Villouvier, Développement de modèles méso-météorologiques, Application
    à l'étude du transport et de la diffusion de polluants à l'échelle régionale.
    Thèse de Docteur Ingénieur, Ecole Centrale de Lyon, (1982).
27 WMO, Guide du système mondial de traitement des données, OMM n⁰ 305, (1983).
28 WMO, Manuel du système mondial de traitement des données, OMM n⁰ 485, (1977).
29 T. Yamada, Numerical simulations of valley ventilation and pollutant trans-
    port, Seventh symposium on turbulence and diffusion, 12-15/11/1985, Boulder,
    (1985).

Regional and Long-range Transport of Air Pollution,
Lectures of a course held at the Joint Research Centre, Ispra, Italy,
15–19 September 1986, S. Sandroni (Ed.), pp. 337–353
© Elsevier Science Publishers B.V., Amsterdam — Printed in The Netherlands

MESOSCALE MODELS INCLUDING TOPOGRAPHY

F. Fiedler

1. INTRODUCTION

   With the advent of larger computers it has become feasible to solve
the partial differential equations, which describe atmospheric mesoscale
flows, numerically. Therefore it has now become a widespread tool to
resolve atmospheric phenomena down to a scale of a few kilometers with
mesoscale atmospheric models. The recent book by Pielke (1984) gives a
wealth of information on the present status of skill and summarizes most
of the inherent problems.

Although many topics are still of an unsolved nature from a theoretical
standpoint, time has come, however, that mesoscale models of the present
generation are able to give a much widened information on processes in
the atmosphere. Additionally to experimentally oriented case studies, with
the help of numerical models a much deeper insight can be gained than was
possible with previous methods.

For describing the dynamics of the lower atmosphere and the diffusion
processes, there are many problems left which need much mare elaborated
methods, especially for the inherent process of plume rise in a complicated
thermally layered boundary layer, the formulation of eddy diffusivities
for elevated and strong point sources as well as for considering effects
of subgrid scale dynamical processes.

   An abundance of uncertainty still exists on the turbulence structure at
higher levels of the boundary layer, e.g. under conditions of convective
rolls or convective cloud cells, within fronts and other nonstationary
conditions. Some of these topics have been refered to in the book by
Nieuwstadt and van Dop (1982).

2. GENERAL PHILOSOPHY OF MESOSCALE MODELING

The literature on experimental studies of atmospheric processes over complex terrain is extremely numerous reflecting the diversity of observed phenomena. In both fields, numerical weather prediction and studies on air pollution, detailed information on these phenomena is the basis for a useful application.

Meteorological conditions are determined by the following mountain effects
a) surface friction (frictional drag)
b) pressure drag
c) gravity waves (wave drag)
d) flow channelling
e) flow decoupling of different layers by thermal stability
f) flow stagnation
g) flow separation in the lee of mountains
h) flow splitting
i) creation of larger horizontal eddies in the lee of mountains
j) secondary circulations due to elevated heat sources
k) exageration of turbulence intensity

It is, however, clear, that in a typical mountanious area it is impossible to measure all the important variables with a satisfactory time and space resolution.

Therefore numerical models will have to serve two purposes
(1) to study the processes created by local and regional scale influences over a much wider field than can be studied experimentally by case studies

(2) to represent a powerful tool for giving an integrating description on the complicated processes which are in most cases only poorly documented by a coarse measurement network.

3. HYDROSTATIC AND NON-HYDROSTATIC MESOSCALE MODELS

In the field of mesoscale modeling mainly two different types of filter conditions have been used. For treating larger mesoscale phenomena hydrostatic models have been developed. Theoretical studies limit their horizontal resolution at the order of about 10 km. Where even smaller scale processes are considered, non-hydrostatic models have to be used.

Over complex terrain frictional drag and pressure drag plays an important role. Frictional drag takes account for the tangential forces along the surfaces where as pressure drag is a consequence of preassure differences on both sides of mountains and valleys. Since hydrostatic models only can take care of hydrostatic pressure drag it is clear that also the dynamics of the smaller scale processes are differently treated in both types of models.

The following discussion on models including topography will mainly relly on the non-hydrostatic model (Karlsruhe Atmospheric Mesoscale Model (KAMM) (Dorwarth, 1986), which has been developed and used in Karlsruhe both for studies on mesoscale processes over irregular terrain as well as for diffusion studies.

For the treatment of mountains in mesoscale models it is now most common to use a coordinate transformation by which the vertical coordinate is transformed to a new coordinate

$$\eta = \frac{H - z}{H - h(x,y)}$$

where H ist the height of the model domain and h is the terrain height. Through this transformation the irregular terrain becomes the lowest coordinate level. Especially boundary conditions can be formulated in a much easier way as is otherwise feasible. In the KAMM-model and additional transformation of the vertical coordinate is used in order to achieve a higher resolution of the lower layers of the model domain.

3.1 BASIC MODEL EQUATIONS

The equations are the equations for momentum-, mass-and heatbalance.

For the momentum-balance the following equations are used:

$$\frac{d}{dt}u = -c_p\Theta\frac{\partial}{\partial x}(\Pi'_d + \Pi'_H) + f(v - v_G\frac{\Theta}{\Theta_G}) - \frac{\partial}{\partial x}\overline{u'u'} - \frac{\partial}{\partial y}\overline{u'v'} - \frac{\partial}{\partial z}\overline{u'w'}$$

$$\frac{d}{dt}v = -c_p\Theta\frac{\partial}{\partial y}(\Pi'_d + \Pi'_H) - f(u - u_G\frac{\Theta}{\Theta_G}) - \frac{\partial}{\partial x}\overline{v'u'} - \frac{\partial}{\partial y}\overline{v'v'} - \frac{\partial}{\partial z}\overline{v'w'}$$

$$\frac{d}{dt}w = -c_p\Theta\frac{\partial}{\partial z}\Pi'_d \qquad\qquad - \frac{\partial}{\partial x}\overline{w'u'} - \frac{\partial}{\partial y}\overline{w'v'} - \frac{\partial}{\partial z}\overline{w'w'}$$

Here u, v and w are the three components of the velocity Θ_v is the virtuel potential temperature, f the coriolis parameter, g the acceleration of gravity and c_p the specific heat of air at constant pressure.

By using the Exner function $\Pi = (P/P_o)^{R/c_P}$ density is eliminated from the equations.

The equation for the energy balance is defined by

$$\frac{d}{dt}\Theta = -\frac{\partial}{\partial x}\overline{u'\Theta'} - \frac{\partial}{\partial y}\overline{v'\Theta'} - \frac{\partial}{\partial z}\overline{w'\Theta'} + S_\Theta .$$

where S_Θ includes all sources and sinks, Finally for the mass balance the simplified equation (shallow convection) is used

$$\vec{\nabla}\cdot\vec{v} = 0$$

In order to incorporate a larger scale driving force for the mesoscale model, the large scale pressure field is subtracted from total pressure which obey is assumed to the relationship

$$\frac{\partial}{\partial x}\Pi_G = \frac{fv_G}{c_p\Theta_G}$$

$$\frac{\partial}{\partial y}\Pi_G = -\frac{fu_G}{c_p\Theta_G}$$

$$\frac{\partial}{\partial z}\Pi_G = -\frac{g}{c_p\Theta_G}$$

This condition means only, that the large scale flow is nearly gestrophic and hydrostatic.

Seperately, mainly for numerical reasons the following relation has been introduced

$$\frac{\partial}{\partial z}\Pi'_H = \frac{g}{c_p}\left(\frac{1}{\Theta_G} - \frac{1}{\Theta}\right)$$

which desbribes a hydrostatic perturbation of pressure.

Finally, the dynamic pressure perturbation in given by

$$\Pi_d' = \Pi - \Pi_G - \Pi_H'$$

With this seperation of pressure the equations of motion have the form

$$\frac{d}{dt}u = -c_p\Theta_v\frac{\partial\Pi}{\partial x} + fv - \frac{\partial}{\partial x}\overline{u'u'} - \frac{\partial}{\partial y}\overline{u'v'} - \frac{\partial}{\partial z}\overline{u'w'}$$

$$\frac{d}{dt}v = -c_p\Theta_v\frac{\partial\Pi}{\partial y} - fu - \frac{\partial}{\partial x}\overline{v'u'} - \frac{\partial}{\partial y}\overline{v'v'} - \frac{\partial}{\partial z}\overline{v'w'}$$

$$\frac{d}{dt}w = -c_p\Theta_v\frac{\partial\Pi}{\partial z} - g - \frac{\partial}{\partial x}\overline{w'u'} - \frac{\partial}{\partial y}\overline{w'v'} - \frac{\partial}{\partial z}\overline{w'w'}$$

3.2 PARAMETERIZATION

The formulation of the closure assumption for the subgrid processes is a key-problem of all models. Depending what purpose the model should serve for the different possible formulations are normally a comprise between model complexity and numerical economy.

In the KAMM-model two versions for the parameterization of subgrid turbulent processes have been used:

a) the gradient assumption
b) second order closure of the Mellor-Yamanda type (Mellor and Yamanda, 1974).

In both parameterization schemes turbulent fluxes are described by

$$-\overline{u'u'} = 2K_M^x\frac{\partial u}{\partial x}$$

$$-\overline{u'v'} = K_M^x\frac{\partial v}{\partial x}$$

$$-\overline{u'w'} = K_M^z\left(\frac{\partial w}{\partial x} + \frac{\partial\eta}{\partial z}\frac{\partial u}{\partial\eta}\right)$$

$$-\overline{u'\Theta'} = K_H^x\frac{\partial\Theta}{\partial x}$$

342

The eddy diffusivities are related via diagnostic equations to the turbulent kinetic energy q^2 and the dimensionless wind shear S_M and the dimensionless temperature gradient S_H including a mixing length l

$$K_M^z = qlS_M$$
$$K_H^z = qlS_H$$

3.3 BOUNDARY CONDITIONS

Special care has to be devoted to the lateral and vertical boundary conditions. At the lower boundary the nonslip condition is used for the horizontal components.

The transformed vertical velocity $\acute{\eta}$ must also vanish at the sursace. When determining $\dot{\eta}$ from

$$\dot{\eta} = u\frac{\partial \eta}{\partial x} + w\frac{\partial \eta}{\partial z}$$
$$= \frac{\partial \eta}{\partial x}\left(u^* - \Delta t c_p \Theta\left(\frac{\partial \Pi_d'}{\partial x} + \frac{\partial \eta}{\partial x}\frac{\partial \Pi_d'}{\partial \eta}\right)\right) + \frac{\partial \eta}{\partial z}\left(w^* - \Delta t c_p \Theta\frac{\partial \eta}{\partial z}\frac{\partial \Pi_d'}{\partial \eta}\right)$$
$$= 0$$

also the horizontal pressure gradient has to be known which is taken from a previous time step.

At the inflow and outflow sides the well-known condition by Orlanski (1976) is used which is solved additionaly

$$\frac{\partial u}{\partial t} = -c\frac{\partial u}{\partial x}$$

Here c is the phase velocity of the waves from the interior of the domain.

In order to avoid reflexion of gravity waves at the upper boundary the damping mechanism described by Durran (1981) is introduced

3.4 TOPOGRAPHY

For applications of the model over mountains the orography of south-western Germany has been applied. Fig. 1 gives an example plot of it.

Presently the model uses a horizontal grid of $\Delta x = \Delta y = 5$ km. It contains in the horizontal direction 45 grid points and in the vertical 24. The vertical resolution is higher in the lowest levels.

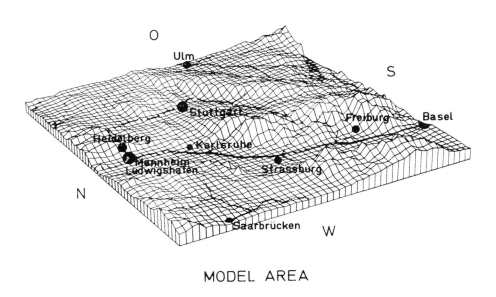

MODEL AREA

Fig. 1 Topography of the model area (Baden-Württemberg)

4. SOME MODELED FEATURES OF THE ATMOSPHERE ABOVE OROGRAPHIC STRUCTURES

a) Flow channeling

One of the most striking dynamical effect of mountains is the channeling of the flow within valleys. In Fig. 2 an example from the Rhine valley is

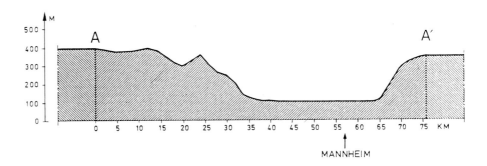

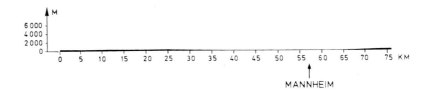

Fig. 2 Cross-section through the Rhine-valley (from Wippermann and Groß, 1981)

presented (Wippermann and Groß, 1981). Although in a one-to-one scale the valley is hardly detectable it has nevertheless a very strong effect on the wind flow. From statistics of a longer time series (Fig. 3) it is seen that geostrophic winds from the west are mainly associated with a southerly valley flow. Geostrophic winds with easterly directions produce with high frequencies a northerly valley flow. This results demonstrate, that the winds within the valley follow primarily the pressure gradient along the valley.

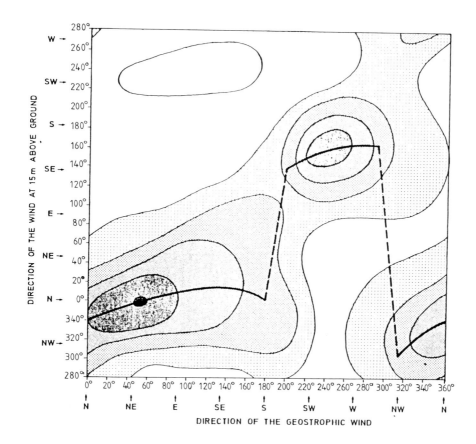

Fig. 3 Distribution of the surface wind in the Rhin-vally as a
function of the large-scale flow (from Wippermann and
Groß, 1981)

Using the statistics of the geostrophic wind as a driving force a wind
roses for the surface wind for every grid-point can be produced which are
in excellent agreement with observations. An example is given in Fig. 4
which shows the wind-rose for Mannheim (Wippermann and Groß, 1981).

This channelling, although of dynamical origin, becomes especially interesting
when thermal effects during the day provide a stronger coupling of the different
layers (Fiedler, 1983). Fig. 5 presents model results for a position in the
middle of the Rhine valley showing the diural increase of the mixed layer

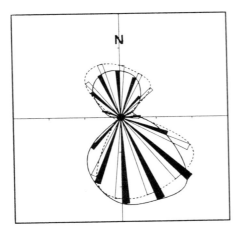

Fig. 4 Wind rose for the surface wind
at Mannheim
(Black bars = model simulations,
White bars = observations)
Simulations for stable stratifications
are shown with a non-hydrostatic model
(upper half) and with a hydrostatic
model (lower half) (from Wippermann
and Groß, 1981).

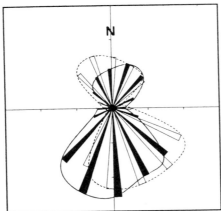

which contains the area of flow channeling (Dorwarth, 1986). A more detailed
result of the flow field in the Rhine valley will be discussed in the next
section.

b) Secondary horizontal circulations in orographic terrain
It is well-known that the largest eddies are always the most effective
ones for the diffusion of masses and other properties. Over horizontally
homogeneous terrain the distance from the ground is the most important scale
determining the eddy sizes.

In flow conditions over hills and valleys additional forces act on the
atmosphere increasing turbulence intensity in wakes of the mountains.

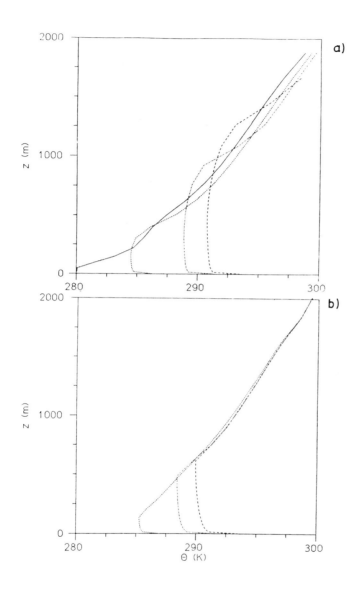

Fig. 5 Convective coupling during the day forces
the flow in upper levels into the direction of the
valley flow (Dorwarth, 1986)

In this way larger eddies with vertical and horizontal rotation axis are
created. Using the three-dimensional version of the mesoscale model for the
area of south-western Germany such eddies have been simulated in the lee of
the black forest.

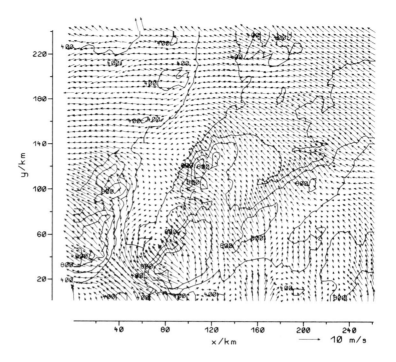

Fig. 6 Simulated surface winds (40 m above ground) for the
area of Baden-Württemberg at March 21, 1985, 12 UTC. Large scale
flow is from south-east (Adrian and Fiedler, 1987)
(this lines are isolines of the topography).

Fig. 6 shows the flow field close to the ground (40 m)
Adrian, Fiedler, 1986).

In this case the mesoscale model has been driven by the output-field of
the model of the German Weather Service, taking care of the larger scale
horizontal variation of the pressure field.

Beside the jet-like flow into the Rhine valley which is known from many
observations as the Möhlin-Jet, in the area of Freiburg a larger horizontal
eddy may be observed which is also assoviated with ascending air. As one can
see from Fig. 7 the general flow characteristics are in reasonable agreement
with surface wind measurements. Also of interests are the splitting points
for the flow created by the mountain especially visible at the southern side

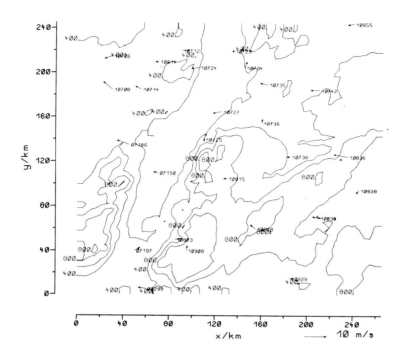

Fig. 7 Observed surface wind conditions at March 21, 1985,
12 UTC

of the Black Forest. This may serve as an example for the significance of
using simulation models with high resolution of the orographic features.
Otherwise it is necessary to use empirical correction factors for the subgrid
scale motion. The classical procedure of introducing an increased value of
roughness length is certainly underestimating these influences.

In applying this type of models the possibility appears on the horizon
to create climatological information e.g. on wind statistics which otherwise
can only be derived by long-term and costly measurements.

c) Diffusion of air-pollutants
As a last exemple, a few results will be shown for the diffusion of trace
gases released into the atmosphere from elevated point sources.

350

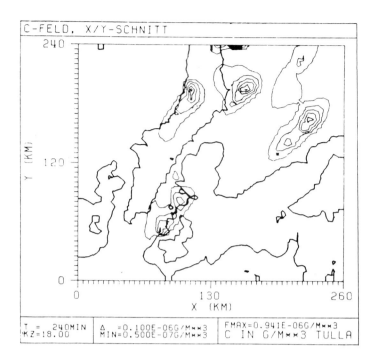

```
C-FELD, X/Y-SCHNITT
```

Fig. 8 Distribution of concentrations of SO$_2$ at a height
of about 600 m above the ground for a south-westerly flow
(March 25, 1985)

Fig. 8 and 9 is taken from a case study of the TULLA measuring compain. It
shows the distribution of concentrations at the ground and in a height about
650 m above ground (constant $\dot{\eta}$ - surface, see model description).

Although treating point sources in grid models is still a problem which
needs improved techniques, the main influence of the turning of wind direction
with height due to topographical pressure disturbances is clearly reflected.

In Fig. 10 and 11 the concentration distribution for easterly flow conditions at two different height levels are presented. We see from these results
again the diverging transport direction created by the flow at 650 m and at
the surface. Especially the surface wind field is deflected to the south in
the region of the Rhine valley being the cause for an appreciable transport
of the pollutants to south.

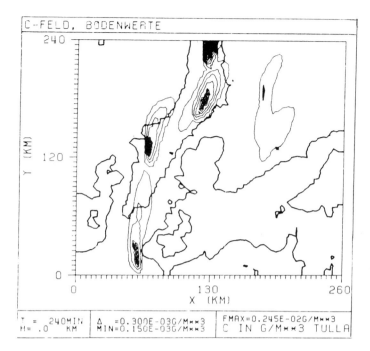

Fig. 9 Distribution of concentrations of SO_2 at the surface for a south-westerly flow (March 25, 1985)

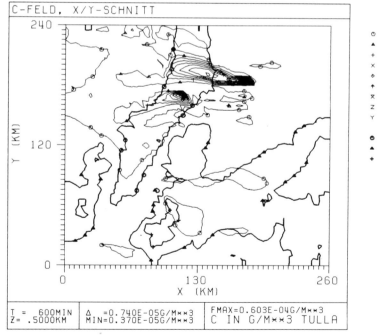

Fig. 10 Distribution of concentrations of SO_2 at a height of about 600 m above the ground for an easterly flow (March 21, 1985)

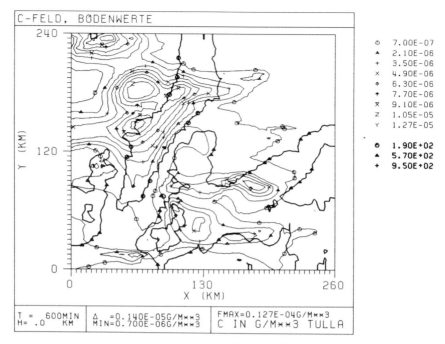

Fig. 11 Distribution of concentration of SO_2 at the ground for on easterly flow (March 21, 1985).

In the context of forest disease and other ecological damages it has often been speculated on the source-receptor relationship either just considering the large scale flow characteristics or only taking surface winds into account. Numerical model results may give here a much better basis for those questions.

5. OUTLOOK

Although many special problems in numerical modeling deserve still an urgent improvement, it has nevertheless reached a level where useful results for interpretation of observations can be gained. Numerical modeling in areas with complex structure of the lower boundary is also the only tool which is able to consider the complicated interaction of the different influences.

Since the influence due to topography are so dominating model results are for many flow characteristics in reasonable agreement with observations.

References

Adrian,G. und F. Fiedler (1987): A three-dimensional model for mesoscale
                          wind forecasts (in preparation).

Dorwarth, G. (1986):   Numerische Berechnung des Druckwiderstandes typi-
                          scher Geländeformen
                          Wiss. Ber. Institut für Meteorologie und Klima-
                          forschung Universität Karlsruhe, Nr. 6.

Fiedler, F. (1983):   Einige Charakteristika der Strömung im Oberrhein-
                          graben
                          Wiss.Ber. Meteorologisches Institut Universität
                          Karlsruhe, Nr. 4, S. 113 - 123.

Mellor, G.L. and Yamada, T. (1974): A hierarchy of turbulence closure models
                          for planetary boundary layers.
                          J. Atmos. Sci. 31, 1791 - 1806.

Nieuwstadt, F.T.M. and H. van Dop (1982): Atmosperic Turbulence and Air
                          Pollution Modelling
                          D. Reidel Publ. Comp., Dortrecht.

Pielke, R.A. (1984):   Mesoscale Meteorological Modelling
                          Academic Press, London.

Wippermann, F. and G. Groß (1981): On the construction of orographically
                          influenced wind roses for given distributions
                          of the large-scale winds.
                          Beitr. Phys.Atm. 54, pp. 492-501.

Regional and Long-range Transport of Air Pollution,
Lectures of a course held at the Joint Research Centre, Ispra, Italy,
15–19 September 1986, S. Sandroni (Ed.), pp. 355–379
© Elsevier Science Publishers B.V., Amsterdam — Printed in The Netherlands

LONG RANGE EPISODIC TRANSPORT MODELLING

H. VAN DOP

1 INTRODUCTION

A long time in the history of mankind the waste disposal in the atmosphere
has been a matter of little concern. The industrial revolution, the population
explosion and the increased economic activity have caused drastic changes in
this respect. An inevitable question nowadays in connection with the production
of air pollution is: what are the environmental consequences. Air pollution
transport and deposition models (ATDM's) provide a means to get an answer to
this question. The public awareness of adverse effects of air pollution
required causal relationships between emissions and its effects. In this way
"the art of atmospheric transport modelling" came into being. When put in a
historical perspective, the first efforts in modelling concerned the regional
air quality in the direct neighbourhood of cities and industrial plants. In the
early sixties the first evidence of air pollution damage due to transport over
longer distances was reported by the Scandinavian countries, which claimed that
the chemical (and consequently biological) composition of their lakes was
affected by emissions of sulphur dioxide in the large industrial centres in
West and Central Europe (ref. 1). Not only the airborne concentrations were a
matter of concern, but also how and where air pollution returned to the earth
surface. A complicating factor was that atmospheric chemical processes should
be taken into account in view of the transport times involved in transport on
the continental scale.

Only very recently it was recognised that the on-going production of air
pollution gradually changes the trace gas composition of the whole troposphere
(ref. 2). This could lead to climatic changes in the next century which,
according to some, may be hardly measurable, and to others, have an apocalyptic
character.

This very brief overview indicates that air pollution transport modelling
covers spatial scales from a few kilometers to the global scale, with
corresponding time scales of minutes to years. Chemical transformation
processes and deposition are necessary ingredients in these models. In this
chapter we emphasise the transport only, the other processes being dealt with
in other chapters.

Almost all air pollution is emitted at, or near the surface of the earth. Transport and chemistry is mainly confined to a thin layer of, say 2-3 km, the atmospheric boundary layer (ABL). The understanding of the physics of the ABL, including cloud formation and precipitation is crucial in the development of transport modelling (and also for the mechanisms which transport (a tiny fraction of) the (chemically converted) emissions up into the background troposphere. As a consequence, transport models should include a reasonable amount of properties of the ABL, such as the three dimensional windfield (with adequate time and space resolution), a description of turbulent motion, including the mixed-layer height, cloud formation and precipitation.

Depending on scale, desired complexity, numerical facilities etc. a model formulation will be chosen. Though a large variety of models exists covering various spatial ranges, two main categories should be distinguished: Eulerian and Lagrangian transport models. Eulerian models describe the dispersion of pollution in a fixed frame of reference (fixed with respect to an observer or to a point on the earth surface). In Lagrangian models the motion of a polluted air parcel is followed from its initial position as it moves along its trajectories. For example $X(t;x_o)$ denotes (one co-ordinate of) the position of an air parcel at time t, which at $t=t_o$ was at x_o.

Of course both descriptions are equivalent, e.g. the wind velocity in the Eulerian frame of reference, $u(x,t)$, is related to the Lagrangian velocity, defined by dX/dt, by

$$\frac{dX(t)}{dt} = u(x=X,t) \; ,$$ (1.1)

where, for simplicity, we have considered one dimension.

The reason why one decides to use a Lagrangian or a Eulerian formulation lies in the application. This will be discussed in the next section, which is devoted to the main concepts in long range transport modelling.

A sometimes underestimated aspect of long range transport modelling is the preparation of the meteorological input variables. These will be discussed in section 3.

Finally in section 4 a review will be given of some important developments in long range transport models.

2 MATHEMATICAL CONCEPTS

Here, a brief review will be given of the mathematical framework of dispersion models. More detailed descriptions can be found e.g. in refs. 3-7. The material which is considered relevant for this chapter is repeated below.

2.1 Conservation equations

The dry atmosphere can be considered as a mixture of ideal gases. The dynamics of atmospheric flow is given by the Navier–Stokes equations, which together with the continuity equation and the energy equation completes the set of atmosheric equations. However, scale and order of magnitude considerations allow several simplifications which reduce the N–S equations to the Boussinesq equations (ref. 8, 9). A crucial result is that the mathematical form of the resulting equations indicates that fluctuations in density need only to be taken into account in combination with the acceleration of gravity. This is commonly but quite loosely expressed in phrases like "the atmosphere can be considered incompressible".

The constant density assumption reduces the continuity equation to

$$\frac{\partial u}{\partial x} + \frac{\partial v}{\partial y} + \frac{\partial w}{\partial z} = 0 \; , \tag{2.1}$$

where (u,v,w) are the wind velocity components in a fixed (Eulerian) frame of reference (conventionally the positive x-axis points eastwards, the y-axis northwards and the z-axis upwards). When χ denotes the concentration of a contaminant (in units of mass per unit volume of fluid) the equation of conservation of mass is simply

$$\frac{\partial \chi}{\partial t} + \frac{\partial u\chi}{\partial x} + \frac{\partial v\chi}{\partial y} + \frac{\partial w\chi}{\partial z} = s \; , \tag{2.2}$$

where s denotes the sources and sinks of contaminant χ.
By putting $\chi = \overline{\chi} + \chi'$, $u = U + u'$ etc. and averaging noting that $\overline{\chi'}$, $\overline{u'}$ etc equal zero, we obtain

$$\frac{\partial \overline{\chi}}{\partial t} + U \frac{\partial \overline{\chi}}{\partial x} + V \frac{\partial \overline{\chi}}{\partial y} + W \frac{\partial \overline{\chi}}{\partial z} = -\frac{\partial}{\partial x} \overline{u'\chi'} - \frac{\partial}{\partial y} \overline{v'\chi'} - \frac{\partial}{\partial z} \overline{w'\chi'} + S \; , \tag{2.3}$$

where also eqn. (2.1) has been used. Eqn. (2.3) is the familiar starting point for numerous approximations. In order to be able to solve eqn. (2.3), assumptions have to be made for the eddy correlation terms at the right of eqn. (2.3). The usual approach is to put

$$- \overline{u'\chi'} = K_x \frac{\partial \overline{\chi}}{\partial x} \; , \quad - \overline{v'\chi'} = K_y \frac{\partial \overline{\chi}}{\partial y} \; \text{and} \quad - \overline{w'\chi'} = K_z \frac{\partial \overline{\chi}}{\partial z} \; , \tag{2.4}$$

a "receipt" which is usually referred to as first order closure or the gradient transfer assumption.
It should be noted that (slightly inconsistent with Eqn. (2.4)) eddy fluxes should be related to gradients in the mean mixing ratio, e.g.,

$-\overline{w'\chi'} = \rho_o K_z \frac{\partial c}{\partial z}$, where the mixing ratio, c equals $\overline{\chi}/\rho_o$ (ρ_o = the mean air concentration). This might be relevant when the vertical dimension of the dispersion domain is so large that the variation of mean density with height, $\partial\rho_o/\partial z$, should be taken into account.

Substituting eqn. (2.4) in eqn. (2.3) yields

$$\frac{\partial\overline{\chi}}{\partial t} + U \frac{\partial\overline{\chi}}{\partial x} + V \frac{\partial\overline{\chi}}{\partial y} + W \frac{\partial\overline{\chi}}{\partial z} = \frac{\partial}{\partial x} (K_x \frac{\partial\overline{\chi}}{\partial x}) + \frac{\partial}{\partial y} (K_y \frac{\partial\overline{\chi}}{\partial y}) + \frac{\partial}{\partial z} (K_z \frac{\partial\overline{\chi}}{\partial z}) + S \ . \qquad (2.5)$$

The empirical coefficients K_x, K_y and K_z determine the strength of the turbulent exchange and may be functions of the co-ordinates in inhomogeneous flows.

2.2 Lagrangian model formulation

In the Lagrangian model formulation the mean concentration of a moving polluted parcel of air is considered. The volume of the parcel is assumed large enough, so that concentration changes due to turbulent exchange through the boundaries can be neglected. It is also assumed that during its travel the distortion of the parcel due to the turbulent motion is small, so that the parcel remains an entity. The mean concentration in the air parcel is then simply given by

$$\frac{d\overline{\chi}}{dt} = S \ , \qquad (2.6)$$

where S, as in eqn. (2.3), contains the source and sink terms. The problem remains to determine the parcel trajectory. This can be done by using the Eulerian-Lagrangian relationship (cf. eqn. (1.1)). In integrated form it reads

$$\underset{\sim}{X}(t) = \underset{\sim o}{X} + \int_o^t \underset{\sim E}{u} (\underset{\sim}{x} = \underset{\sim}{X}(t'), \ t') \ dt' \ , \qquad (2.6a)$$

where $\underset{\sim}{X}(t)$ denotes the 3-D parcel position; the initial position is denoted by $\underset{\sim o}{X}$ and $\underset{\sim E}{u}$ is the Eulerian wind velocity field.

Usually an air parcel is considered which is adjacent to the earth's surface and which has a fixed height, h. Since the wind velocity changes with height the mean transport wind should be determined from a vertical average of the wind velocity over the parcel height h. In practice, the wind velocity is obtained from radiosonde data or from dynamic models of the wind field. The 925 mbar level wind data (~ 800 m altitude) are considered representative for the transport wind in trajectory models.

The solution of the equation (2.6) for a sufficient number of initial positions provides a fast and relatively easy picture of atmospheric dispersion.

Trajectory models can be modified such that they may account for a variable parcel height, or include horizontal diffusion. The Lagrangian approach has been successfully applied in the EMEP study on long range transport and deposition of sulphur dioxide and sulphate on the European continent (ref. 10).

2.3 Eulerian model formulation

The master equation for a number of applications in atmospheric dispersion calculations is eqn. (2.5). In a few cases the equation has analytical solutions for which we refer to Pasquill and Smith or Sutton (refs. 4 and 3). In mesoscale and long range transport applications non-homogeneity and non-stationarity of wind and turbulence make that only numerical solutions will provide solutions to eqn. (2.5). Nevertheless we shall briefly discuss some current mathematical simplifications of the transport equation. First, over (more or less) flat terrain vertical transport of diffusion will be dominated by diffusion and so the fourth term at the left of eqn. (2.5) may be omitted. Also in moderate wind conditions it is assumed that the horizontal advection dominates the horizontal turbulent diffusion. With these assumptions eqn. (2.5) reduces to

$$\frac{\partial \overline{\chi}}{\partial t} + U \frac{\partial \overline{\chi}}{\partial x} + V \frac{\partial \overline{\chi}}{\partial y} = \frac{\partial}{\partial z} (K_z \frac{\partial \overline{\chi}}{\partial z}) + S \ . \qquad (2.7)$$

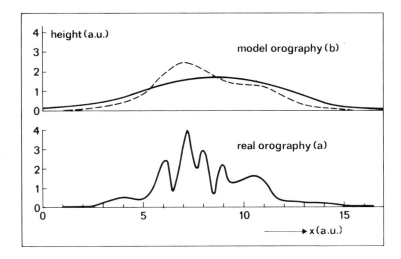

Fig. 1. Sketch of variability in terrain height (a) and the representation in a discrete grid (b). The model topography depends on the grid size:
———— gridsize = 6 a.u., – – – – 2 a.u.

Besides, Van Dop et al. (ref. 11) show that in the necessary discretization of eqn. (2.5) the horizontal diffusion will in most cases have sub-grid scale dimensions (when the natural variability of the wind-field is well-represented at the selected grid-size), and therefore, can be neglected.

Some emphasis should be given to atmospheric modelling in complex orography (see also Fig. 1). Then vertical mean velocities can no longer be neglected. In case the wind velocity input to eqn. (2.5) is obtained only from wind observations, we stand for the formidable task to transform these data to the three dimensional wind field with the spatial resolution corresponding to the selected grid size of the transport model. When the data are obtained from a dynamical mesoscale model the situation is hardly better, though we are more or less forced to use the corresponding gridsize and topography of the dynamical model. It is nevertheless clear that in an area with complex topography the use of a dynamical model in combination with the transport model offers a distinct advantage and, provided that computer capacity is sufficient, this method should be preferred.

Eulerian models have the advantage over Lagrangian models that they offer more flexibility to incorporate meteorological properties, such as the wind velocity variation with height, vertical exchange processes and the treatment of deposition and emissions.

Additionally, in Eulerian models (arbitrarily) small grid sizes can be used, which taking into account chemical processes may be of advantage, if not required. (Note that in Lagrangian models the parcel size may be chosen small too, but then one has to include interparcel exchange processes. This can all be done, but the Lagrangian formulation looses then its major attractivity: its simplicity.)

In summary Eulerian models seem to provide the only viable general approach to atmospheric transport modelling, including deposition and transformation processes.

2.4 Vertically integrated models

An approach which deserves separate attention is the use of the vertical ABL structure to define layers in the dispersion model. The normal numerical procedure is to select a vertical (and horizontal) grid size which is small enough to resolve a sufficient amount of details of the atmospheric structure. This usually leads to the choice of an equidistant grid, or a grid of which the gridsize is gradually increased at higher altitudes. The number of gridpoints required in the vertical direction is of the order of 10-20. This can be rather demanding with respect to computer capacity and processing and possibly inefficient, the more so since it may be doubted whether this detail in vertical structure is really needed in long range transport modelling.

In the ABL we distinguish a surface layer, which is a part of the day-time or night-time boundary layer. Above these layers is the free troposphere. The interaction between the free troposphere and the former layers is weak, at least in the absence of clouds. The thickness of the layers varies spatially and temporally. In many circumstances, however, the heights of those layers are well-defined. An attractive alternative, therefore, could be to describe the transport in two or three layers only, considering layer averaged concentrations only. This approach is adopted e.g. in the SAI, KNMI and RIVM model (refs. 12-14) and is rigorously treated by Lamb (ref. 15). Because of its potential importance a summary of this approach is presented here.

For simplicity we describe an atmospheric boundary layer where only one layer (the mixed-layer) varies spatially and temporally (cf. Fig. (2)). We define a layer average value of say $\phi(x,y,z,t)$ by brackets,

$$\langle\phi\rangle = \frac{1}{h} \int_o^h \phi(x,y,z,t)dz \ . \tag{2.8}$$

From this definition follows that

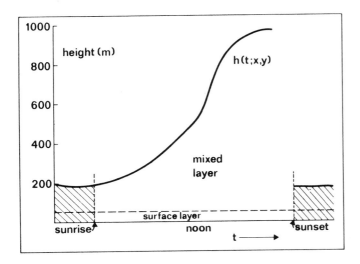

Fig. 2. The day-time and nocturnal boundary layer height at location (x,y) as a function of time. The hashed area represents the (stable) nocturnal boundary layer. The dashed line represents the (constant) surface layer height.

$$\frac{\partial \langle \phi \rangle}{\partial t} = \langle \frac{\partial \phi}{\partial x} \rangle + (\ \phi(x,y,h,t) - \langle \phi \rangle\ )\ \frac{\partial \ell n h}{\partial t}\ , \tag{2.8a}$$

$$\frac{\partial \langle \phi \rangle}{\partial x} = \langle \frac{\partial \phi}{\partial x} \rangle + (\ \phi(x,y,h,t) - \langle \phi \rangle\ )\ \frac{\partial \ell n h}{\partial x}\ , \tag{2.8b}$$

and a similar expression for $\partial \langle \phi \rangle / \partial y$.

We apply these operations to the conservation equations (2.1 and 2.2). Linearity yields

$$\langle \frac{\partial u}{\partial x} \rangle + \langle \frac{\partial v}{\partial y} \rangle + \langle \frac{\partial w}{\partial z} \rangle = 0 \text{ and} \tag{2.9}$$

$$\langle \frac{\partial \chi}{\partial t} \rangle + \langle \frac{\partial u \chi}{\partial x} \rangle + \langle \frac{\partial v \chi}{\partial y} \rangle + \langle \frac{\partial w \chi}{\partial z} \rangle = \langle s \rangle\ . \tag{2.10}$$

Using eqns. (2.8) the continuity equation (2.9) can be rewritten as

$$\frac{\partial \langle u \rangle}{\partial x} + \frac{\partial \langle v \rangle}{\partial y} + \frac{w_h}{h} = (u_h - \langle u \rangle)\ \frac{\partial \ell n h}{\partial x} + (v_h - \langle v \rangle)\ \frac{\partial \ell n h}{\partial y}\ , \tag{2.11}$$

where the subscript h refers to the value at h.
Also eqn. (2.10) can be rewritten as

$$\frac{\partial \langle \chi \rangle}{\partial t} + \frac{\partial \langle u \chi \rangle}{\partial x} + \frac{\partial \langle v \chi \rangle}{\partial y} + \langle u \chi \rangle\ \frac{\partial \ell n h}{\partial x} + \langle v \chi \rangle\ \frac{\partial \ell n h}{\partial y} + \langle \chi \rangle\ \frac{\partial \ell n h}{\partial t} =$$

$$\chi_h\ \frac{\partial \ell n h}{\partial t} + (u \chi)_h\ \frac{\partial \ell n h}{\partial x} + (v \chi)_h\ \frac{\partial \ell n h}{\partial y} - \frac{(w \chi)_h}{h} + \langle s \rangle\ . \tag{2.12}$$

Here we define the fluctuations as

$$\langle u' \chi' \rangle \equiv \langle u \chi \rangle - \langle u \rangle \langle \chi \rangle \text{ etc.} \tag{2.13}$$

Substituting this in (2.12) and substracting from the result eqn. (2.11) multiplied by $\langle \chi \rangle$, we obtain

$$\frac{\partial \langle \chi \rangle}{\partial t} + \langle u \rangle\ \frac{\partial \langle \chi \rangle}{\partial x} + \langle v \rangle\ \frac{\partial \langle \chi \rangle}{\partial y} = \tag{2.14}$$

$$-\frac{\partial}{\partial x} \langle u' \chi' \rangle - \frac{\partial}{\partial y} \langle v' \chi' \rangle + \frac{\Delta \chi}{h} (\frac{\partial h}{\partial t} + u_h\ \frac{\partial h}{\partial x} + v_h\ \frac{\partial h}{\partial y} - w_h) + \langle s \rangle\ ,$$

where $\Delta \chi = \chi_h - \langle \chi \rangle$.

When we define an entrainment velocity w_e as

$$w_e \equiv \frac{\partial h}{\partial t} + u_h \frac{\partial h}{\partial x} + v_h \frac{\partial h}{\partial y} - w_h \qquad (2.15)$$

(see Deardorff (ref. 16)), we finally obtain

$$\frac{\partial \langle \chi \rangle}{\partial t} + \langle u \rangle \frac{\partial \langle \chi \rangle}{\partial x} + \langle v \rangle \frac{\partial \langle \chi \rangle}{\partial y} = \frac{\Delta \chi}{h} \cdot w_e + \langle s \rangle , \qquad (2.16)$$

where we have neglected again the fluctuating contribution to the horizontal transport.

When we have to consider more than one layer we have for each layer an equation similar to (2.16) (presumably not more than two or three). This is considerable progress compared with eqn. (2.5) which has to be solved for each grid cell in the vertical domain.

Note that the spatial variation in the boundary layer height and the (layer averaged) wind field are not independent. They must obey eqn. (2.11).

2.5 Numerical methods

Solutions of the transport equation can be obtained by numerical methods only. Often practical limitations (computer facility, run-time) determine which method should be followed. This does not take away that the primary requirement for a numerical method should be that errors introduced by the method should be smaller than the other uncertainties in the long range transport problem.

The basic equation is a parabolic partial differential equation (eqn. (2.5)). In case more components are involved, $\chi_1 \cdots \chi_n$, we have a set of n of these equations which are coupled by chemical reactions. The numerical implications of the chemistry will not be discussed here so that this section is devoted to the numerical analysis of eqn. (2.6). There is, however, one important feature of numerical chemical schemes which should be borne in mind: they cannot cope with negative concentrations. This has an important feed-back on numerical schemes for advection, since they usually create (small) negative concentration values. In the discussion of various advection schemes this aspect will therefore be emphasized.

The transport equation should be completed with an initial field, $\overline{\chi}(x,y,z,t=0) = \overline{\chi}_0$ and with boundary conditions, which define the in- and outflow over the model boundaries.

The sources are contained in the term S at the right of eqn. (2.5). They consist of surface sources, which are estimated total emissions over relatively large areas, and point sources. In emission inventories surface sources are usually presented on a grid with a size of comparable dimensions with the numerical grid size. These sources can be easily implemented in transport models. The lateral dimensions of releases from point sources, however, may be sub-grid scale for quite a long time. Implementation of these sources can only be carried out accurately when the "plume" dimension has become at least of the same order as the numerical grid-size. This problem may be tackled by approximating the concentration fields of point sources by analytical Gaussian fields close to the source (cf. ref. 17), in the initial stage of the dispersion.

The most frequently used way to solve eqn. (2.5) was introduced by Yanenko (ref. 18) and Marchuk (ref. 19). It consists of splitting the process in a advection step and a diffusion step according to

$$\frac{\partial \overline{\chi}}{\partial t} + U \frac{\partial \overline{\chi}}{\partial x} + V \frac{\partial \overline{\chi}}{\partial y} = 0 \text{ and} \tag{2.17a}$$

$$\frac{\partial \overline{\chi}}{\partial t} = \frac{\partial}{\partial z} \left(K_z \frac{\partial \overline{\chi}}{\partial z} \right) + S \ . \tag{2.17b}$$

When χ is chemically reactive a third equation is added which describes this. Here we discuss the numerical solution of (2.17) only.

The advantage of the splitting is that we are now able to apply numerical solution methods which comply with the different mathematical nature of eqns. (2.17a and b). The (parabolic) equation (2.17b) has a diffusive character. Its numerical solution poses no difficulties for which we refer to the literature (refs. 19-21). The remaining part of this section is devoted to the solution of the advection eqn. (2.17a).

The ideal advection scheme should have the following properties.
- It should be mass conserving.
- The accuracy must be sufficient, i.e. the contaminant must be transported with the right speed (and in the right direction).
- The scheme should yield positive concentrations everywhere.
- The numerical diffusion should be much smaller than the real atmospheric diffusion.
- Process time and memory requirements should be low.

It is obvious from the large number of publications still appearing on this subject that the ideal scheme has not yet been formulated, though computer capacity increases by an order of magnitude every ten years. We will therefore discuss a few selected methods which all have their pros and cons.

(a) <u>The pseudo spectral method</u>. The advection equation may be splitted again, so that subsequently the advection in respectively the x- and y-direction:

$$\frac{\partial \overline{\chi}}{\partial t} + U \frac{\partial \overline{\chi}}{\partial x} = 0 \text{ and} \qquad (2.18a)$$

$$\frac{\partial \overline{\chi}}{\partial t} + V \frac{\partial \overline{\chi}}{\partial y} = 0 . \qquad (2.18b)$$

By Fourier transforming eqns. (2.18) two ordinary differential equations result which can easily be solved numerically. The solution is transformed back by the inverse transformation. It is an accurate method which can be applied to a course grid and it can cope with sharp concentrations gradients. An inherent assumption in the method is that it requires periodic boundary conditions. This makes that precautions should be taken to prevent spurious inflow. There are two basic solutions of this problem. The first consists of damping the concentrations near the edges so that concentrations values at the boundaries are always small (ref. 22, 23) and the second is obtained by modifying the numerical concentration field (by substracting an analytical field) such that the flux over the boundaries is virtually zero. The method generates (small) negative concentrations which makes it not very suitable in combination with chemistry.

(b) <u>The second moment method</u>. By considering the evolution of moments of a concentration distribution it is possible to displace a rectangular concentration distribution exactly (ref. 24, 25). Concentrations remain

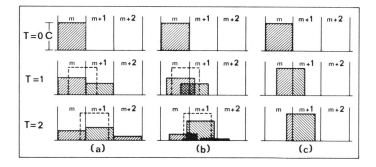

Fig. 3. An initially rectangular concentration distribution advected to the right in three sequential time steps. Dashed distribution corresponds to transport by continuum advection. (a) Simple finite-difference. (b) Difference scheme using reconstruction with first moment of concentration distribution. (c) Difference scheme using reconstruction with first and second moments. (ref. 24).

strictly non-negative. However, the method is rather "expensive" with respect to computer time. An illustration of how the method works is given in Fig. 3.

(c) The "chapeau function" method. This method is recommended by Chock (ref. 26). The concentration field is expanded in orthogonal functions (chapeau functions). The method is cheap and applicable on arbitrary rectangular grids. The method, however, is fairly dispersive and generates negative concentrations of considerable magnitude.

(d) Smolarkiewicz' scheme. Essentially it is an upwind scheme to which a convection procedure is applied. In the upwind scheme numerical diffusion occurs with a magnitude that is characterised by the diffusion coefficient $K_n \sim \frac{1}{2} \|u\| \Delta x (1-\lambda)$, where λ is the Courant number ($\lambda = \|u\| \Delta t / \Delta x$). Smolarkiewicz introduces an artificial "antidiffusion" velocity $\tilde{u} = - (K_n/\chi) \partial\chi/\partial x$ which counteracts the numerical diffusion (ref. 27). The method is non-negative, the computer time and memory requirements are reasonable and implementation is simple.

3 METEOROLOGICAL DATA

In the introduction the paramount importance of meteorological data of the atmospheric boundary layer was already indicated. The major features are the transport by the mean atmospheric flow, the turbulence, and regarding deposition, clouds and precipitation.

It appears that exchange processes at the surface of the earth play an important role in relation to processes in the boundary layer. Therefore, at the end of this section also some attention will be paid to some geophysical properties.

3.1 The wind field

Wind data can be obtained either from routine synoptic observations, or from a dynamical model. The former data can be made available easily but have the disadvantage that most data are ground-based and only over land sufficiently dense, while upper air data are spatially and temporally too sparse to give a reasonably detailed analysis of the wind field for mesoscale and long-range transport models. Moreover, it requires considerable effort to derive reasonable wind fields from observations, which are often affected by the local situation (ref. 28-30), and therefore not very representative for a (grid square) average wind velocity field.

An alternative is offered by generating wind data from regional meteorological models (ref. 31) or limited area models. These models are primarily designed to be used as regional weather forecast models. Since they calculate meteorological fields with a spatial and temporal resolution which fit with

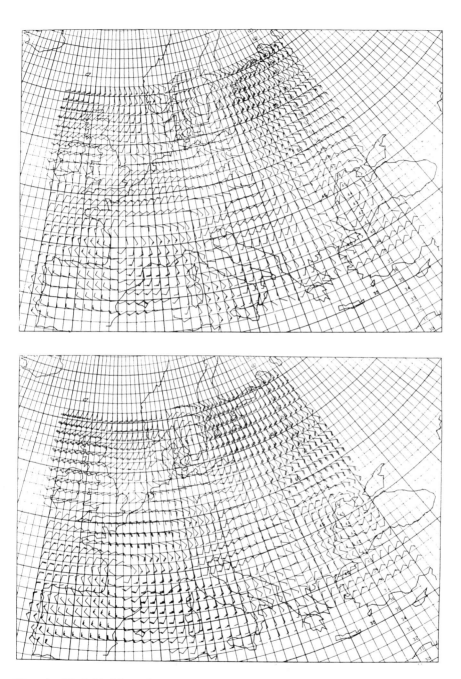

Fig. 4. Wind field analysis over Europe on 26 May, 1986 (a) 1000 mbar (b) 850 mbar. The data are derived from the ECMWF model.

those of mesoscale and long range transport models, they are very suitable as
input for dispersion modelling on this scale. For example the limited area
model (LAM) which is currently used at KNMI is derived from the ECMWF (European
Centre for Medium Range Weather Forecasting). There are 7 layers below 3000 m
altitude in the model and the parametrization of boundary layer processes is
simple. 1000 and 850 mbar windfields can be easily produced with the ECMWF
computer facility (Fig. 4). Moreover, these models contain simple
parametrizations of precipitation (see below) and an extension to the
prediction of other meteorological phenomena (cloud cover) is expected in the
near future.

3.2 Precipitation

An important element in long range transport modelling is the precipitation:
crudely half of the atmospheric pollutants are deposited on the surface by
precipitation. The process of cloud formation and precipitation cannot yet be
fully described by numerical meteorological models. Intertwined in this
process is the air pollution, which is absorbed by (or evaporates from) cloud
or rain droplets, or which serves as condensation nuclei for the formation of
water droplets when the pollution consists of particulate matter. The physical
processes involved are discussed in another chapter. Here, the discussion is
limited to how precipitation may be quantitavily described, that is, how
representative amounts of precipitation can be estimated from observations or
model data.
First should be noted that precipitation, be it from single convective clouds
or originating from large frontal systems, is highly intermittent and spatially
strongly varying, when it is compared with the density in space and time of the
synoptic rain collector network. The spatial resolution amounts to 50-100 km
over Europe (n.b. no observations over sea!), which largely exceeds the size of
convective storms or the width of rain bands and cells in frontal systems. The
information gathered by the synoptic network is therefore not much more than a
random sample. A second problem is that amounts of precipitations are reported
as accumulated sums over 6 or 12 hours. This is clearly unsufficient for a
reliable three hourly rain analysis on a grid of say 80x80 km. On the other
hand regional meteorological models are able to provide precipitation data with
the desired resolution in time and space though their results should be
considered with care. An advantage of models is that they also predict
precipitation over sea.
In Fig. 5 an experimental precipitation analysis is shown, which is a mixture
of LAM precipitation data and observations. The 6 and 12 hourly rain data were
divided over preceding 6 hour intervals using the synoptic weather code

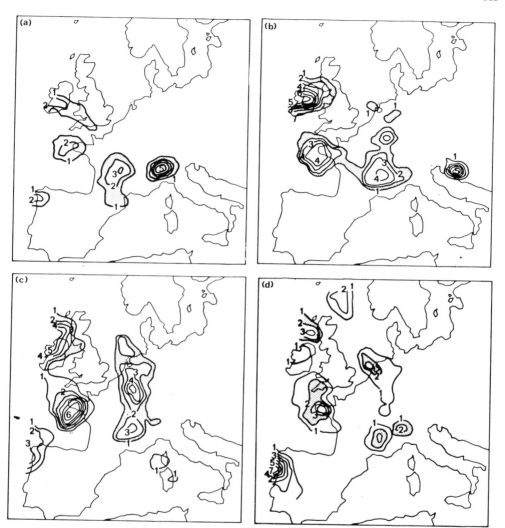

Fig. 5. Precipitation field analysis based on 6, 12 and 24 h synoptic data, on 3 May, 1986 12 GMT (a), 12 (b) and 4 May 00 (c) and 06 (d) GMT.

according to a procedure developed by Scheele (ref. 32). The results of Fig. 5 have a spatial resolution of ~ 80x80 km.

It should be noted that these data are average values over grid cells. A more detailed analysis of the precipitation data over The Netherlands (cf. Fig. 6) shows that the precipitation is confined to a much smaller region and with a higher intensity than indicated in Fig. 5.

One may hope to avail of high resolution rain data from (interlinked) radar networks in the future.

3.3 Atmospheric turbulence

All movements which are not contained in the average 3-dimensional windfield are caused by atmospheric turbulence. Though the transport in the horizontal direction, parallel to the surface, is dominated by the mean windfield, the transport in vertical direction is mainly the result of turbulent atmospheric motions. The intensity of these motions is closely related to the mean wind shear and the atmospheric stability. These matters are amply discussed in Dr. Smith's chapter. We shall review those topics which are relevant for long range transport modelling.

In the lower troposphere we may distinguish a few layers: the surface layer, the atmospheric boundary layer (in which the surface layer is included), and the free troposphere.

The surface layer covers the range $z_o \ll z \lesssim \min (2\|L\|, 0.1h)$, where z_o is the surface roughness, h the mixed-layer depth and L the Obukhov length. The turbulence in this layer is determined by the friction velocity (u_*) and L only. There is no wind turning with height in this layer. Windspeed, temperature and concentration profile are governed by similarity laws for which the governing parameters are u_*, θ_* ($=H_o/u_*$, where H_o equals the surface heat flux) and v_d. The deposition velocity v_d is specific for the concentration and the nature of the surface. A typical range of the surface layer height is 20-200 m.

During day-time a mixed-layer develops which is characterised by a near neutral potential temperature gradient ($\partial\theta/\partial z \simeq 0$). In horizontally homogeneous and almost cloud free conditions the dynamics of this layer can be formulated (ref. 33). The development of the mixed-layer height, h(t) is then roughly known. We may recall that this is an important parameter in vertically integrated transport models (see section 2.4). The growth rate of h is expressed in H_o, L and γ, where γ is atmospheric lapse rate (dθ/dz) above the mixed layer. A convenient empirical formula is (ref. 13)

$$\frac{dh}{dt} = \frac{H_o(t)}{\gamma h(1-c_3 L/h)} [1 + c_1 + c_2 (L/h)^2] , \qquad (3.1)$$

3 mei 08 UT - 4 mei 08 UT

0 1 3 10 15 mm

KNMI

Fig. 6. Microanalysis of precipitation over The Netherlands at the same period as in Fig. 5. (The numbers denote the values in 0.1 mm.)

where c_1–c_3 are 0.4, 31 and 8.7 respectively.

In a moist boundary-layer cloud formation will occur just below the mixed layer height. Condensation processes enhance the atmospheric instability and pollution, which normally will be confined in the mixed layer, may be entrained (locally!) to much higher altitudes in towering cumulus clouds, and thus be transported into the free troposphere. A typical value of the daily maximum height at moderate latitude is 1600 m.

During night-time a radiation inversion develops with a quasi equilibrium height given by

$$h = c \ \frac{u_*/f}{1 + 1.9 \ h/L} \tag{3.2}$$

(ref. 34), with a typical value of $\sim$ 200 m. Within this layer vertical motions are strongly damped so that the vertical dispersion will be small. It is determined by the wind (direction) shear and the local temperature gradient $d\theta/dz$, usually notated as $(\frac{g}{T} \frac{d\theta}{dz})^{\frac{1}{2}} \equiv N$, the Brunt-Vaissala frequency. It should, however, be noted that buoyant plumes of (large) point sources may penetrate the nocturnal inversion layer, so that pollution is directly injected in the (slightly stable) free atmosphere.

Above the nocturnal and day-time boundary layer is, what we indicate here as free of back-ground troposphere. For modelling purposes in this capping layer some modellers define a cloud-layer, or reservoir layer (ref. 15 and 14). At these altitudes it may be assumed that the influence of the earth surface is small: the mean flow is geostrophic and turbulence levels are generally low.

3.4 Geophysical data

The most important parameters to describe the turbulence characteristics of the ABL are z_0, u_*, H_0, $d\theta/dz$ and in addition to the downward flux of pollution, v_d. When these parameters are known most other can be derived (see e.g. ref. 35). Of these parameters z_0, H_0 and v_g depend on the nature of the surface. For the determination of the sensible heat flux, H_0, for instance it is necessary to estimate the net (short wave) radiation, and then, from an energy balance consideration how this radiation energy is converted into sensible, latent and soil heat flux. This partition, but also the roughness length z_0 and the surface resistance can be related to the terrain characteristics (see tables I and II). In ref. 36 a method is described to determine these characteristics from land maps for areas of $\sim$ 10x20 km^2. Satellite survey of the earth surface, however, offers a more convenient method

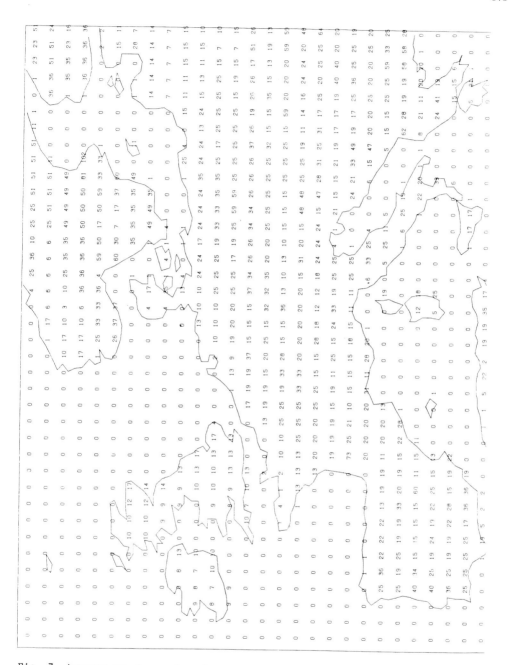

Fig. 7. Average roughness length on squares of 1° x 1° [in cm]. Data are based on the classification given in Table I.

to determine global surface characteristics[*]. The spatial resolution of most data is 1° x 1°. An example of satellite data-based determination (ref. 37) of the roughness length is given in Fig. 7. The same data in combination with the data of table II may yield a similar pattern for the deposition velocity over the European continent (and any other region) for various pollutants.

TABLE I
Roughness length for some terrain categories, according to Wieringa (ref. 30)

Terrain category	z_o[m]
Water surface	0.0005
Open field	0.03
Field with scattered trees anedges	0.25
Roads/railways	0.50
Forest	1.00
Built-up areas	2.00

TABLE II
Terrain dependent surface resistances [SM^{-1}] of PAN, ozone, NO and NO_2.

type	PAN	O_3 day	O_3 night	NO	NO_2 day	NO_2 night
water	10000	2500		10000	5000	
cropland, arable land	150	200	600	1000	300	1000
meadows, grassland	300	200	300	3000	300	1000
permanent crops	150	200	600	1000	300	1000
buit up areas	500	500	500	1000	1000	1000
forest	150	150	1200	2000	150	400
bare soil	300	200	200	1000	1000	
wet land	300	2500		10000	5000	

---

[*] A more elaborate list of roughness data can be found in the chapter on deposition processes.

4 CURRENT DEVELOPMENTS IN LONG RANGE TRANSPORT MODELLING

4.1 The EMEP model

The EMEP (European Monitoring and Evaluation Programme) model is a (Lagrangian) trajectory model which describes the long term average air concentrations and total deposition of SO_2 and SO_4^{2-} over Europe (ref. 10). It was one of the first operational models and was initially used for the evaluation of the acidification of the Scandinavian lakes. The meteorological data are based on the 850 mbar radiosonde observations. The spatial resolution is 127x127 km^2. Chemical transformation rate and dry and wet deposition are assumed to be linearly proportional to the concentrations. A constant ABL height of 1000 m is assumed in the routine model version. Notwithstanding its simplicity and shortcomings the model proved to be extremely useful after extensive tests and validations. A research version which includes (many) other compounds and a complex chemical scheme is under development (ref. 38).

4.2 The EPA regional oxidant model

This Eulerian grid model, which is developed at the Environmental Protection Agency (EPA), (ref. 15) has the objective to guide the formulation of regional emission control strategies in the United States. The model contains three "dynamic" layers, as discussed in section 2.4. The horizontal domain is ~ 1000x1000 km^2 and the grid size approximately 18 km (2500x3 grid points). The model aims at including all relevant chemical and physical processes related to transport, transformation and deposition of air pollution, such as full photochemistry (including slow reactions), night-time chemistry of the products and precursors of photochemical reactions, cumulus cloud physics and chemistry, mesoscale vertical motion, sub-grid scale chemistry and emissions from natural sources of volatile organic compounds (VOC's), NO_x and stratospheric ozone intrusions.

4.3 The NCAR acid deposition modelling project

This project (ref. 39) is funded by the US EPA and the National Science Foundation (NSF). Its principal task is to develop a Eulerian regional acid deposition model which is suitable for assessing source-receptor relationships. The prototype model contains 15 layers (at fixed heights) and has a horizontal resolution of 80 km (1700x15 grid points). It covers an area of approximately 3000x3000 km^2. The project is designed for the development of "credible" models, based on recent progress and state of the art knowledge of mesoscale meteorology, tropospheric chemistry and advanced computing.

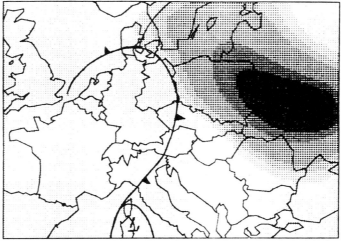

I-131 CONCENTRATION (BQ/M3); H=500M; T1/2=2 DAYS
DATE: 86   4 27

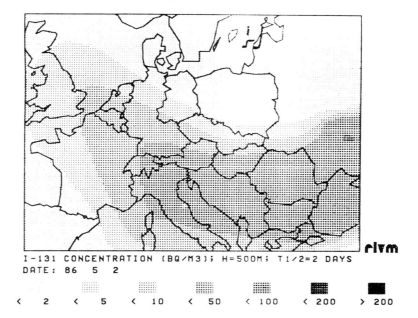

I-131 CONCENTRATION (BQ/M3); H=500M; T1/2=2 DAYS
DATE: 86   5   2

< 2   < 5   < 10   < 50   < 100   < 200   > 200

Fig. 8. Airborne I-131 concentrations in Bqm^{-3} on 27 April, 1986 (a) and 2 May, 1986 (b).

4.5 The PHOXA project

The two previous projects pertain primarily to application at the American continent. The PHOXA (Photochemical Oxidant and Acid Deposition Model Application) project was initiated by the German and Dutch Government. The primary goal is to be able to develop different control strategy options in Europe. It attempts to apply a number of existing transport and deposition models. Among these are

- a two-layered trajectory model including chemistry;
- a regional transport model developed by SAI based on their episodic photo oxidant model (ref. 12);
- the transport and deposition of acidifying pollutants—model (TADAP) developed by ERT, a very sophisticated transport model including a detailed analysis of dry and wet deposition of acidifying pollutants.

4.6 RIVM/KNMI co-operation on dispersion modelling

Two Dutch institutes (the Institute for public health and environmental hygiene (RIVM) and KNMI) are combining their research and experience (refs. 13 and 14) on mesoscale and long range transport modelling.

The objectives are to develop and make operational
- a two-layered trajectory model including full chemistry;
- a Eulerian (vertically integrated) transport and deposition model including simple linear chemistry and deposition, to be applied in Europe;
- an episodic photo oxidant model and an acid deposition model, including full chemistry. The transport model is based on the above mentioned model;
- a (partially prognostic) model to be used for the dispersion of accidental releases of toxic or hazardous material (calamities at nuclear or chemical plants).

At KNMI the emphasis is on the meteorological preprocessing, i.e. the preparation of meteorological fields and the prognostic aspects. The more operational activities are carried out by RIVM. The first two objectives have been achieved. An illustration of the capacity of the Eulerian model was the analysis of transport and deposition of the Chernobyl nuclear accident (ref. 40). In Fig. 8, a pattern of calculated air concentration of I^{137} is compared with measurements at various sites in Europe (ref. 41).

REFERENCES

1 A. Eliassen, The OECD study of long range transport of air pollutants: long range transpoet modeling, Atmos. Environ., 12 (1978) 479–488.

2   Global tropospheric chemistry: a plan for action. Report of the US National Academy of Science (1985).

3   W.A.L. Sutton, On the equation of diffusion in a turbulent medium, Proc. Roy. Soc., A 182 (1943) 48.

4   F. Pasquill and F.B. Smith, Atmospheric Diffusion, Ellis Horwood Ltd., 1983.

5   A.S. Monin and A.M. Yaglom, Statistical FLuid Mechanics, Vol I, MIT, Cambridge (Mass.), 1971.

6   F.T.M. Nieuwstadt and H. van Dop, Atmospheric Turbulence Modelling, Reidel, 1982.

7   H. van Dop, Atmospheric Distribution of Pollutants and Modelling of Air Pollution Dispersion in O. Hutzinger (Ed.), The Handbook of Environmental Chemistry, 4A, Springer, Berlin, 1985, pp. 107–147.

8   E.A. Spiegel and G. Veronis, On the Boussinesq approximation for a compressible fluid. Astrophys. J. 131 (1960 442–447.

9   Donaldson, C. du P., Construction of a dynamic model of the production of atmospheric turbulence and the dispersal of atmospheric pollutants, in D.A. Hangen (Ed.), Workshop on Micrometeorology, A.M.S., Boston, 1973.

10 A. Eliassen, A review of long-range transport modeling, J. Appl. Meteor., 19 (1980) 231–240.

11 H. van Dop, B.J. de Haan and C. Engeldal, The KNMI mesoscale air pollution model, Royal Netherlands Meteorological Institute, Scientific Report W.R. 82–6, 1982.

12 H. Meinl and P.J.H. Builtjes, Photochemical Oxidant and Acid Deposition Model Applications (PHOXA), Dornier, Friedrichshafen, 1984.

13 H. van Dop and B.J. de Haan, Mesoscale air pollution dispersion modelling, Atm. Environment, 17 (1983) 1449–1456.

14 N.D. van Egmond and H. Kesseboom, Mesoscale Air Pollution Dispersion Models – I Eulerian Grid model, Atmos. Environment, 17, 1983, 257–265.

15 R.G. Lamb, A regional Scale (1000 km) Model of Photochemical Air Pollution, Part I: Theoretical Formulation. EPA-600/3-83-035, Environmental Protection Agency, Research Triangle Park (NC), 1982.

16 J.W. Deardorff and E.W. Peterson, The boundary layer growth equation with Reynolds averaging, J. Atmos. Sci., 37, 1980, 1405–1409.

17 P. Karamchandani and L.K. Peters, Analysis of the error associated with grid representation of point sources, Atmospheric Envirionment, 17 (1983) 927–933.

18 N.N. Yanenko, The Method of Fractional Steps, Pringer, New York, 1971.

19 R.D. Richtmyer and K.W. Morton, Difference Methods for Initial-Value Problems, Interscience Publishers, New York, 1967.

20 P.J. Roache, Computational Fluid Dynamics, Hermosa, Albequerque, 1976.

21 R. Peyret and T.D. Taylor, Computational Methods for Fluid Flow, Springer Verlag, New York, 1983.

22 O. Christensen and L.P. Prahm, A pseudospectral method for dispersion of atmospheric pollutants, J. Appl. Meteor, 15 (1976) 1284–1294.

23 D. Gottlieb and S.A. Orszag, Numerical Analysis of Spectral Methods, SIAM, Philadelphia, 1979.

24 B.A. Egan and J.R. Mahoney, Numerical modelling of advection and diffusion of urban area source pollutants, J. Appl. Meteor., 11 (1971) 312–322.

25 L.B. Pedersen and L.P. Prahm, A method for numerical solution of the advection equation, Tellus XXVI (1974) 594–602.

26 D.P. Chock, A comparison of numerical schemes for solving the advection equation, Part II, Atmospheric Environment, 19 (1985) 571–586.

27 P.K. Smolarkiewicz, A fully multidimensional positive definite advection transport algorithm with small implicit diffusion, J. Comp. Phys., 54 (1984) 325–362.

28 H. van Dop, B.J. de Haan and G.J. Cats, Meteorological input for a three dimensional medium range air quality model, Proc. 11th NATO/CCMA International Technical Meeting on Air Pollution Modeling and its Application, Amsterdam, 24–27 November (1980b) 64–72.

29 J. Wieringa and P.J.M. van der Veer, Internal KNMI Report No. V-278 (in Dutch), 1976.

30 J. Wieringa, Estimation of mesoscale and local roughness for atmospheric transport modeling, Proc. 11th NATO/CCMS International Technical Meeting on Air Pollution Modeling and its Application, Amsterdam, 24-27 November (1980a) 279-295.

31 R.A. Anthes and T.T. Warner, Development of hydrodynamic models suitable for air pollution and other mesometeorological studies, Mon. Wea. Rev, 106 (1978) 1045-1078.

32 M.P. Scheele, private communication.

33 H. Tennekes, A model for the dynamics of the inversion above a convective boundary layer, J. Atmos. Sci., 32 (1973) 992-995.

34 F.T.M. Nieuwstadt, The nocturnal boundary layer, theory and experiments (thesis), KNMI, De Bilt, 1981.

35 A.P. van Ulden and A.A.M. Holtslag, Estimation of Atmospheric Boundary Layer Parameters for Diffusion Applications, J. Climate Appl. Meteor., 22 (1985) 1196-1207.

36 H. van Dop, Terrain classification and derived meteorological parameters for interregional transport models, Atmospheric Environment, 17, 6 (1983) 1099-1105.

37 M.F. Wilson and A. Henderson Sellers, A global archive of land cover and soils data for use in general circulation models, J. Climatology, 5 (1985) 119-143.

38 A. Eliassen, O. Hov, I.S.A. Isaksen, J. Saltbones and F. Stordal, A Lagrangian long-range transport model with atmospheric boundary layer chemistry, Journal of Applied Meteorology, 21 (1982) 1645-1661.

39 The NCAR Regional Acid Deposition Model, NCAR Technical Note TN-256+STR, Boulder, Colorado, 1985.

40 Updated background information on the nuclear reactor accident in Chernobyl, USSR, WHO Report, Copenhagen (1986).

41 F.A.A.M. de Leeuw, R.M. van Aalst, H.J. van Rheineck-Leyssins, H. Kesseboom, N.D. van Egmond, M.P. Scheele, H. van Dop and A.P. van Ulden, The Chernobyl Accident: Reconstructing the radioactivity concentration and deposition patterns over Europe. Submitted to Science.

Regional and Long-range Transport of Air Pollution,
Lectures of a course held at the Joint Research Centre, Ispra, Italy,
15–19 September 1986, S. Sandroni (Ed.), pp. 381–389
© Elsevier Science Publishers B.V., Amsterdam — Printed in The Netherlands

STATISTICAL LONG RANGE TRANSPORT MODELS

W. KLUG

1. INTRODUCTION

For many applicational purposes long term averages of air pollutant concentrations and depositions are either required or of interest. Long term averages are considered to be seasonal or annual mean values of the mentioned quantities. One way of obtaining these long term values is, of course, to run a detailed transport and diffusion model hour after hour and compute from the hourly output data the required averages. It is obvious that this procedure requires large amounts of detailed input data on the one hand and excessive computer time on the other. The other method one can think of is to use a relatively simple model with averaged input data and mean values of parameters such as deposition velocity and others. The parameter values can be optimised by comparing the model results with observations and adjusting the parameters by using least-square-methods. Such a model is called a 'statistical' model, it is semi-empirical and its justification is the performance of the model in describing the observed values of concentrations and depositions. However, by this very nature this kind of model has - in spite of its usefulness and its possibility to give insight into the processes that govern observations - also drawbacks. The most important being that the parameters determined for one special situation of environmental conditions (e.g. concentrations of other pollutants) must not hold if applied to other such conditions. But, they are useful, easy to handle and to understand. In the following, two examples of this type of model are given.

2. THE WET-DRY-MODEL

F.B. Smith (1981) and nearly at the same time, but independently, Venkatram et al. (1982) have published similar work on statistical long-range transport modelling. Their work was based on the earlier work of Rodhe and Grandell (1972), but is was generalized and also applied.

The basic idea of this type of model is that the atmosphere can be subdivided at any moment in time in regions on a synoptic scale where there is no precipitation and others where precipitation occurs almost everywhere. This division seems to be very appropriate since most of the air pollutants behave differently whether precipitation is present or not. In order to simplify the terminology we call the first part of the atmosphere 'dry' and the second 'wet', without implying the original meaning of these words.

As we meteorologists know the synoptic features of the atmosphere are in constant motion, which means that the wet and dry areas are also moving and developing in time. Furthermore, a parcel of polluted air in the atmospheric boundary layer will become subject of these conditions and it is also possible - since the precipitation clouds move with air flows mainly outside the atmospheric boundary layer - that such a parcel can move from a dry region into a wet and vice versa. The model we are now constructing has the following features: We are considering sulfur in the atmosphere in two chemical states, one of which is SO_2 and the other SO_4. We assume that the polluted air has traveled far enough so that it is well mixed in the atmosphere boundary layer with height h. Furthermore, there are two removal processes working on SO_2 and SO_4, being wet and dry removal. The sulfur is released as SO_2 into the atmosphere and can be transformed during its transport into SO_4. The transformation rate is called α and has the dimension $[sec^{-1}]$. If we denote SO_2 with q and SO_4 with Q and the subscript D refers to dry regions, W to wet regions, then four equations can be established which describe the behaviour of the sulfur admixtures in a Lagrangian frame work

$$\frac{dq_D}{dt} = s_w q_w - s_D q_D - \frac{v_g}{h} q_D - \alpha_D q_D \qquad (1)$$

$$\frac{dq_w}{dt} = s_D q_D - s_w q_w - \frac{v_g}{h} q_w - \alpha_w q_w - A q_w \qquad (2)$$

$$\frac{dQ_D}{dt} = s_w Q_w - s_D Q_D + \alpha_D q_D \qquad (3)$$

$$\frac{dQ_w}{dt} = s_D Q_D - s_w Q_w + \alpha_w q_w - A Q_w \qquad (4)$$

In the above equations V_g stands for the dry deposition velocity of SO_2; the parameters S_W, S_D for the reciprocal characteristic time scales connected with leaving a wet region and entering a dry one or vice versa; the rate of wet removal is denoted by A and will be discussed later. Following Venkatram (1986) we can draw the following interaction diagram:

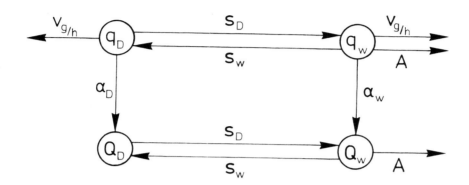

A few comments must be made on the wet removal coefficient and the characteristic time scales $(S_{D,W})^{-1}$. The rate of removal, A , of the pollutant is assumed to be linearly dependent on the rate of precipitation λ with a coefficient of proportionality Λ .

$$A = \Lambda \cdot \lambda \tag{5}$$

A has dimensions of $[sec^{-1}]$, $\Lambda [mm^{-1}]$ and λ therefore $[mm/sec]$. Λ is often called the washout coefficient.
If we denote

$$S_D = \frac{1}{\tau_D} \quad \text{and} \quad S_W = \frac{1}{\tau_W} \tag{6}$$

it is clear that τ_D and τ_W are the characteristic time scales for residence in the dry and wet states, respectively. Although these time scales refer to a Lagrangian frame of reference it was shown by Hamrud and Rodhe (1981) that there is little difference in the statistics of precipitation when Lagrangian and Eulerian systems are compared. Therefore, it was possible to estimate from pre-cipitation observations at fixed points the magnitude of τ_D and τ_W . Whereas τ_D is estimated to be in the order of 70 h, τ_W is only in the order of 7 h. V_g , the dry deposition velocity of SO_2, has a value of 0.8 cm sec^{-1} and h , the height of the mixed layer can be estimated to 1000 m. The rate of conversion of SO_2 to SO_4, α , is thought to be typically about 1% per hour or $\alpha = 2.8 \cdot 10^{-6}\ sec^{-1}$. In the above model A is assumed to be equal for SO_2 and SO_4, but it is of course possible to introduce an A_q and a different A_Q .

The equations (1) through (4) form a system of linear first-order simultaneous differential equations which can be solved by conventional methods if the parameters are constant. For details see the original paper by Smith (1981). If, however, the parameters vary along a trajectory numerical solutions of the equations have to be obtained.

Another solution is of interest when no differentiation is made between wet and dry areas, in other words the probability of precipitation is equally likely at all times. Then the above set of equations simplifies to two equations:

$$\frac{dq}{dt} = - \left(\frac{v_g}{h} + \alpha + A \right) q \tag{7}$$

$$\frac{dQ}{dt} = \alpha q - A Q \tag{8}$$

Two more comments are necessary. First, it should be noted that the equations (1) through (4) are written in a Lagrangian frame of reference, which means that we consider the rate of change of concentrations of an air parcel which is moved by atmospheric motions. We do not find therefore advection nor diffusion terms in these equations. The simplest approach to take care of advection and diffusion is to assume straight trajectories from the source to the receptor, where the horizontal dispersion H can be written as (Venkatram 1986)

$$H = \frac{f_\Theta}{2 \pi r u} \tag{9}$$

where f_θ is the relative frequency with which the wind blows from the source to the receptor, r is the source-receptor distance and u is the mean wind speed in this sector. Note that the vertical diffusion has been eliminated by assuming a well mixed boundary layer with a constant concentration in the vertical.

The second remark refers to the emissions. The source term has been omitted for simplicity in the above equations. It is obvious that the emissions have to be taken into account either by an initial concentration or by converting the emissions into additional concentrations en route.

Finally, it should be mentioned that Smith (personal communication) added to the equations (3) and (4) a dry deposition term for SO_4.

3. THE BOX-MODEL

Klug (1982) describes a simple Eulerian Box-Model which contains the same physical processes as in the Wet-Dry-Model but is formulated in a fixed coordinate system. It is assumed that the box has horizontal dimensions l (order of 50-150 km) and vertical dimensions h, the height of the mixed layer. The concentration q is uniform within the box. Advection is taken care of by an advected concentration q_0 on the upwind side of the box with mean wind u. The source is an uniform area source of strength S. The removal processes included are dry deposition, wet deposition and chemical (first order) transformation.

The rate equation of the concentration in the box reads as follows:

$$\frac{\delta q}{\delta t} = \frac{S}{h} - \frac{u}{l}\left(q - q_0\right) - \left(\frac{V_g}{h} + A + \alpha\right)q \tag{10}$$

It is interesting to study the non-dimensional form of eq. (10), which can give us some insight in the order of magnitude of the different terms:

$$\underset{\text{I}}{\frac{l}{u \cdot q}\frac{\delta q}{\delta t}} = \underset{\text{II}}{\frac{S \cdot l}{q \cdot u \cdot h}} - \underset{}{\left(1 - \frac{q_0}{q}\right)} - \underset{\text{III}}{\frac{l \cdot V_g}{h \cdot u}} - \underset{\text{IV}}{\frac{A \cdot l}{u}} - \underset{\text{V}}{\frac{\alpha l}{u}} \tag{11}$$

using typical values for SO_2 in an industrial area (Ruhr area), we put $l = 50$ km, averaged emissions over 50 km x 50 km amount to a value of $S = 5 \cdot 10^{-6}$ g SO_2/ (m^2 sec). Further typical values are $q = 50$ µg SO_2/m^3, $h = 1000$ m, $u = 5$ m sec^{-1}, $V_g = 0.01$ m sec^{-1}, $\alpha = 0.02/h$ and $A = 0.1/h$. The advection concentration q_0 is assumed to be $q_0 = 0.29$ q.

With these values inserted into equation (12) we obtain the following magnitudes of the different terms.

I = 1   II = 0.8   III = 0.1   IV = 0.5   V = 0.06

One can also note, that always I > 0, II $\lessgtr$ 0 depending on the advected concentration $q_0/q \gtrless 1$ and the last three terms are always negative because of their nature as removal terms.

We shall now rewrite equation (10) into the two equations necessary for dealing with SO_2 and SO_4.

$$\frac{\delta q}{\delta t} = \left(1 - \beta\right)\frac{S}{h} - \frac{u}{l}\left(q - q_0\right) - \left(\frac{V_{g2}}{h} + A_2 + \alpha\right)q \tag{12}$$

$$\frac{\delta Q}{\delta t} = \frac{3}{2} \left(\beta \frac{S}{h} + \alpha q \right) - \frac{u}{l} \left(Q - Q_o \right) - \left(\frac{V_{g4}}{h} + A_4 \right) Q \tag{13}$$

where β is the fraction of SO_2 emitted directly as sulphate, V_{g2} and V_{g4} the respective dry deposition velocities of SO_2 and SO_4 and A_2 and A_4 the wet deposition rates multiplied with the rain probability. The factor 1.5 in eq. (13) comes from the differences in molecular weight between SO_2 and SO_4.

When applying this model the wind field is described by one precipitation-wind rose over the whole of Europe with eight equal sectors. The differential equations (12) and (13) are integrated numerically for each sector separately until a steady state is reached.

Klug and Lüpkes (1985) compared the results of 10 different long term interregional air pollution models. The emission inventory was that published by EMEP and is based on the year 1978. The region for which SO_2 and SO_4 concentrations, wet and dry deposition were calculated covers middle Europe excluding a part of the Mediterranean Sea and Southern France. The meteorological input data used was a uniform precipitation-wind rose for the whole area and the values for $A = 10^{-4}$ sec^{-1}, $V_{g2} = 0.8$ cm sec^{-1}, $V_{g4} = 0.2$ cm sec^{-1}, $\alpha = 2.8 \cdot 10^{-6}$ sec^{-1}, $\beta = 0.05$ where chosen. A few results of this comparison are shown in Fig. 1-3, where the Smith- and Klug-model results are compared. It is obvious that the resulting concentration and deposition patterns are closely related, which is true for the results of nearly all models, which were compared. A detailed quantitative comparison will be given in the course "Model Evaluation procedures for Long Term Average Models".

4. REFERENCES

- Klug, W. (1982), Physical Transport or the Problem how to model Air Pollution, in 'Air Pollution by Nitrogen Oxides', Elsevier Scientific Publishing Company, Amsterdam, p. 243-248.

- Klug, W. and Lüpkes, C. (1985), A Comparison between Long Term Interregional Air Pollution Models, Final Report for Umweltbundesamt Berlin.

- Rodhe, H. and Grandell, J. (1972), On the Removal Time of Aerosol Particals from the Atmosphere by Precipitation, Scavenging, Tellus 24, p. 442-454.

- Smith, F.B. (1981), The Significance of Wet and Dry Synoptic Regions on Long-Range Transport of Pollution and its Depositions, Atmospheric Environment 15, p. 67-98.

- Venkatram, A., Ley, B.E. and Wong. S.Y. (1982), A Statistical Model to Estimate long-term Concentrations of Pollutants associated with long-range Transport, Atmospheric Environment 16, p. 249-257.

- Venkatram, A. (1986), Statistical long-range Transport Models, Atmospheric Environment 20, p. 1317-1324.

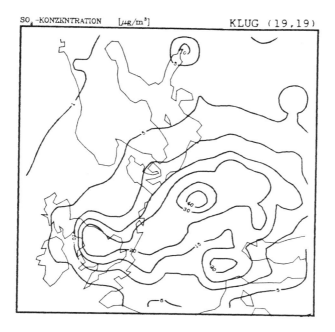

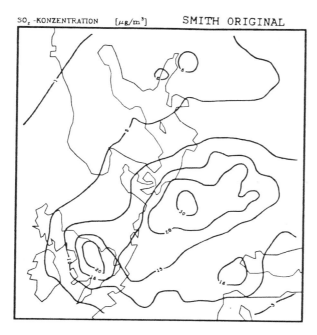

Fig. 1: SO$_2$ concentrations in units of SO$_2$

388

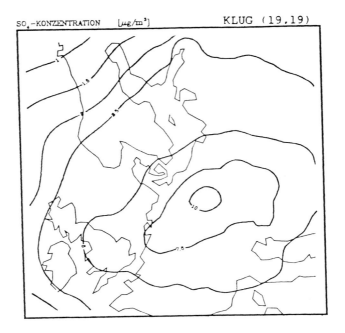

Fig. 2: SO_4 concentrations in units of SO_4

S-TOTALE DEPOSITION [g/m²/a]   KLUG (19,19)

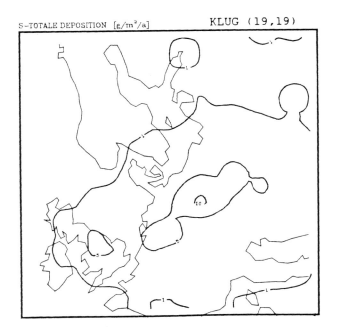

S-TOTALE DEPOSITION [g/m²/a]   SMITH ORIGINAL

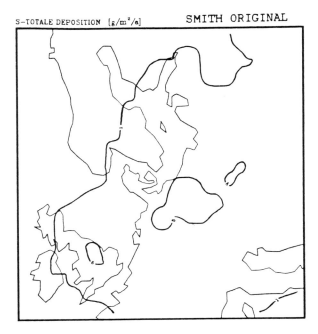

Fig. 3: total depositions of S

Regional and Long-range Transport of Air Pollution,
Lectures of a course held at the Joint Research Centre, Ispra, Italy,
15–19 September 1986, S. Sandroni (Ed.), pp. 391–412
© Elsevier Science Publishers B.V., Amsterdam — Printed in The Netherlands

MODELS FOR PHOTOCHEMICAL PROCESSES

Øystein Hov

1 INTRODUCTION

Ozone and other photochemical oxidants are natural constituents of the atmosphere. About 90% of the ozone in the atmosphere is found in the stratosphere. It absorbs most of the solar ultraviolet radiation before the troposphere is reached, where about 10% of the atmospheric content of ozone is found. The main sources for tropospheric ozone are influxes from the stratosphere and photochemical production involving nitrogen oxides, hydrocarbons and carbon monoxide from natural and anthropogenic sources. Surface removal is the main sink for tropospheric ozone.

Up to about 1970 it was thought that photochemical air pollution was concentrated mainly to some urban areas in the US, while the abundance of tropospheric ozone elsewhere was mainly a result of the combined action of processes driven by nature and not by man. In the fall of 1971, a photochemical pollution episode in Europe was reported from the Netherlands (TNO, 1971), and high concentrations of ozone in air coming from the European continent, were observed in Britain during June and July, 1971 (Atkins et al., 1972). The first high ozone concentrations in Scandinavia were measured on the west coast of Sweden in March 1972 and the source areas were thought to be North-Central Europe (Grennfelt, 1973).

Today it is well established that enhanced ozone concentrations in the troposphere can be found on many different scales in time and space.

In the atmospheric boundary layer, that is the lowest part of the troposphere where the atmosphere is under the direct influence of the ground through the exhange of heat, momentum, moisture and other gaseous and particulate material, enhanced ozone concentrations may be found

(i)   in the plumes from refineries, petrochemical industry and power
        plants typically 100 km away from main sources, or a few hours of
        transportation
(ii)   in plumes from urban areas typically 100 km away
(iii)   in air transported over long distances (1000 km or more, over
        several days) and in air masses found in stagnant anticyclones.

Furthermore, there is growing evidence that there has been an increase in the concentration of ozone in the middle troposphere over Europe over the past 15 years. The upward trend in Europe is mainly judged from a reliable record of ozone sonde observations over Hohenpeinßenberg in Bavaria (Attmannspacher et al., 1984). The ozone sonde records indicate a 50-70% increase in the concentration of ozone above the boundary layer from 1967-1982. Other evidence of increasing ozone over Europe is reviewed in a recent report on photooxidants-precurser relationships (Hov et al., 1986).

2 MATHEMATICAL FORMULATION OF AN AIR QUALITY MODEL

Mass conservation of each chemical species is a common requirement in air quality modelling. This can be written as

$$\frac{\partial c_i}{\partial t} + \nabla \cdot \vec{v} c_i = \nabla \cdot D_i \nabla c_i + R_i \tag{1}$$

where c_i denotes the concentration of species i, $\vec{v}$ is the wind field (stochastic variable), D_i molecular diffusion coefficient for species i, while R_i includes the effects of atmospheric transformation, removal and emission. The ensemble averaged atmospheric diffusion equation is usually applied:

$$\frac{\partial \langle c_i \rangle}{\partial t} + \nabla \cdot \vec{v} \langle c_i \rangle = \nabla \cdot (K \cdot \nabla \langle c_i \rangle) + R_i (\langle c_1 \rangle), \ \ldots, \ \langle c_p \rangle),$$
$$i = 1, \ \ldots, \ p \tag{2}$$

where $\langle c_i \rangle$ is the ensemble averaged concentration, $\vec{v}$ is the deterministic wind velocity and K the tensor introduced to approximate the turbulent transport term by the mean concentration gradient, p is the number of components (see e.g. Seinfeld, 1975).

In Table 1 is given an overview of the factors which make up R_i. Eq. (2) is often taken as the general form of the continuity equation for a chemical species. In practical applications, less general forms of the equation are used depending on the problem to be analysed, by reducing the dimensionality to two or one, by transformation into a trajectory model, etc. A review is published by Johnson, 1983. In practical applications, the mathematical formulation is a compromise based on many factors: The quality and detail of the meteorological data, emission fields, data describing chemical transformation, availability of computational and economical resources, and availability of air quality data against which the model can be compared.

TABLE 1
Overview of factors in the R_i-term in eq. (2).

Factor	Comments
Emissions	Anthropogenic or natural
Removal	Surface dry deposition, precipitation scavenging, aerosol interaction chemical reactions.
Transformation	Photochemical or dark reactions, aerosol reactions.

The numerical solution of eq. (2) or some siplified form of eq. (2) is not straight forward. The component of the equation system which describes the chemical transformation is in general non linear, and exhibits stiff properties. This means that there is a wide range of chemical decay times involved. Special techniques have to be applied to keep the computer cost down and obtain solutions with acceptable numerical accuracy. A so called quasi-steady state approximation technique has been developed by Hesstvedt et al., 1978, which cuts down the computer cost for the integration of kinetic equations, but which depends on other methods (e.g. Gear, 1971) for cross checks to assess the numerical accuracy.

Operator splitting techniques can be used to solve particular versions of eq. (2). It is then possible to apply special numerical procedures to solve the various parts of the equation, where the basic elements are those due to transport and those due to chemistry (see McRae et al., 1982, Hov, 1983).

3 FORMATION OF PHOTOCHEMICAL OXIDANTS

Photochemical oxidants, notably O_3 and PAN, are formed when a mixture of organic gases, nitrogen oxides and air is exposed to solar radiation.

NO_2, NO and O_3 take part in a cyclic set of reactions:

$$NO_2 + h\nu \xrightarrow{k_1} NO + O(^3P) \qquad (R1)$$

$$O(^3P) + O_2 + M \rightarrow O_3 + M \qquad (R2)$$

$$NO + O_3 \rightarrow NO_2 + O_2 \qquad (R3)$$

M denotes an air molecule. The dissociation rate coefficients are important in a photochemical model, since the solar energy drives the chemistry. In general,

$$k_i(z,t) = \int_{\lambda_i}^{\lambda_2} \emptyset_i(\lambda,z)\sigma_i(\lambda,z)F(t,\lambda,z)d\lambda \qquad (3)$$

where λ_1 and λ_2 denote the wavelength interval of absorption for species i, $\emptyset_i(\lambda,z)$ denotes the wavelength and height dependent quantum yield, $\sigma_i(\lambda,z)$ the wavelength and height dependent absorption crossection, and $F(t,\lambda,z)$ denotes actinic irrandiance (solar flux) which depends on the time of day, wavelength and height. Dissociation rate coefficients are usually calculated theoretically. Several parameterisation schemes are in use, e.g. Luther and Gelinas, 1976, Isaksen et al., 1976 and Schere and Demerjian, 1977. Absorption by O_3 and O_2, Rayleigh scattering, reflection by the surface or clouds, and absorption and scattering by aerosols affect the solar flux in the atmosphere.

Reaction (R3) followed by (R1) and (R2) does not affect the ozone concentration. NO may be converted to NO_2 without any loss of O_3, however:

$$HO_2 + NO \rightarrow NO_2 + OH \qquad (R4)$$
$$RO_2 + NO \rightarrow NO_2 + RO \qquad (R5)$$

which followed by (R1) and (R2), lead to the formation of O_3. RO_2 denotes an peroxyalkyl radical (e.g. CH_3O_2). Hydroperoxy and peroxyalkyl radicals arise from the oxidation of hydrocarbons, which occurs through reactions of the form

$$HC_i + OH \rightarrow products \qquad (R6)$$

This reaction is usually the rate determining step in a chain of reactions where the hydrocarbon (HC_i) eventually is decomposed into stable end products (CO, CO_2) or intermediate species are formed, like aldehydes. Several ractions of the types (R4) or (R5) may occur during the decomposition, depending on the number of carbon and hydrogen atoms in the original molecule.

Hydrocarbons containing double bonds (olefins) also react with ozone. This is an important point to bear in mind in air quality modelling covering several days, since O_3-hydrocarbon reactions dominate at night when hydroxyl vanishes.

The hydroxyl radical is assumed to drive the major part of the hydrocarbon degradation. It is important to keep track of its production and loss mechanisms. The major production in polluted air comes from the degradation of hydrocarbons or intermediate species like formaldehyde. In clean air the most important source of hydroxyl is through the reactions

$$O_3 + h\nu \rightarrow O(^1D) + O_2 \qquad \lambda < 3100A \qquad (R7)$$

followed by

$$O(^1D) + H_2O \rightarrow 2OH \qquad (R8)$$

Important loss of hydroxyl and hydroperoxy radicals takes place through

$$NO_2 + OH \rightarrow HNO_3 \qquad (R9)$$
$$SO_2 + OH \rightarrow HSO_3 \qquad (R10)$$
$$HO_2 + CH_3O_2 \rightarrow CH_3O_2H + O_2 \qquad (R11)$$
$$HO_2 + HO_2 \rightarrow H_2O_2 \qquad (R12)$$
$$HO_2 + NO_2 \rightarrow HO_2NO_2 \qquad (R13)$$

Photolysis of CH_3O_2H, H_2O_2 and HO_2NO_2 and thermal decomposition of HO_2NO_2 also lead to formation of OH or HO_2 in which case these species only serve as a temporary storage for radicals. The species are water soluble, however, and the reactions (R11) - (R13) followed by removal of the products in water droplets, serve as radical sinks.

TABLE 2
Ratios of paraffins, olefins and aromatics to acetylene on a ppbC basis as measured in clean air in the northern hemisphere (Penkett, 1982) and in the Lincoln Tunnel, New York (Killus and Whitten, 1983).

Ratio (on a ppbC basis)	Lincoln Tunnel	Clean air, northern hemisphere
paraffins/acetylene	6.8	7.4
olefins/acetylene	3.2	1.0*
aromatics/acetylene	3.9	1.6

* This ratio is 0.0 if only anthropogenic sources of olefins are considered.

3.1 Hydrocarbons in atmospheric photochemistry

The emissions of organic material into the atmosphere are very complex. The mass of natural emissions dominates on a global basis, but it is generally believed that natural organic emissions play a minor role in the formation of pollutants which affect the air quality close to the ground (Dimitriades, 1981). Attention will therefore be paid here to the chemical transformation of anthropogenic hydrocarbon emissions.

Vehicle traffic is a dominating source of many hydrocarbons. The hydrocarbon composition measured in road tunnels or close to busy roads, is representative of the composition of the emissions. The large spread in reactivity of the anthropogenic hydrocarbons give rise to a change in the concentration distribution with time or distance from the source. In Table 2 is shown the average ratios of groups of hydrocarbons to acetylene as measured in the Lincoln Tunnel in New York and in clean air in the remote troposphere far away from anthropogenic sources. The numbers can be viewed on the background of the reactivity distribution of anthropogenic hydrocarbons illustrated in Figure 1. Acetylene is a slowly reacting tracer of automobile exhaust.

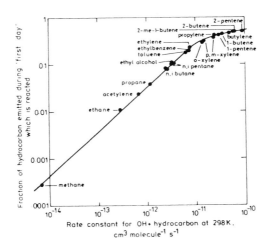

Fig. 1. The fraction of each hydrocarbon emitted into an Eulerian urban box model for London which was reacted in a day as a function of the OH reaction rate coefficient. The continuity equation (2) was simplified to

$$\frac{dc}{dt} = R - (c-c_b)(\frac{1}{h}\frac{dh}{dt} + \frac{v}{L})$$

where c_b is the background concentration, h the mixing height, v mean wind (2.4 m/s) and L horizontal dimension of the model box (50 km) (Derwent and Hov, 1979).

Table 2 shows that the relative importance of slowly reacting hydrocarbons (paraffins) increases with distance from the source, while the most reactive hydrocarbans (olefins) are almost depleted far away from the sources.

An exact description of the chemical degradation of the volatile organic species in the atmosphere is not possible because of the complexity of the reaction mechanisms involved and the large number of different organic species present. The chemical formulation must be simplified at the same time as the essential features of the system must be maintained. Sufficient detail is needed so that the model performance can be validated and the predictions interpreted in a useful way.

Several methods have been used to simplify the model formulation for hydrocarbon transformation. The classification of these models following of that of Leone and Seinfeld (1985) is illustrated in Fig. 2. The most straightforward procedure is the explicit representation of the chemical degradation of each individual hydrocarbon thought to play a significant role in photochemical ozone formation. To provide a realistic description a large number of species and reactions need to be included (e.g. Leone and Seinfeld, 12 primary organics, 215 reactions; Derwent and Hov (1979), 37 primary organics, 300 reactions).

In the surrogate mechanisms, (Dodge 1977, Atkinson et al., 1982) the organic species in a particular class, e.g. alkenes, are represented by one or more members in that class, with more or less parameterisation of the chemical pathways. Another method is to lump the hydrocarbons by classes according to a common basis such as molecular structure or reactivity class, and using a generalised mechanism for each class.

Two techniques for lumping reactivity class have been used. Molar weighted averaging has been used to obtain the rate constants for the generalised reactions, the contribution of each member of the group being determined by its initial concentration and the rate constants applicable to that species (Falls and Seinfeld, 1978, McRae and Seinfeld, 1983), a reactivity weighting technique has also been used in which the initial concentration of a representative species is adjusted so that its rate of reaction with OH, using an unmodified k_{OH}, is the sum of the k_{OH} [HC] terms for the individual species (Penner and Walton, 1982). Since the composition of the hydrocarbon in each class will change with time, the lumped reaction rate parameter will therefore change with time in a way that is difficult to follow in the model.

Lumping of molecular structure is used in the Carbon Bond Mechanism (CBM) of Killus and Whitten (1983), this is a generalised mechanism in which carbon atoms of similar bonding are treated similarly, regardless of the molecules in which they occur. In the latest version (CBM-X), 8 types of carbon atoms are treated, two each of carbonyl, olefin and aromatic species, alkyl carbon atoms and CO. The CBM approach allows a rather straightforward split of individual hydrocarbons into carbon bond classes, and the range of reaction rate parameters to be averaged in each bond is much narrower than for a lumped mechanism.

398

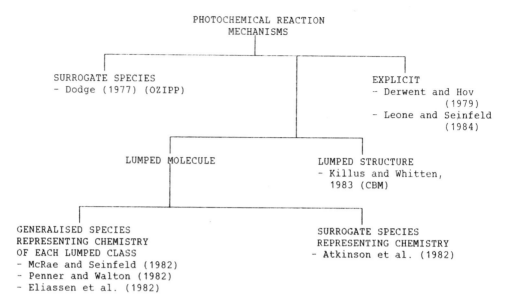

PHOTOCHEMICAL REACTION
MECHANISMS

SURROGATE SPECIES
- Dodge (1977) (OZIPP)

EXPLICIT
- Derwent and Hov
(1979)
- Leone and Seinfeld
(1984)

LUMPED MOLECULE

LUMPED STRUCTURE
- Killus and Whitten,
1983 (CBM)

GENERALISED SPECIES
REPRESENTING CHEMISTRY
OF EACH LUMPED CLASS
- McRae and Seinfeld (1982)
- Penner and Walton (1982)
- Eliassen et al. (1982)

SURROGATE SPECIES
REPRESENTING CHEMISTRY
- Atkinson et al. (1982)

Fig. 2: Classification of Photochemical Reaction Mechanisms for Ozone
Formation.

4 APPLICATION OF 7 DIFFERENT CHEMICAL MECHANISMS IN THE MODELLING OF REGIONAL
   OXIDANT FORMATION

It is of interest to consider the question - if different chemical mechanisms were introduced into a model for calculation of photochemical oxidants
on a regional scale (e.g. for Europe and covering an episode of 5 d), would the
results look very different if the meteorology, boundary and initial conditions, emissions and other parameters were kept the same? This question has
been investigated using a very simple model formulation:

$$\frac{dc}{dt} = \frac{F}{h} + p - Lc \qquad (4)$$

where c is the concentration, F emissions, h mixing height, p gas phase chemical production and Lc gas phase chemical loss. No ground removal or wet chemical interaction is allowed. The mixing height was assumed to be constant equal
to 1000 m. As a starting point was taken the average hydrocarbon emissions for
the UK as estimated by Derwent and Hov (1979) for 35 different hydrocarbon species. The description of the decomposition of these hydrocarbons involved about
160 different chemical species and more than 300 reactions. The model was integrated for 5 days with constant emissions in time, and the NOx and HC-emissions
were identical on a mass basis for all the chemical mechanisms applied. The
other mechanisms investigated were

(i)   The scheme published by Rodhe, Crutzen and Vanderpol (1981) involving 15 species and 19 reactions,

(ii)   The chemistry from the extended EMEP-model (somewhat modified from the version published by Eliassen et al., 1982) involving 40 species and about 60 reactions,

(iii)   The chemistry used in the publication by Sverdrup and Hov (1984) involving 79 species and about 160 reactions,

(iv)   An updated version of the chemistry published by Derwent and Hov (1979) with about 160 species and more than 300 reactions,

(v)   The mechanism published by Atkinson et al. (1982) stated to be suitable for incorporation into models of the urban environment (i.e. not tested for regional models) involving 54 species and about 80 chemical reactions,

(vi)   The CBM-III mechanism involving 36 species and 75 reactions (Killus and Whitten, 1983), and used primarily in urban applications,

(vii)   The CBM-X mechanism involving 65 species and 146 reactions.

The same reaction rate and photodissociation rate coefficients were used in all models for similar processes. In Figs. 3 and 4 the development of the concentration of OH and O_3 with time is shown for the different mechanisms. The results are similar on the first day, while later they diverge. The CBM-III mechanism is intended for use in urban smog conditions and not in multiday simulations. Mechanism 1 is seen to give results very different from the other ones. The differences in OH have important implications for the prediction of hydrocarbon decay rates and the conversion of SO_2 and NOx to acid gases and aerosols.

In Fig. 5 the development of O_3 is shown for cases when the same OH-profile was used in all mechanisms, equal to the EMEP-result (Fig. 3). The range of the results is considerably narrower, and this shows that the formulation of the OH-chemistry is very important in explaining the differences apparent in Fig. 4.

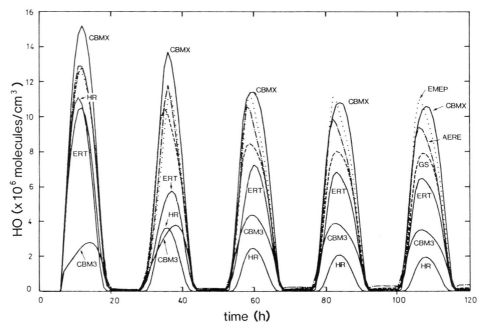

Fig. 3: Calculation of the concentration of hydroxyl with time for 7 different chemical mechanisms in a box model with constant emissions of NO_x and VOC, no ground removal or ventilation and a constant mixing height (1000 m).

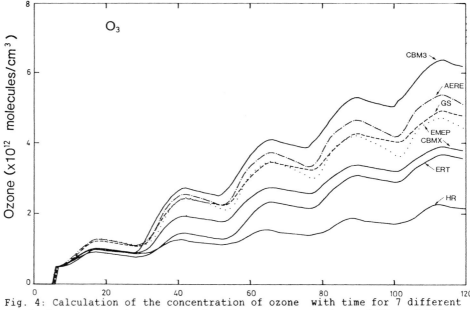

Fig. 4: Calculation of the concentration of ozone with time for 7 different chemical mechanisms in a box model with constant emissions of NO_x and VOC, no ground removal or ventilation and a constant mixing height (1000 m).

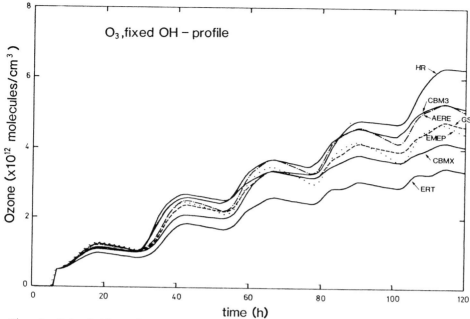

Fig. 5: Calculation of the concentration of ozone calculated with the same OH-profile equal to the "EMEP"-curve in all chemical mechanisms in a box model with constant emissions of NO_x and VOC, no ground removal or ventilation and a constant mixing height (1000 m).

5 MODELS FOR THE FORMATION OF PHOTOCHEMICAL OXIDANTS

The spatial and temporal scales of dispersion and transformation of NOx, HC, O_3 and other photochemical oxidants, are given in Table 3, and if the atmospheric fate of a pollutant is to be computed by a model, it is required that the temporal and spatial domains are at least as indicated there. If the generation of ozone is analysed in a model covering an area extending a few kilometers and over a few hours or less, the computational result will be strongly dependent on the boundary and initial conditions.

The chemical species which affect the air quality through their presence in the atmospheric boundary layer (ABL), are in general removed from the ABL on a time scale comparable to the typical time length between ABL break-down situations. This occurs on the average every 2-4 d at European latitudes (Smith and Carson, 1977).

TABLE 3

Characterization of dispersion and transformation of chemical species in
terms of spatial and temporal scales.

Species	Scales for dispersion and transformation		Comments
	Space	Time	
O_3	1000 km	5 d	
PAN	1000 km	5 d	NO_2 is, together with PAN, a dominating
NO_2	Some 100 km	2 d	oxidized nitrogen species in long-range transported air
HNO_3 (NO_3^-)	1000 km		
CO	10000 km	30 d	Significant anthropogenic influence on the tropospheric CO-budget
HCHO	100 km	1 d	
CH_4	Global	10 y	
NMHC	10-1000 km	1-100 h	Nonmethane hydrocarbons

The design of an air quality model is based on the problem to be analysed.
The calculation of street canyon concentrations of a primary pollutant like CO,
requires different chemical, physical and meteorological considerations compared to the modelling of the generation, transport and loss of ozone or
sulphate in the ABL.

5.1 Model for long-term changes in ozone in the lower troposphere

The mixing time within the hemisphere from the pole to the equator, or from
the ground to the tropopause, is typically a few months, while the interhemi-
spheric mixing time is 1-2 years. To analyse changes in the composition of the
lower troposphere over a time period of many years, a global model is required
since the changes of interest occur on a temporal scale which is longer than
the typical time for exchange within and even between the hemispheres. As an
approximation complete mixing in the zonal direction may be assumed, since the
time period required to obtain good zonal mixing is 2-3 weeks, reducing the
dimensionality of the model to two. Such a model has been developed at the
University of Oslo (Isaksen, 1979, Isaksen et al., 1985): A two-dimensional,
meridional model with 19 x 20 grid points, i.e. every 10 degrees latitude from
pole to pole, and every 250 m in the lower troposphere and 1 km grid distance
from 3 km through the upper troposphere to 15 km. A surrogate chemical
mechanism is used in the model, where the degradation of ethane, n-butane,
ethene, propene, m-xylene, methane and terpenes is described in an explicit
way. All hydrocarbon emissions are represented in terms of these seven species.

The continuity equation in two dimensions is solved for each species, using fields for the mean and diffusive motion derived from wind observations. The performance of the model has been evaluated by comparing the prediction of the global distribution of ozone and other trace gases with measurements. The model performance may also be evaluated through parallel computation of the distribution of an inert tracer with model where the representation of tropospheric transport is better (like GCMs). Using this global 2-d model of the troposphere, Isaksen (1979) concluded that the increase in anthropogenic emissions of hydrocarbons and nitrogen oxides may have given rise to a 30-40 per cent increase in the concentration of ozone in the lower troposphere at mid-latitudes in the northern hemisphere during the last 25 years. Such a model is suitable to find out how slow changes in the emissions of anthropogenic and natural emissions of NOx, HC including CH_4, SO_2 and CO may influence the composition of the troposphere (Isaksen and Hov, 1986).

5.2 Models for long-range transport of photochemical oxidants (regional episodes)

In Europe there is no consensus with respect to the time interval over which the ozone concentration should be averaged in order to get the best measure of the potential for environmental damage. In the U.S. it is required by law to control the maximum one hour average ozone concentration. There is therefore an emphasis in Northern America on developing complex 3-dimensional Eulerian models to calculate hourly ozone values on a regional scale. Several projects to develop models to simulate oxidant formation, transport and control strategies are in progress: Transport and deposition of acidifying pollutants (TADAP), Regional Oxidant Model (ROM) (Lamb and Novak, 1984) and the NCAR model (ADMP, 1985).

The PHOXA-project in Europe has the same general purpose as the U.S. model projects. Emphasis is put on calculating the distribution of photochemical oxidants in Europe during episodes of a few days' length (van Ham and Builtjes, 1985).

In Europe, the modelling of photochemical oxidants on a regional scale in the atmospheric boundary layer has also been carried out using the Lagrangian trajectory model developed within EMEP (European Monitoring and Evaluation Programme) and adding a rather comprehensive description of the photochemistry of photooxidants (Eliassen et al., 1982; Hov et al., 1985).

In a model of atmospheric transport and chemistry, the following aspects are involved (after Pasquill and Smith, 1983):

(i)   source characteristics: the mean rate of emission and the variation
         with time and in space, source height, buoyancy of the effluents;
(ii)   advection: determination of appropriate advecting winds;
(iii)   dispersion: lateral and vertical dispersion, influence of mixing
         height;
(iv)   composition changes: chemical transformation;
(v)   loss processes out of the mixing layer: description of such processes,
         ultimate return of material back into the boundary layer;
(vi)   dry deposition process: the loss of pollution by absorption, sedimen-
         tation and impaction to the underlying surface;
(vii)   wet deposition process: the removal of pollution by precipitation by
         uptake into cloud droplets or into precipitation falling through the
         sub-cloud mixing layer;
(viii)   the fate of the pollution once it has reached the surface.

Models may be subdivided as Eulerian or Lagrangian. In the first case, the
concentrations and deposition are estimated on a grid basis, while in the
Lagrangian case a succession of trajectories and the chemical development of
individual air parcels is followed in time.

If an Eulerian grid is chosen, an immediate problem is the one of undesired
computational dispersion associated with numerical integration of the advection
equation. This problem can be reduced by characterizing the mass distribution
within each grid element by its first and second moments and assume conservation of these moments or by using spectral or pseudospectral methods (for
review see Eliassen, 1980).

Eulerian models are useful when many parameters are specified on a grid
basis or when concentration fields in a grid are required.

Lagrangian models are in general simpler in concept and source-receptor
relationships are more easily arrived at than in Eulerian models. Vertical
resolution is difficult to describe in a Lagrangian model, as well as horizontal diffusion. Lagrangian models may be most appropriate for calculating
average concentrations over more than 12-24 h, in which case the swinging of
the trajectories due to changes in the synoptic situation often dominates over
the effect of horizontal diffusion.

PHOXA-model. The modelling area for the PHOXA-model is shown in Fig. 6. It
covers most of the heavily industrialized areas in Europe north of the Alps.
The grid cells have a dimension of approximately 30x30 km^2 depending on the
location. An emission database has been set up for SO_2, CO, NOx, NH_3 and hydrocarbons split into reactivity classes, and with information about point sources
and area sources. A condensed version of the CBM-X mechanism is used to
describe chemical transformation. Wind data and mixing heights are prepared by

The Royal Dutch Meteorological Institute. Further details about the PHOXA-model are given by van Ham and Builtjes (1985) and Builtjes (1985).

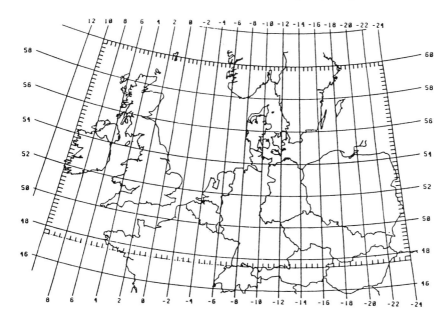

Fig. 6. The PHOXA modelling area.

TADAP-model. The TADAP (Transport and Deposition of Acidifying Pollutants) model is a highly complex episodic acid deposition model developed under the sponsorship of the Ontario Ministry of the Environment, the Canadian Atmospheric Environment Service and the German Umweltbundesamt. TADAP is developed under a contract with Environmental Research and Technology in the U.S. It consists of a number of modules simulating transport, diffusion, chemical transformation in the gas phase and in the aqueous phase, cloud physics and wet and dry deposition.

NCAR-model. The NCAR acid deposition modelling project (ADMP) is funded by EPA, the National Science Foundation (NSF), the National Research Council (NRC) and NCAR. The objective is to develop an Eulerian regional acid deposition model suitable for assessing source-receptor relationships in North America. The initial task is to construct a frame including a mesoscale meteorological model and a new model for transport and chemical transformation (ADMP, 1985). There are efforts being done to prepare for a transfer of the model to Europe in the frame of Eurotrac.

The extended EMEP-model with chemistry. In Europe, the modelling of photochemical air pollution on a regional scale in the atmospheric boundary layer
has been carried out using the Lagrangian trajectory model developed within
EMEP and adding a rather comprehensive description of the photochemical generation of oxidants. The grid is shown in Fig. 7. The grid element size is 150x150
km^2 at $60^0 N$ latitude.

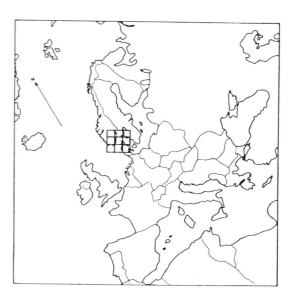

Fig. 7. The EMEP modelling area.

The chemical mechanism is a surrogate mechanism where all anthropogenic
emissions of hydrocarbons are represented by ethane, n-butane, ethene, propene,
m-xylene and formaldehyde. The photodissociation coefficients are calculated
taking into account the degree of cloud cover as judged from information about
relative humidity. Approximately 85 per cent of the computer power is used to
run the chemical part of the model. A number of episodes where high ozone concentrations have been recorded in Southern Scandinavia have been interpreted
using this model (Eliassen et al., 1982; Hov et al., 1985).

Other long-range transport models. Carmichael et al. (1986) developed a 3-d
grid model for regional-scale transport, chemistry and deposition of acidifying
compounds, photochemical oxidants and their precursors. There is also a
description of the exchange of air between the atmospheric boundary layer and
the free troposphere, and in-cloud and below-cloud wet removal and chemistry. A
schematic diagram of the chemical processes treated in the model is shown in

Fig. 8. A very comprehensive oxidant model is developed at EPA for the north-eastern part of the U.S., the regional oxidant model (ROM), Lamb and Novak (1984). This is a 3-d Eulerian model with a fine grid of less than 20x20 km^2.

5.3 Comparing long-range transport models

The chemical formulations in the TADAP, NCAR and PHOXA-models have been developed along similar lines: From what is believed to be a very complete and comprehensive description of the photochemical development in polluted air, a condensed mechanism is deviced to be applied in the model. These chemical schemes generally are of the same complexity as the one used in the EMEP-model: Typically 100 reactions and 40-60 species, and for the hydrocarbon emission inventory, information is required about how to split the emissions into 5-10 different species.

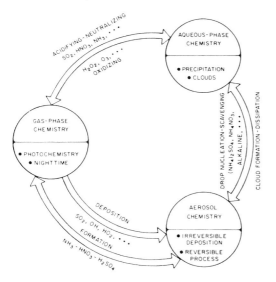

Fig. 8. Schematic representation of the chemical processes treated in the model of Charmichael et al. (1986).

The question of averaging time for the computed concentrations is an important one. The comprehensive Eulerian models with high resolution grid systems (about 30 km x 30 km in PHOXA, 18 km x 18 km in the ROM-model) are designed to calculate hourly averaged ozone concentrations. The calculated concentration fields will always be the product of a large body of information which sometimes may be accurate on a 18 or 30 km scale and on an hourly basis, but which in many cases at best is fairly accurate when averaged over longer times and lengths. Many of the input data are uncertain and represent averages over long time periods and for areas larger than one grid cell, which makes it impossible to claim that "true" hourly ozone concentrations can be calculated in a 18 or 30 km grid.

If a trajectory model with similar chemistry, emissions, and meteorology as used in PHOXA is developed and run in parallel with the 3-dimensional PHOXA model-calculation, insight can be gained in the success and shortcomings of the trajectory model compared to the Eulerian model. A trajectory model can be run at lower cost than a 3-d Eulerian model, and the comparison with the results from a more complex model will help when judging the value of the longer time-series of results from the simpler model. This is a valid statement also when it is taken into account that there may be significant errors in our understanding of emissions and the description of meteorological and chemical processes, and when it is remembered that ambient air measurements for model validation are scarce.

The line of simplicity starting with a complex PHOXA-type Eulerian model can go via a trajectory model using the same grid, chemistry and meteorology as the Eulerian model to the extent that it is applicable (usually no vertical resolution in a trajectory model, horizontal diffusion cannot be described).
Further simplification can be introduced by applying a coarser grid, as in the extended EMEP-model. It is difficult to state a priori that this simplification will give less acceptable results than a trajectory model on a finer grid. The reason for this is that some representative transport height has to be defined, and averaging the emissions over fairly large grid cells reduces the chances of the very large errors in the description of emissions along a given trajectory. Also, averaging the emissions over large grid cells is an indirect way of parameterising horizontal diffusion in a trajectory model.

A rather long step down the line of simplification of a PHOXA-type model is taken when replacing the chemistry with a reaction very similar to the first order decay parameterisation found in most models for the calculation of acid deposition on a monthly or annual basis:

$$HC + OH \rightarrow p\ Ox + RCHO + q\ OH$$

where hydrocarbons are represented by a single lumped molecule HC, OH is taken to have a fixed concentration (e.g. 1×10^6 molecules/cm^3), p and q are some constants to be determined once and for all, Ox oxidants (O_3), RCHO aldehyde. The particular form of the condensed reaction for oxidant formation has been suggested by van Aalst (1985). There is no reason to believe that this will work for the formation of oxidants as well as it does for sulphur and possibly also for the formation of NO_2 and nitrate on a regional scale averaged over long time periods (one month or more). Hydrocarbon decomposition in the atmosphere involves processes covering a wide time span (reactivity factor), the products are different for different HCs (stoichiometry factor), and some HCs are more common than others (inventory factor). For sulphur and nitrogen, the time determining step after all only involves SO_2 and NO_2. The formation of O_3 and other photochemical oxidants is also quite nonlinear. The rate of ozone formation diminishes as the ozone concentration increases, and the differential equations with time which arise from a set of reactions describing photochemical smog, are in general highly nonlinear. Using the simple reaction above with a prescribed OH-concentration, the nonlinearity would be overlooked.

There may also be a meteorological reason why long term average concentrations (e.g. monthly means) of ozone calculated in this way may be much less reliable than the mean concentration obtained from a PHOXA-type model or a trajectory model where calculations actually have been carried out using a chemical formulation where the time development day after day of the chemical species such as OH, O_3, H_2O_2 and others is calculated on the basis of information about emissions, solar radiation and other meteorological and physical processes. Oxidant episodes are often associated with high temperatures, strong insolation and light winds and slow subsidence (persistent inversions). The meteorological circulations that give rise to such conditions can vary significantly from year to year, because the occurrence of the "extreme" weather conditions which favour photochemical oxidant episodes in the ABL, is extremely sensitive to small circulation changes. This is in some contrast to the formation of acidic substances which is more related to the "normal" weather pattern of migrating cyclones.

It is interesting to note that there is a nonlinear relationship between changes in mean temperature and the corresponding change in the probability of extreme temperatures, with small changes in mean temperature sometimes resulting in relatively large changes in event probabilities (Mearns et al., 1984).

A different approach to simplify the chemical description of the formation of photochemical oxidants was taken by Derwent and Hov (1980). For each hydrocarbon, the inventory, reactivity and stoichiometry factors were established and used to define its potential contribution to the formation of ozone, PAN and other photochemical oxidants. This leads to a much simpler set of differential equations than the systems resulting from the complete set of chemical reactions. This model approach has later been used in the UK to estimate the potential for formation of ozone and other photochemical oxidants from the emissions in any one country and transported to any other European country (Derwent, private communication, 1985).

410

REFERENCES

1 ADMP, The NCAR Eulerian regional acid deposition model. Acid deposition modeling project. NCAR/TN-256, NCAR Boulder, CO, 1985.
2 D.H.F. Atkins, R.A. Cox and A.E.J. Eggleton, Photochemical ozone and sulfuric acid aerosol formation in the atmosphere over southern England, Nature, 235 (1972) 372-376.
3 R. Atkinson, A. Lloyd and L. Winges, An updated chemical mechanism for hydrocarbons/NOx/SO$_2$ photo oxidation suitable for inclusion in atmospheric simulation models, Atmos. Environ., 16 (1982) 1341-1355.
4 W. Attmannspacher, R. Hartmannsgruber and P. Lang, Langzeittendenzen des Ozons der Atmosphäre aufgrund der 1967 begonnenen Ozonmeßreihen am Meteorologischen Observatorium Hohenpeißenberg. Meteorol. Rdsch., 37 (1984) 193-199.
5 R.M. van Aalst, A method to include oxidants in a NOx long range transport model. Paper presented at the Technical meeting on Atmospheric computations for Assessment of Acidification in Europe, Warsaw, Poland, September 4-5, 1985.
6 P.J.H. Builtjes, The PHOXA-project, photochemical oxidants and acid deposition model application. Proceedings of COST 611-workshop at RIVM, Bilthoven, The Netherlands, 23-25 September, 1985.
7 G.R. Carmichael, L.K. Peters and T. Kitada, A second generation model for regional-scale transport, chemistry, deposition. Atmos. Environ., 20 (1986) 173-188.
8 R.G. Derwent and Ø. Hov, A simplified numerical method for estimating the potential for photochemical air pollution formation in the United Kingdom. AERE R-9682. HMSO, London, 1980.
9 R.G. Derwent and Ø. Hov, Computer modelling studies of photochemical air pollution formation in North West Europe. AERE R-9434, HMSO, London, 1979.
10 B. Dimitriades, The role of natural organics in photochemical air pollution. Issues and research needs. JAPCA 31 (1981) 229-235.
11 M.C. Dodge, Combined use of modelling techniques and Smog Chamber Data to derive ozone - precursor relationships, US Environmental Protection Agency Report, PA-600/3-77-001a (1977) 881-889.
12 A. Eliassen, A review of long-range transport modeling, J. Appl. Met., 19 (1980) 231-240.
13 A. Eliassen, Ø. Hov, I.S.A. Isaksen, J. Saltbones and F. Stordal, A lagrangian long-range transport model with atmospheric boundary layer chemistry, J. Appl. Met., 21 (1982) 1645-1661.
14 A.H. Falls and J.H. Seinfeld, Continued development of a kinetic mechanism for photochemical smog, Envir. Sci. Technol., 12 (1978) 1398-1406.
15 C.W. Gear, The automatic integration of ordinary differential equations, Comm. A.C.M., 14 (1971) 176-179.
16 P. Grennfelt, Measurement of ozone in Gothenburg January 1972-August 1973 and studies of co-variations between ozone and other air pollutants, IVL-Report B 221 (Swedish Environmental Research Institute, Gothenburg, Sweden), 1973.
17 J. van Ham and P.J.H. Builtjes, Study on photochemical oxidants and precursors. Phase IV. Long-range transport phenomena. Draft final report, part II, TNO, Apeldoorn, The Netherlands, 1985.
18 E. Hesstvedt, Ø. Hov and I.S.A. Isaksen, Quasi-steady state approximation in air pollution modelling: Comparison of two numerical schemes for oxidant prediction. Int. J. Chem. Kinet., 10 (1978) 971-994.
19 Ø. Hov, Numerical solution of a simplified form of the diffusion equation for chemically reactive atmospheric species. Atmos. Environ. 17 (1983) 551-562.
20 Ø. Hov, F. Stordal and A. Eliassen, Photochemical oxidant control strategies in Europe: A 19 days' case study using a Lagrangian trajectory model. NILU TR 5/85, Box 130, N-2001 Lillestrøm, Norway, 1985.

21 Ø. Hov, K.H. Becker, P. Builtjes, R.A. Cox and D. Kley, COST-611 task force on the photooxidants-precursor relationship in Europe. Report to be published by CEC, Brussels, 1986.

22 I.S.A. Isaksen, Transport and distribution of pollutants in the troposphere. Proc. WMO No. 538, Geneva, 1979, pp. 347-358.

23 I.S.A. Isaksen, Ø. Hov, S.A. Penkett and A. Semb, Model analysis of the measured concentration of organic gases in the Norwegian Arctic. J. Atm. Chem., 3 (1985) 3-27.

24 I.S.A. Isaksen and Ø. Hov, Calculations of trends in the tropospheric concentration of O_3, OH, CO, CH_4 and NOx. Tellus, in the press, 1986.

25 I.S.A. Isaksen, K.H. Midtbø, J. Sunde and P.J. Crutzen, A simplified method to include molecular scattering and reflection in calculation of photon fluxes and photodissociation rates. Geophysica Norvegica, 31 (1976) 11-26.

26 W.B. Johnson, Interregional exchanges of air pollution: Model types and applications. JAPCA 33 (1983) 563-574.

27 J.P. Killus and G.Z. Whitten, A new carbon bond mechanism for air quality monitoring, US Environmental Protection Agency Report No. EPA-600/3/-83-041, 1983.

28 R.G. Lamb and J.H. Novak, U.S. EPA Regional oxidant model for the transport of photochemical oxidants and their precursors. Proc. EPA-600/9-84-006, 1984.

29 P.A. Leighton, Photochemistry of air pollution, Academic Press, New York, 1961.

30 J.A. Leone and J.H. Seinfeld, Comparative analysis of chemical reaction mechanisms for photochemical smog, Atmos. Environ., 19 (1985) 437-465.

31 F.M. Luther and R.J. Gelinas, Effects of molecular multiple scattering and surface albedo on atmospheric photodissociation rates, J. Geophys. Res., 81 (1976) 1125-1132.

32 G.J. McRae, W.R. Goodin and J.H. Seinfeld, Development of a second-generation mathematical model for urban air pollution - I. Model formulation. Atmos. Environ., 16 (1982) 679-696.

33 G.J. McRae and J.H. Seinfeld, Development of a second generation mathematical model for urban pollution - II Model performance and evaluation, Atmos. Environ., 17 (1983) 501-523.

34 L.O. Mearns, R.W. Katz and S.H. Schneider, Extreme high-temperature events: Changes in their probabilities with changes in mean temperature. J. Clim. Appl. Met., 23 (1984) 1601-1613.

35 F. Pasquill and F.B. Smith, Atmospheric diffusion, John Wiley & Sons, New York, 1983, 437 pp.

36 S.A. Penkett, Non-methane organics in the remote troposphere, in E.D. Goldberg, (ed.) Atmospheric chemistry, Dahlem Konferenzen 1982. Springer Verlag, Berlin, pp. 329-355.

37 J.E. Penner and J.J. Walton, Air Quality Model Update, Lawrence Livermore Laboratory UCID - 19300. Lawrence Livermore National Laboratory, University of California, Livermore, California, 1982.

38 J.T. Peterson, Calculated actinic fluxes (290-700 mm) for air pollution photochemistry application. U.S. Environmental Protection Agency Report EPA-600/4-76-025, 1976.

39 PHOXA, Photochemical oxidant and acid deposition model application "(PHOXA)", UBA, FRG and Ministry of Housing, Physical Planning and Environment, the Netherlands, 1984.

40 H. Rodhe, P.J. Crutzen and A. Vanderpol, Formation of sulfuric and nitric acid in the atmosphere during long-range transport, Tellus, 33 (1981) 132-141.

41 K.L. Schere and K.L. Demerjian, Calculation of selected photolytic rate constants over a diurnal range. U.S. Environmental Protection Agency Report EPA-600/4-77-015, 1977.

42 J.H. Seinfeld, Air Pollution: Physical and chemical Fundamentals, McGraw-Hill, Inc., New York, 1975, 523 pp.

43 F.B. Smith and D.J. Carson, Some thoughts on the specification of the boundary layer relevant to numerical modelling, Boundary Layer Met., 12 (1977) 307-330.

44 G.M. Sverdrup and Ø. Hov, Modelling study of the potential importance of heterogeneous surface reactions for NOx transformations in plumes, Atmos. Environ., 18 (1984) 2753-2760.
45 TNO, Analysis of the smog situation in Holland, Report G500 (Institute of Public Health Engineering TNO, Delft, Netherlands, in Dutch), 1971.

Regional and Long-range Transport of Air Pollution,
Lectures of a course held at the Joint Research Centre, Ispra, Italy,
15–19 September 1986, S. Sandroni (Ed.), pp. 413–435
© Elsevier Science Publishers B.V., Amsterdam — Printed in The Netherlands

MODELS FOR DEPOSITION PROCESSES

H. van Dop

1 INTRODUCTION

High concentrations in air are no longer the most imminent danger of
releases of waste in the atmosphere. On the one hand the attention has shifted
towards the gradual but steady increase of atmospheric trace gases such as
ozone, methane and carbondioxide (ref. 1, 2), and on the other hand to the
deposition of the major air pollutants such as sulphur dioxide, nitrogen
oxides, sulphates, nitrates and volatile organic compounds. The first signs
that the continuous deposition of acidic material could have adverse effects on
the ecosystem came from the Scandinavian countries in the early seventies
(ref. 3). There, the fish die-back was ascribed to a gradual increase of the
acidity of many lakes, a direct consequence of deposition of sulphate and
nitrate in a weakly buffered aquaous environment. The matter got more and
world-wide attention when a few years later Canadians reported similar
phenomena in their lakes (ref. 4) and when from the European continent the
first reports occured on the deplorable state of the forests (ref. 5, 6),
allegedly a consequence of acid deposition.
Since then an enormous research effort has been undertaken to understand and
eventually to describe quantitavely the chain of processes leading from
emissions to deposition and effects on terrestrial and aquaous ecosystems.

Important tools in this evaluation are atmospheric transport and deposition
models. These models are in fact drastic mathematical simplifications of an
enormous amount of physical and chemical processes going on in the lowest
layers of the atmosphere, the atmospheric boundary layer (ABL). In a previous
chapter mathematical modelling was considered with the emphasis on atmospheric
transport. Here, we will emphasize how the deposition is treated in these
models.
First, dry deposition of gases and particles is discussed. Gravitational
settling occurs mainly in the vicinity of sources and is relatively unimportant
in mesoscale and long range transport modelling. Therefore, this process will
be neglected. Also detailed physical and chemical mechanisms underlying dry
deposition will not be treated here: the reader is referred to Drs. Beilke's

and Fowlers' contributions. Only that background information is presented which is needed to understand the mathematical formulation in ATDM's. Some emphasis will be given to currently used values of deposition velocities for various surfaces and chemical compounds.

In the third and final section the wet deposition processes will be considered. By definition these processes comprise the total surface flux of pollutants in the aquaous phase (rain or fog droplets). A full description of these processes does not exist as yet and many details are still uncertain, if not unknown. Here, only some of the main ideas will be presented. Moreover, wet deposition can only be included in transport models in a lumped way, since a complete description of the process would surpass the present computer capacity. The section will be concluded with some remarks on precipitation analysis.

2. DRY DEPOSITION

2.1 Introduction

We summarise the main results of the theory for dry deposition. The equation of conservation of a contaminant is given by

$$\frac{\partial \overline{\chi}}{\partial t} + U \frac{\partial \overline{\chi}}{\partial x} + V \frac{\partial \overline{\chi}}{\partial y} + W \frac{\partial \overline{\chi}}{\partial z} = - \frac{\partial}{\partial x} (\overline{u'\chi'}) - \frac{\partial}{\partial y} (\overline{v'\chi'}) - \frac{\partial}{\partial z} (\overline{w'\chi'})$$

$$+ D \left\{ \frac{\partial^2 \overline{\chi}}{\partial x^2} + \frac{\partial^2 \overline{\chi}}{\partial y^2} + \frac{\partial^2 \overline{\chi}}{\partial z^2} \right\} + \overline{S} \qquad (2.1)$$

(cf. Eq. 2.3 in chapter on long range episodic transport modelling). The molecular diffusivity term is retained here. We make the following assumptions:
- horizontal homogeneity (note that this also implies W=0)
- steady state
- absence of sources and (chemical) sinks.

Then eqn. (2.1) reduces to

$$0 = - \frac{\partial}{\partial z} (\overline{w'\chi'}) + D \frac{\partial^2 \overline{\chi}}{\partial z^2} . \qquad (2.2)$$

When eqn. (2.2) is integrated from o-z, bearing in mind that at height z in the surface layer the turbulent transport largely exceeds the molecular transport and that at the surface turbulence vanishes, we obtain

$$\overline{w'\chi'} = - D \left(\frac{\partial \overline{\chi}}{\partial z}\right)_o \equiv F , \qquad (2.3)$$

which shows that for a particular surface and compound the turbulent flux, F, in the surface layer is constant. This and the observation that the mean concentration varies with height has led, in analogy with Ohm's law, to the

concept of a deposition velocity, V_d: the difference between concentration
values at different heights $\overline{\Delta\chi}$ equals the product of the surface resistance
r_d, ($\equiv V_d^{-1}$) and the (constant) flux F: or

$$\overline{\Delta\chi} = r_d * F .$$ (2.4)

Usually one chooses the difference between an arbitraty level z (in the surface
layer) and the surface. When it is assumed that at the surface $\overline{\chi}=0$ we obtain

$$F = V_d \overline{\chi} .$$ (2.5)

We have repeated this analysis here to indicate the rather restrictive nature
of eqn. (2.5). Nevertheless it is the starting point of most formulations of
dry deposition in numerical modelling.

2.2 The deposition velocity of gases

When the implication that $\overline{\chi}=0$ at the surface is not satisfied (e.g. for
water vapour) eqn. (2.5) should be replaced by

$$F = V_d (\overline{\chi} - \overline{\chi}_o) ,$$ (2.6)

(ref. 7), an expression which is only useful when $\overline{\chi}_o$ can be determined.

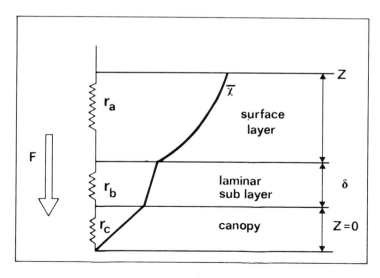

Fig. 1. Surface deposition velocity (or resistance) concept. F. denotes the
constant downward flux. The thickness of the laminar sublayer is δ.

However, for most air pollution components it can be assumed that $\overline{\chi_o}=0$.
The analogy with Ohm's law can be extended by assuming that the total surface
resistance consists of a series of separate resistances, the aerodynamic
resistance, r_a, the laminar resistance, r_b, and the foliar or canopy resistance
r_c, so that

$$r_d \equiv 1/V_d = r_a + r_b + r_c ,\qquad\qquad (2.7a)$$

or,

$$V_d = \frac{1}{r_a + r_b + r_c} .\qquad\qquad (2.7b)$$

The resistances are schematically shown in Fig. 1. The aerodynamic resistance
is determined by the turbulent transport properties. In the surface layer these
can be expressed in the Obukhov length, L, the friction velocity u_* and the
surface roughness z_o. Thus

$$r_a = [\ln(z/z_o) - \psi_h(z/L)] / ku_* ,\qquad\qquad (2.8)$$

where k = 0.4, the Von Karman constant.
The stability function ψ_h, corresponding with the turbulent transport of heat
(ref. 8) is also assumed to be representative for the transport of a passive
contaminant. It is given in a suitable approximation by

$$\psi_h = - 5\ z/L \qquad\qquad\qquad z/L > 0 \quad \text{and}$$

$$\psi_h = \exp \{0.598 + 0.390 \ln (-z/L) - 0.09 [\ln (-z/L)]^2\} \qquad z/L < 0 \qquad (2.9)$$

(ref. 8).

The laminar resistance can be written as

$$r_b = (a\ Re_*^{\ m}\ Sc^n + b) / u_* \qquad\qquad (2.10)$$

(ref. 7, 10), where a($\approx$0.17), b(=3), m(=0.5) and n(=2/3) are constants and Re_*
is the roughness Reynolds number, $u\ z_o/\nu$, and Sc = ν/D the Schmidt number. (The
Stanton number is defined as B $\equiv (u_*\ r_b)^{-1}$).

Simple parameterisations of (2.10) are

$$r_b = \frac{2}{k} (\frac{\kappa}{D})^{2/3} / u_* \qquad\qquad (2.11)$$

(ref. 9), where κ is the thermal diffusivity of air, or

$$r_b = \frac{2.6}{k} / u_* \, , \qquad\qquad (2.12)$$

(ref. 11, 12).

The surface resistance r_s depends on the state and nature of the vegetation and on the chemical properties of the contaminant. Its value should be determined from experiments. For sulphur dioxide some values are listed in table I. Similar measurements for other gases are scarce. One may, however, obtain r_s-values for other gases by scaling them to the values in table I according to their reactivity and solubility relative to SO_2. For instance reasonable values for the surface resistance of NO_2, O_3 and NO are respectively 250, 250 and 2000 s m^{-1}. The surface resistance depends strongly on the surface moisture. This is for instance parameterised by Onderdelinden et al. (ref. 13) as

$$r_s = r_{sd} \, (0.12 + 0.29 \, \exp[-t_w/6] + 0.59 \, (1 - \exp[-t_d/6])) \, , \qquad (2.13)$$

TABLE I

SO_2 surface resistances and surface roughness - mid summer

land type	surface resistance (s/m) insolation (W/m²)			surface roughness (cm)
	>400	200-400	0-200	
urban	1000	1000	1000	1000	100
agriculture	70	120	200	500	25
range	100	140	200	500	5
decidious forest	60	130	300	1000	100
coniferous forest	150	240	400	1000	100
forest/swamp mix	70	140	300	1000	100
water	0	0	0	0	*
swamp	50	60	75	100	15
mix ag/range	100	140	200	500	10

* Water surface roughness calculated from friction velocity (adapted from ref. 9).

where r_{sd} is the "dry" resistance and t_w and t_d are the number of "dry" and "wet" hours in the preceding period.

An alternative to the procedure outlined above is to rely completely on (atmospheric) measurements of the deposition velocity with the assumptions that most data pertain to a height of ~1 m. These values include the canopy, laminar and turbulent resistance up to the measurement height. The deposition velocity at other levels can then easily be obtained by applying (2.7) at z and a reference height and eliminating r_b+r_c:

$$V_d(z) = [1/V_g + r_a(z) - r_a(z_r)]^{-1} , \qquad\qquad ((2.14)$$

where V_g is the empirical deposition velocity (ref. 14). Some values of V_g are listed in table II.

TABLE II
Deposition velocity, V_g [cm s^{-1}] for SO_2 and various terrain categories.

terrain category	deposition velocity
water surface	1.1
open field (range)	0.7
open field with scattered trees and hedges	1.0
forest	1.5
built up areas	1.5

2.3 The deposition velocity of particles

The mechanisms which are responsible for the deposition of particulate material are different from those of gases. We consider only deposition of the smaller particles (smaller than ~ 20 μm), since gravitational settling is not very important in long range transport modelling. In the laminar layer the diffusivity of particles is governed by Brownian motion and therefore decreases with increasing particle size. The resistance of this layer for particles in the range 0.05-1 μm is high. When particles smaller than 0.2 μm touch the surface their small momentum is overcome by molecular forces and they tend to stick to the surface, resulting in a low value for r_c. For particles in the range 0.2-2 μm the distinction between the laminar and canopy layer does not appear to be very useful. Nevertheless the deposition velocity concept can still be applied. In Fig. 2 a compilation is given of deposition velocities of aerosols over a vegetation surface, as a function of particle size. Based on the above considerations a simple parameterisation was derived (ref. 11),

$$V_d = \frac{1}{r_a + r_s} \, ,$$

where the laminar and canopy resistance are combined in r_s. Based on observations r_s can be evaluated from

$$r_s = 500 \ u_*^{-1} \qquad\qquad\qquad L > 0$$

$$= 500 \ u_*^{-1} \ / \ (1 + (\frac{300}{-L})^{2/3}) \qquad L < 0 \qquad\qquad (2.15)$$

$$= 1.1 \ 10^3 \ u_*^{-1} \ (h/-L)^{-2/3} \qquad L < 0, \quad h/L < -30 \ .$$

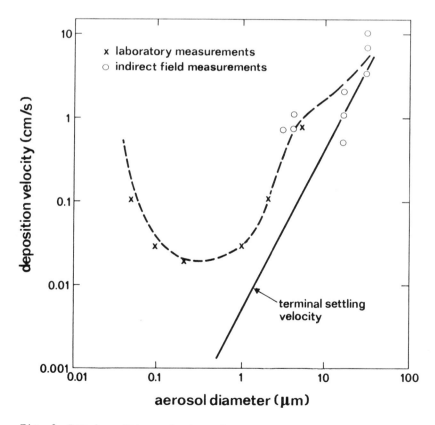

Fig. 2. Dry deposition velocity of aerosol as a function of diameter, over land surfaces ($1 < z_o < 20$ cm) (adapted from ref. 51).

TABLE III

Land cover classes (from Wilson and Henderson-Sellers, ref. 16) and associated roughness lengths, z_o (from Wieringa (ref. 52).

type	z_o(m)	type	z_o(m)
open water	0.0002	thorn shrub	--
inland water	0.005	temperature meadow and	
bog or marsh	0.03	permanent pasture	0.07
ice	--	temperature rough grazing	0.1
paddy rice	--	tropical grassland plus shrub	0.24
mangrove	--	tropical pasture	--
dense needleleaf evergreen		rough grazing plus shrub	0.24
forest	1.0	pasture plus tree	0.24
open needleleaf evergreen		semi-arid rough grazing	--
woodland	0.5	tropical savanna	--
dense mixed forest	1.0	pasture plus shrub	--
open mixed woodland	0.5	arable cropland	0.17
evergreen broadleaf woodland	0.5	dry farm arable	0.17
evergreen broadleaf cropland	0.17	nursery and market gardening	--
evergreen broadleaf shrub	--	cane sugar	--
open deciduous needleleaf		maize	0.24
woodland	--	cotton	--
dense deciduous needleleaf		coffee	--
forest	--	vineyard	0.35
dense evergreen broadleaf		irrigated cropland	0.17
forest	--	tea	--
dense decituous broadleaf		equatorial rainforest	--
forest	1.0	equatorial tree crop	--
open deciduous broadleaf		tropical broadleaf forest	--
woodland	0.5	tundra	0.03
deciduous tree crop	--	dwarf shrub	0.17
open tropical woodland	--	sand desert and barren land	--
woodland plus shrub	0.35	scrub desert and semi-desert	--
dense drought deciduous forest	--	semi-desert and scattered trees	--
open drought deciduous woodland	--	urban	--
deciduous shrub	--		

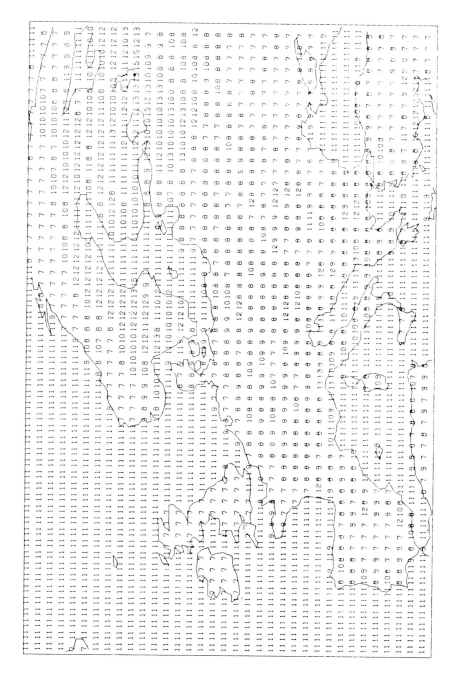

Fig. 3. Average deposition velocity on grid squares of 1° x 1° based on table
II and on the land use data in Henderson Sellers (ref. 16), see table III.

Sheih et al. (ref. 12), however, adopt for rough surfaces a (much smaller) uniform value for r_s of 10^2 s m^{-1}.

2.4 Land use

In the previous section it appeared that the surface type plays an important role in the determination of the deposition velocities of gases and particles. For relatively small regions it is feasible to make a classification of terrain features by hand (ref. 12, 15), based on readily-available maps. In general a global (satellite-based) land-use classification will be more appropriate. In Wilson and Henderson-Sellers (ref. 16) some database are described which contain a variety of land cover and soil data. The data pertain to a 1° x 1° longitude latitude grid. They distinguish 52 different terrain types. Further, four additional factors are tabulated:

- the height of the surface elements
- the density of the surface elements
- the seasonal variation of land cover
- the soil moisture.

Two classes of vegetation are assigned to each grid square: the primary class which covers more than 50% and the secondary class hich covers 25-50% of the surface. Some soil properties are also included in the data base. A list of land cover classes is given in table III. Based on these data and on the data of table II it is possible to evaluate mean deposition velocities for the major air pollution components. This is illustrated in Fig. 3.

3 WET DEPOSITION

3.1 Introduction

By wet deposition we indicate all processes which involve the impact on the earth surface of pollution contained in condensed (or liquid) atmospheric water vapour. This is usually in the form of precipitation (rain, snow of hail), but also fog may contribute to the deposition process.

Initially pollution is injected in air. From the gaseous phase it has to be transferred to the liquid phase. Once in the liquid phase the pathway of pollution is strongly intertwained with that of atmospheric moisture. The understanding of the wet deposition processes cannot be complete without a detailed knowledge of atmospheric water vapour cycle.

Precipitation droplets, which often originate from melted snow, have a diameter of 100-5000 μm. It is formed in clouds from cloud droplets which have a typical diameter of 1-50 μm. The water content of clouds usually varies between 0.1 and 1 g m^{-3}. The precipitation is formed in clouds by accretion or coalescence processes, where the enhanced codensation on small aerosol particles (pollution!) may play a role. In Fig. 4 the various phases and transfer processes of water vapour are indicated. Note the cyclic or reversible characters of the transfer processes. Condensation occurs when air masses are forced to move to higher altitudes so that their temperature decreases. This may be the case in the vicinity of a front (i.e. the separation plane between air masses of different origin and composition) or in convective systems which originate from local updrafts during daytime when solar radiation is moderate or strong. Both cases usually lead to a marked difference in precipitation intensity. Frontal precipitation is typical a few mm per hour whereas in showers precipitation intensity may be tens of mm per hour.

The above considerations imply that the atmospheric water vapour cycle is obvious strongly coupled to atmospheric dynamics and thermodynamics.

In this chapter we shal emphasize the wet removal processes only, and presume that all additional atmospheric data, such as flow field, temperature, cloud cover, cloud water content droplet size spectrum, precipitation etc. are known.

Because different removal mechanisms play a role for small (cloud) and large (rain) droplets, distinction is made between in-cloud scavenging and sub-cloud

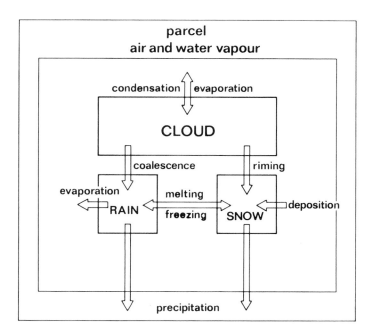

Fig. 4. Various phases of water in the atmosphere.

scavenging of air pollution. Also the nature of air pollution (gas of aerosol) plays a role in this respect. Before we give a few applications we summarize in the next sections the major transfer mechanisms of airborne pollution.

3.2 Below cloud scavenging

Here we consider the interaction of aerosols and gases with rain droplets, which, by their size have a fall velocity, V_r. The fall velocity is a function of droplet radius R, and can be conveniently expressed as

$$V_r = 27.3 - 1026 \, R + 348 \, R^2 \qquad (0.05 \leqslant R < 0.7 \text{ mm})$$

and (3.1)

$$V_r = -156 - 613 \, R + 123 \, R^2 \qquad (0.7 \leqslant R < 2.9 \text{ mm})$$

(R in mm and V_r in cm s^{-1}) (ref. 17). Note that V_r, being a downward velocity, is taken negative.

Not only the fall velocity depends on the raindrop size and for the evaluation of scavenging processes average quantities have to be considered. When in general $\psi(R)$ is a raindrop size dependent function, we obtain the average of ψ by

$$\langle \psi \rangle = {}_o\!\int^\infty \psi(R) \, f(R) \, dR \, , \qquad (3.2)$$

where f(R) is the probability density function of rain drops. For practical applications the Marshall and Palmer distribution is used. It reads

$$f(R) = \alpha \, \exp^{-\alpha R} \, , \qquad (3.3)$$

with $\alpha = 8.2 \, J^{-0.21}$ mm^{-1} and J the rainfall rate (mm h^{-1}).

So is for example the liquid (rain) water fraction in a unit volume of air given by

$$r_f = \frac{4\pi N}{3} \, {}_o\!\int^\infty R^3 \, f(R) \, dR \, , \qquad (3.4)$$

where N is the number of raindrops in the unit volume; the rainfall rate J is

$$J = -\frac{4\pi N}{3} \, {}_o\!\int^\infty R^3 \, V_r(R) \, f(R) \, dR \qquad (3.5)$$

and the average concentration in the aquaous phase (per unit volume (air +

water)) of a pollutant, c_w, is related to the concentration in the raindrops, $\hat{c}_w(R)$ by

$$c_w = \frac{4\pi N}{3} \int_o^\infty R^3 \, \hat{c}_w(R) \, f(R) \, dR \; . \qquad (3.6)$$

Finally note that the mixed concentration of an ensemble of raindrops falling through a unit horizontal area (rain collector) is given by

$$\hat{C}_w = - \frac{\int_o^\infty R^3 \, V_r(R) \, \hat{c}_w(R) \, f(R) \, dR}{\int_o^\infty R^3 \, V_r(R) \, f(R) \, dR} \; . \qquad (3.6a)$$

With the above expressions in mind we shall consider in some more detail the aerosol and gas scavenging processes for a raindrop of a specific size R only, knowing that the results can be generalised by assuming or measuring a raindrop spectrum and carrying out the averaging procedure (eqn. 3.2).

We shall first consider the scavenging of aerosols. When a raindrop (size R and velocity V_r) falls through an aerosol polluted layer[*], it encompasses a volume per unit of time of $\pi R^2 V_r$. When the aerosol concentration in air (mol m^{-3}) is denoted by c_a, then this volume contains $\pi R^2 v_r c_a$ aerosol mass. We define now the concept of collection efficiency, $E(R,a)$, as the probability that an aerosol particle (radius a) is caught by a raindrop (radius R). This probability covers all the microphysical processes by which aerosol particles may enter a raindrop. Due to the relatively large speed of most raindrops with respect to the aerosol the dominant transfer mechanisms are impaction, interception and Brownian diffusion (ref. 18, 19, 20). The mass transfer rate for one raindrop is thus

$$\pi R^2 \, V_r(R) \, E(R,a) \, c_a \; .$$

The mass transfer rate per unit volume is given by (see eqn. 3.2)

$$\partial c_w / \partial t = \{-\pi N \int_o^\infty R^2 \, V_r(R) \, E(R,a) \, f(R) \, dR\} \, c_a \; . \qquad (3.7)$$

The minus sign is introduced to make $\partial c_w / \partial t$, being the rate of change of aerosol concentration in the liquid phase, positive.

---

[*] It is assumed that the layer is uniformly polluted.

Note that in eqn (3.7) E depends on the aerosol radius and that with a variable
aerosol size a similar procedure has to be followed as indicated in eqn. (3.2)
to determine the transfer rate averaged over all particle sizes. If we define a
transfer coefficient (wash-out coefficient) as

$$\Lambda = -\pi N_o \int_o^\infty R^2 \, V_r(R) \, E(R,a) \, f(R) \, dR \tag{3.8}$$

we may write, noting that by definition

$$\partial c_a / \partial t = -\partial c_w / \partial t \, ,$$

$$\tag{3.9}$$

$$\partial c_a / \partial t = -\Lambda \, c_a \, ,$$

which described the decrease in air concentrations during a precipitation
event. Its solution is simply

$$c_a = c_a(o) \, e^{-\Lambda t} \, . \tag{3.10}$$

It follows from eqn. (3.8) that for a uniform rain distribution
$(r(R) = \delta(R-r_o))$ the scavenging coefficient is related to the rainfall rate
(cf. eqn. 3.5) by

$$\Lambda = \frac{3E}{4r_o} \cdot J \, . \tag{3.11}$$

In practical applications the somewhat more general expression

$$\Lambda = \Lambda_w \, J^n \tag{3.12}$$

is often used (see section 3.3), where Λ_w and n are determined from empirical
of theoretical studies.

The derivation of eqns. (3.9-3.12) was for illustrative purposes kept as simple
as possible. Complicating factors in aerosol scavenging are that (i) one has to
consider scavenging by snow and hail as well, (ii) at high humidities
hygroscopic aerosols tend to grow by nucleation processes, thus changing their
diameter (and capture properties). The latter process can be conveniently
parameterised by reevaluating the collection efficientcy E(R,a) (see Fig. 5).
Further we have assumed that the transfer process is irreversible, an
assumption which is violated when rain evaporates and the dissolved aerosol
becomes airborne again. This may lead to a significant redistribution of
aerosol (ref. 21). For more details on aerosol scavenging we refer to the

excellent review of Hales (ref. 20).

We now turn to the below cloud scavenging of gases. The essential difference with the considerations above is that the transfer mechnisma is reversible: gases may both be absorbed by raindrops and desorb from them. Similar to the deposition process formulation (see section 2.1 eqn. (2.6)), these processes can be expressed as a flux, F, (in moles $m^{-2} s^{-1}$)

$$F = K (c_a - H\hat{c}_w) \; , \tag{3.13}$$

where K is an "exchange velocity" between the gas phase and the aquaous phase and H Henry's laws constant, which is determined by the ratio of equilibrium concentration in the air and raindrop respectively away from the interface. The solubility is inversely proportional to H. The inverse of the "exchange

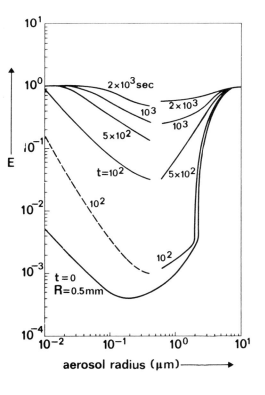

Fig. 5. Collection efficiency for raindrops (R = 0.5 mm) and nucleating aerosols, as a function of initial aerosol radius (after Slinn (ref. 53)).

428

velocity", the exchange resistance can be decomposed as a series of an atmospheric resistance, $1/k_a$, and an aquaous resistance, H/k_w, so that

$$\frac{1}{K} = \frac{1}{k_a} + \frac{H}{k_w} \ . \tag{3.14}$$

In (3.14) k_a can be estimated from

$$k_a = \frac{\rho D}{2R} \ Sh \ , \tag{3.15}$$

where Sh is the Sherwoodnumber,

$$Sh = 2 + 0.6 \ Re^{\frac{1}{2}} \ Sc^{1/3} \ , \tag{3.16}$$

with Re the raindrop Reynoldsnumber (Re = $-2R \ V_r/\nu$) and Sc the Schmidt Number (Sc = ν/D).
The aquaous resistance can be expressed as

$$k_w \sim D_w/R \ , \tag{3.17}$$

where D_w is the molecular diffusivity in water. The constant of proportionality depends on the microdynamics of the raindrop. For many gases the solubility is sufficient to neglect the third term in eqn. (3.14) ($H \rightarrow 0$) so that K is determined by the gasphase resistance only.
The rate of change of concentration in air (per unit volume) can be inferred from eqn. (3.13) by integrating over the droplet surface and the raindrop size distribution:

$$\frac{\partial c_a}{\partial t} = - \ 4\pi N_o \int^\infty R^2 \ K(R) \ (c_a - H \ \hat{c}_w(R)) \ f(R) \ dR \ . \tag{3.18}$$

Starting from eqn. (3.18) it is possible under some simplifying assumptions to obtain a gas scavenging coefficient (ref. 20, 22). When the solubility of the gas in water is high the term $H \ \hat{C}_w$ may be neglected in eqn. (3.18) and we get

$$\frac{\partial c_a}{\partial t} = - \ \Lambda \ C_a \ , \tag{3.19}$$

with

$$\Lambda = - \ 4\pi N_o \int^\infty R^2 \ K(R) \ f(R) \ dR \ , \tag{3.20}$$

the scavenging coefficient formulation.

When the timescale of the transfer is much smaller than other relevant
timescales we may assume steady state conditions, or

$$c_a = H \, \hat{c}_w = H \, \hat{C}_w \, , \qquad (3.21)$$

which relates the air concentration directly to the concentration of the
species in the rainwater collector, $\hat{C}_w$. In many practical applications the
ratio of $\hat{C}_w / c_a$ is referred to as the scavenging ratio, ζ, which in this
particular case equals H^{-1}.

With a precipitation rate J the surface flux is $\hat{C}_w \, J$ (mol m^{-2} s^{-1}). When we
assume a uniform distribution of pollutant up to a (constant) altitude h the
rate of change of concentration in air can then be written as

$$\frac{\partial c_a}{\partial t} = - \frac{\hat{C}_w J}{h} = - \frac{\zeta J}{h} \cdot c_a \, , \qquad (3.22)$$

which defines again a scavenging coefficient

$$\Lambda = \frac{\zeta J}{h} \, . \qquad (3.23)$$

Apart from the transfer into the aquaous phase a pollutant may be subject to
chemical transformation by the presence of other trace compounds in the rain
(or cloud) drop. The rate of change of a pollutant due to chemical reactions
can be formulated as

$$\frac{\partial c_w}{\partial t} = - k \, c_w \, , \qquad (3.24)$$

where k is the reaction coefficient. The addition of chemical transformation
further complicates the wet deposition process because it feeds back on the
transfer rates as well. Usually some limiting cases are considered. For example
rapid (irreversible) chemical reactions make the transfer mechanism also
irreversible. On the other hand slow (compared with the transfer rate)
reactions make that transfer and chemistry may be treated separately. For
further details we refer to ref. (20) and the references therein.

3.3 In-cloud scavenging

In fact we deal with the same processes here. There are, however, some
differences. Firstly the cloud drop size spectrum differs essentially from the
raindrop spectrum. Other differences are that in a cloud environment more or
less continuously condensation and evaporisation processes occur and that
relative velocity differences between airborne pollution and cloud droplets are

smaller than in rain scavenging. This makes that interception and impaction are
less important transfer mechanisms. The nucleation process is here the dominant
transport mechanism but also phoretic effects (i.e. relative motion induced by
temperature-, concentration- or electrical gradients) might be important. The
incomplete knowledge of dynamical and physical processes in clouds hampers an
accurate and detailed description of the various transfer mechanisms.

In most practical applications transfer rates are expressed in bulk form
(cf. eqns 3.19 and 3.22), defining an average scavenging coefficient or
scavenging ratio. For a more extensive treatment of this subject the reader is
referred to ref. (20, 23-26).

3.4 Scavenging coefficients and scavenging ratios

In this section some data will be presented on scavenging coefficients and
ratios, for some of the more important air pollutants. Data for sulphur
dioxide, nitric oxides and acid and sulphate and nitrate aerosol are given. The
data are taken from a review article by Van Aalst and Diederen (ref. 50).

The SO_2 solubility in water is low and depends on both temperature and
acidity of the raindrop. The loss of atmospheric SO_2 by wet deposition is not
very important in many cases. Scavenging coefficients and ratios are summarised
in Table IVa,b.

TABLE IVa
Scavenging coefficients (s^{-1}) for sulphur dioxide. The rainfall rate J is in
mm hr^{-1}.

$\Lambda * 10^{-5}$	reference	remarks
~1	27	literature review
0.4 - 6	28	" "
17 $J^{0.6}$	29	laboratory measurements
2.85 J	30	non-frontal rain (UK)
2.6	31	field measurements (UK)
2.6 J	32	" " "

TABLE IVb

Scavenging ratios for sulphur dioxide. P = precipitation amount (mm).

$\zeta * 10^5$	reference	remarks
1.6	28	single rainstorm (P = 0.3)
16	28	" " (P = 11-20 mm)
0.1	31	field measurements
0.2	33	" "
5.3	33	single rain events

The solubility of NO_2 in water is low. Airborne NO and NO_2 concentrations hardly decrease after precipitation. Therefore, direct uptake of NO and NO_2 by precipitation and cloud droplets is often neglected. For NO a scavenging coefficient was determined of 10^{-5} J s^{-1} (ref 32), where J is the rainfall rate (mm h^{-1}). Some data for NO_2 are presented in table V.

Contrary to the oxides, nitric acid is highly soluble and rapidly scavenged by cloud and rain droplets (cf. Table VI). Even by snow nitric acid is scavenged.

TABLE V

Scavenging coefficients for NO_2 (s^{-1}). The rainfall rate J is in mm hr^{-1}.

$\Lambda * 10^{-5}$	reference	remarks
4 $J^{0.6}$	29	rain chamber
2.2 J	32	non frontal rain

TABLE VI

Scavenging coefficients for HNO_3. The rainfall rate is in mm h^{-1}.

$\Lambda * 10^{-5}$	reference	remarks
20.000	22	in-cloud scavenging
10	22	rain (J = 1)
40	22	rain (J = 15)
25	47	snow

The scavenging ratio for snow is estimated to be $5 \ 10^3$.

Finally we present some data for aerosol scavenging. Though most experimental data pertain to sulphate aerosol or aerosol mixtures, they are thought to be generally valid for any aerosol, the major parameter being particle size only. The data are summarized in Tables VIIa and b.

TABLE VIIa

Scavenging coefficients for sulphate and nitrate aerosol Λ (in s^{-1}). The rainfall rate J is in mm h^{-1}.

$\Lambda \ 10^{-5}$	reference	remarks
10–100	34	in-cloud (theory and experiments)
1–100	34	below cloud
~ 100	35	in-cloud (observations)
~ 10	27	below cloud
1–100	36	review
2–10	37	single precipitation events. values depend on J and R
$35 \ J^{0.78}$	38	recommended to be used in transport models
24 J	38	" " " " " " "
5	39	snow

TABLE VIIb

Scavenging ratios for nitrate and sulphate aerosol. The precipitation amount P
is in mm.

$\zeta * 10^5$	reference	remarks
1 - 10	40	–
0.01 - 0.2	42	values depend on R
4 - 10	43	sulphate
8.1 - 11	44	field study (Can)
8.3 - 13	45	field study (Nor)
19 $P^{-0.88}$	48	–
1 - 10	46	sulphate
8	47	snow
0.8	49	unrimed snow
8	49	rimed snow

REFERENCES

1 Global tropospheric chemistry: a plan for action. Report of the US National Academy of Science (1985).
2 W. Seiler, European experiment on transport and transformation of environmentally relevant trace constituents in the troposphere (EUROTRAC): a project proposal, Fraunhofer Institut, Garmisch Partenkirchen, 1986.
3 OECD: The OECD programme on long range transport of air pollutants. Measurements and findings, OECD, Environment Directorate, 2 Rue Pascal 75775, Paris Cedex 16, France, 1977.
4 United States – Canada: Memorandum of intent on transboundary air pollution: strategies, development and implementation. Interim Report, February, 1981.
5 W.H. Smith, Air pollution and forests, Springer Verlag, NY, 1981.
6 B. Ulrich, R. Mayer, T.K. Khanna, Chemical changes due to acid precipitation in a loss-derived soil in Central Europe, Soil Sci., 130 (1981) 193.
7 J.A. Businger, Evaluation of the Accuracy with which dry deposition can be measured with current micrometeorological techniques. Submitted to J. Climate and Appl. Meteor. (1986).
8 A.J. Dyer and B.B. Hicks, Flux-gradient relationships in the constant flux layer, Quart. J. R. Met. Soc., 96 (1970) 715-721.
9 M.L. Wesely and J.D. Shannon, Improved estimates of sulphate dry deposition in eastern North America. Environ. Progress, 3 (1984) 78-81.
10 C.R. Owen and W.R. Thompson, Heat transfer across rough surfaces, J. Fluid Mech., 15 (1963) 321-334.
11 The NCAR Regional Acid Deposition Model, NCAR Technical Note TN-256+STR, Boulder, Colorado, 1985.
12 C.M. Sheih, M.L. Wesely and B.B. Hicks, Estimated dry deposition velocities of sulfur over the eastern U.S. and surrounding region, Atmos. Environ, 13 (1979) 1361-1368.
13 D. Onderdelinden (private communication)

14 H. van Dop en B.J. de Haan, Mesoscale air pollution dispersion modelling, Atm. Environment, 17 (1983) 1449–1456.
15 H. van Dop, Terrain classification and derived meteorological parameters for interregional transport models, Atmospheric Environment, 17 (1983) 1099–1105.
16 M.F. Wilson and A. Henderson Sellers, A global archive of land cover and soils data for use in general circulation models, J. Climatology, 5 (1985) 119.
17 A.N. Dingle and Y. Lee, Terminal fallspeed of raindrops, J. Appl. Met., 11 (1972) 877.
18 A.N. Dingle and Y. Lee, An analysis of in-cloud scavenging, J. Appl. Met., 12 (1973) 1295.
19 G. Hidy, Removal of gaseous and particulate pollutants, In Chemistry of the Atmosphere, S.I. Rasoule, ed. Plenum, NY, 1973.
20 J.M. Hales, The mathematical characterisation of precipitation scavenging in O. Hutzinger (Ed.) The Handbook of Environmental Chemistry 4A, Springer, Berlin, 1985 pp. 108–222.
21 A. Tremblay and Henry Leighton, A three dimensional cloud chemistry model, J. Climate and Appl. Meteor., 25 (1986) 652–671.
22 S.Z. Levine and S.E. Schwartz, In and below-cloud scavenging of nitric acid vapor, Atmos. Env., 16 (1982) 1725.
23 R.C. Easter and J.M. Hales, PLUVIUS, A generalized one-dimensional model of reactive pollutant behavior, including dry deposition precipitation formation, and wet removal, PNL-4046 ED 2 Pacific Northwest Laboratory (1984), (available through NTIS).
24 R.C. Easter and J.M. Hales, Mechanistic evaluation of precipitation scavenging using a one-dimensional reactive storm model, Final report to Electric Power Research Institute RP-2022-1 (1982).
25 J.M. Hales, A generalized multidimensional model for precipitation scavenging and atmospheric chemistry, Atm. Env. In Press (1985).
26 D.A. Hegg, S.A. Rutledge and P.V. Hobbs, A numerical model for sulfur chemistry in warm-frontal rainbands, J. Geophys. Res., 89 (1984) 7133.
27 J.A. Garland, The dry deposition of sulphur dioxide to land and water surfaces, Proc. Roy. Soc. Londen A., 354 (1977) 245–268.
28 T.A. McMahon and P.J. Denison, Empirical atmospheric deposition parameters – a survey, Atmos. Environ., 13 (1979) 571–585.
29 S. Beilke, Laboratory investigations of washout of trace gases, Proc. Symp. on Precipitation Scavenging USAEC Symp. Services No. 22 (1970) 261.
30 P.R. Maul, Preliminary estimates of the washout coefficients for sulphur dioxide using data from an East Midlands ground level monitoring network, Atmospheric Environment, 12 (1978) 2515.
31 T.D. Davies, Sulphur dioxide precipitation scavenging, Atmospheric Environment, 17 (1983) 797.
32 A. Martin, Estimated washout coefficients for sulphur dioxide, nitric oxide, nitrogen dioxide and ozone, Atmospheric Environment, 18 (1984) 1955.
33 T.D. Davies, Precipitation scavenging of sulphur dioxide in an industrial area, Atmospheric Environment, 10 (1976) 879.
34 K.P. Makhon'ko, Simplified theoretical notion of contaminant removal by precipitation from the atmosphere, Tellus, 19 (1967) 467.
35 H. Sievering, C.C. van Valing, E.W. Barrett and R.F. Pueschel, Cloud scavenging of aerosol sulphur: Two case studies, Atmospheric Environment, 18 (1984) 2685.
36 T.A. McMahon and P.J. Denison, Empirical atmospheric deposition parameters – a survey, Atmospheric Environment, 13 (1979) 571.
37 H.M. Davenport and L.K. Peters, Field studies of atmospheric particulate concentration changes during precipitation, Atmospheric Environment, 12 (1978) 997.
38 B.C. Scott, Theoretical estimates of the scavenging coefficient for soluble aerosol particles as a function of precipitation type, rate and altitude, Atmospheric Environment, 16 (1982a) 1753.

39 B.J. Huebert and C.H. Robert, The dry deposition of nitric acid to grass, J. Geophys. Res., 90, D1 (1985) 2085.

40 W.G.N. Slinn, Precipitation scavenging, In: Atmospheric Sciences and Power Production, D. Randerson (ed.), US DOE Technical Information Center, Oak Ridge, 1982.

41 W.G.N. Slinn, L. Hasse, B.B. Hicks, A.W. Hogan, D. Lal, P.S. Liss, K.O. Munnick, G.A. Sehmel and O. Vittori, Some aspects of the transfer of atmospheric trace constituents past the air-sea interface, Review paper, Atmospheric Environment, 12 (1978) 2055.

42 G.R. Carmichael and M. Reda, Non-isothermal SO_2 absorption by water droplets – The effect of precipitation intensity, sulphate aerosol scavenging and aquaous S(IV) oxidation, Atmospheric Environment, 16 (1982) 2905.

43 S.E. Lindberg and G.M. Lovett, Field measurements of particle dry deposition rates to foliage and inert surfaces in a forest canopy, Environ. Sci. Technol., 19 (1985) 2381

44 L.A. Barrie and J. Neustadter, The dependence of sulphate scavenging ratios on meteorological variables, In: Precipitation scavenging, dry deposition and resuspension (Pruppacher et al., (eds.)), Elsevier Science Publ., New York, 1983.

45 D.M. Whelpdale, Rep. EMEP/CCC 5/81, Norwegian Institute for Air Research Lillestrøm, Norway, 1981.

46 B.L. Niemann, Scavenging ratios for exeptional wet sulphate episodes in eastern North America from three event networks, In: Precipitation scavenging, dry deposition and resuspension (Pruppacher et al., (eds.)), Elsevier Science Publ., New York, 1983.

47 B.J. Huebert, F.C. Fehsenfeld, R.B. Norton and D. Slbritton, The scavenging of nitric acid vapor by snow, In: Precipitation scavenging, dry deposition and resuspension (Pruppacher et al., (eds.)), Elsevier Science Publ., New York, 1983.

48 J.M. Hales and M.T. Dana, Regional scale deposition of sulphur dioxide by precipitation scavenging, Atmospheric Environment, 13 (1979a) 1121.

49 B.C. Scott, J. Appl. Meteorol., 20 (1981) 619.

50 R.M. van Aalst and H.S.M.A. Diederen, Removal and transformation processes in the atmosphere with respect to SO_2 in S. Zwerver and J. van Ham (Eds.), Interregional air pollution modelling, the state of the art, Plenum, New York, 1985, 83–147.

51 H. Sievering, Profile measurements of particle mass transfer at the air-water interface, Atmospheric Environment, 15 (1981) 123.

52 J. Wieringa, Windklimaat van Nederland, Staatsuitgeverij, Den Haag (1983).

53 W.G.N. Slinn, Some approximations for the wet and dry removal of particles and gases form the atmosphere. Water, Air and Soil Pollution, 7 (1977) 513.

Regional and Long-range Transport of Air Pollution,
Lectures of a course held at the Joint Research Centre, Ispra, Italy,
15–19 September 1986, S. Sandroni (Ed.), pp. 437–465
© Elsevier Science Publishers B.V., Amsterdam — Printed in The Netherlands

LONG-RANGE (CAPTEX) AND COMPLEX TERRAIN (ASCOT)
PERFLUOROCARBON TRACER STUDIES

J.L. HEFFTER, T. YAMADA, AND R.N. DIETZ

1   INTRODUCTION

The verification of computer models for predicting pollutant concentrations and depositions, both in long-range (several thousand kilometers) and complex terrain (mountains and valleys) applications is essential and can be performed in field experiments with the use of the previously described perfluorocarbon tracer (PFT) technology (ref. 1).

A major long-range field experiment, the Cross Appalachian Tracer Experiment (CAPTEX '83), was conducted for model evaluation purposes (ref. 2). During CAPTEX '83, about 3000 3- to 6-hour ground-level measurements were taken at 80 sites in northeast U.S. and southeast Canada at distances out to 1100 km from two sources at Dayton, Ohio and Sudbury, Ontario. In addition, about 1200 upper-air samples were obtained from seven aircraft sampling along various flight paths over the CAPTEX '83 ground sampling network. A description of the experimental PFT sampling results and how they were used to evaluate the NOAA - Air Resources Laboratory Branching Atmospheric Trajectory (BAT) long-range transport and dispersion model will be given in the next section of this paper.

In the Atmospheric Studies in Complex Terrain (ASCOT) field experiments, also supported by the U.S. Department of Energy, atmospheric transport and diffusion of airborne materials over complex terrain have been investigated. Extensive observations were made during the summer of 1980 over the Geysers area in northern California (ref. 3). High concentrations of pollutants were expected to occur during nocturnal periods since turbulent mixing was suppressed, and the drainage flows that frequently developed were shallow. The objectives of the experiments, most of which were conducted during the nighttime, were to investigate the structure of nocturnal drainage flows and the behavior of tracer gases released over the sloping terrain.

In the third section of the paper, the ASCOT data, including ground-level PFT concentrations, are used to test the performance of a three-dimensional hydrodynamic wind model and a random-particle statistical diffusion model. The determination of tracer concentrations aloft using previously described (ref. 1) vertical atmospheric sampling cables (VASCs) will be demonstrated.

This work was performed under the auspices of the United States Department of Energy under Contract No. DE-AC02-76CH00016.

2 VALIDATION OF LONG-RANGE ATMOSPHERIC TRANSPORT MODELS

The adverse effects of air pollutants on man and the environment during
their long-range transport and dispersion continues to be of major concern
(e.g., acid rain, arctic haze, accidental pollutant releases). Air pollution
models have been developed to simulate atmospheric and chemical processes out
to distances of several thousand kilometers from pollutant sources. Verification of these model calculations is essential in efforts to establish the
uncertainty associated with model simulations.

2.1 Review of several long-range tracer experiments

One of the most credible methods for air pollution model evaluation and
verification is from field experiments in which a known amount of tracer is
released from a source, or several sources, and concurrent measurements of
concentration and deposition are made at various downwind distances and locations. The first major long-range experiment of this type was over a 6-week
period during the winter and spring of 1974 from a source near Idaho Falls
(Table 1). Samples were collected in the midwest U.S. along an arc about
1500 km downwind. Due either to passage over the mountains or to greater dispersion during winter, little tracer was measured along the sampling arc (ref.
4). A reanalysis of the data (ref. 5) showed that the low concentrations could
have been due to the enhanced dispersive effects of wind sheer.

TABLE 1

Major long-range field experiments.

Year	Name (Release site)	Maximum sampling distance (km)
1974	Long-range (Idaho Falls)	1500
1980	Oklahoma (Norman)	600
1982-83	ACURATE (Savannah River)	1100
1983	CAPTEX (Dayton, OH & Sudbury, Ont.)	1100
1987	ANATEX (Glasgow, MT & St. Cloud, MN)	3000

A second experiment was successfully conducted in July 1980 from a source at
Norman, Oklahoma, with two releases that included two perfluorocarbon tracers
specifically developed for use in long-range field experiments (ref. 6). The
main objectives were to test the release, sampling, and analysis techniques of

the new tracers to distances of at least 600 km from the source where the farthest sampling arc was located, and to demonstrate the capability to perform this type of experiment at a reasonable cost.

Two more recent long-range field experiments have been conducted for the explicit purpose of model evaluation and model calculation verification. The Atlantic Coast Unique Regional Atmospheric Tracer Experiment (ACURATE) (ref. 7) made use of a tracer of opportunity from the Savannah River Plant, South Carolina, and the Cross Appalachian Tracer Experiment (CAPTEX '83) (ref. 2) used the newly developed perfluorocarbon tracers released from Dayton, Ohio, and Sudbury, Ontario.

A new long-range tracer experiment, ANATEX (Across North America Tracer Experiment), is planned for early 1987 (ref. 8). Two new perfluorocarbon tracers will be released, one each from Glasgow, Montana, and St. Cloud, Minnesota, with surface sampling out to 3000 km from the Glasgow source. This section will focus on CAPTEX and will describe the experiment and summarize the sampling results.

The National Oceanic and Atmospheric Administration Air Resources Laboratory began the development of computer long-range transport and dispersion models in the mid 1970's (ref. 9) and has continued development since then (ref. 5, 10). A revised version of one of the latest models, the Branching Atmospheric Trajectory (BAT) model (ref. 11) is outlined. A statistical long-range model evaluation technique is described and applied to the revised BAT model using the results from CAPTEX '83.

2.2 CAPTEX '83

CAPTEX '83 consisted of seven tracer releases, five from Dayton, Ohio, and two from Sudbury, Ontario, during mid-September through October 1983. One of the newly developed tracers, perfluoromethylcyclohexane (PMCH, C_7F_{14}), proven reliable in the previous Oklahoma experiment, was released over a 3-hour period from the surface as a gas in each of the seven tests. PMCH is inert and can be measured at its background concentration of 4 parts in 10^{15} parts of air which is equivalent to 4 femtoliters/liter (fL/L). The CAPTEX source sites and surface sampling network in the northeast U.S. and southeast Canada are shown in Fig. 1. The source release points at Dayton and Sudbury are designated as "R." The 86 sampling sites, shown by circles, were arrayed in arcs at approximately 100 km intervals from 300 km to 1100 km from the Ohio release point. Regular network rawinsonde stations ("N") and ten additional stations ("E"), established by the Electric Power Research Institute (EPRI) for enhanced spatial coverage, took two special soundings a day (06 GMT and 18 GMT) following each release, as long as tracer remained over the network, in addition to the regular 00 GMT and 12 GMT schedule.

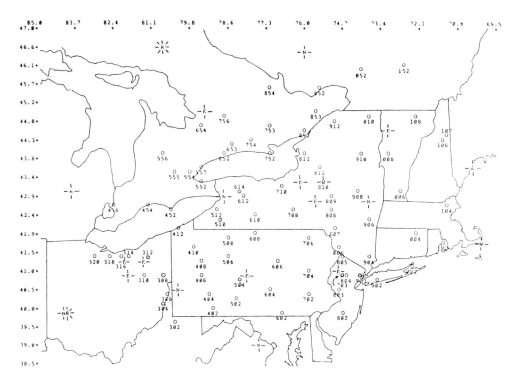

Fig. 1. CAPTEX '83 release sites (R), surface sampling sites (circles with numbers), network rawinsondes (N), and EPRI rawinsondes (E).

A tracer release was scheduled only when the winds were forecast to transport the tracer over the sampling network. Automatic sequential samplers, each containing 23 sampling tubes, were operated at each sampling site by volunteers from the National Weather Service Cooperative Observer Network in the U.S. and at government-operated monitoring stations in Canada. After each release, samplers were programmed to collect six consecutive air samples from 3 to 6 hours long, depending on location with respect to the release point. Approximately 3000 surface air samples were scheduled to be collected during CAPTEX. Quality-assured concentrations were reported for 75% of these samples. As an example of sampling results, Fig. 2 shows the patterns of the maximum surface tracer concentrations in fL/L measured at each sampling site after two separate releases. Values in parentheses may be low due to maximum concentrations occurring at the sites before sampling started. The pattern shown in Fig. 2a for release #5 from Sudbury is associated with strong, vertically well-mixed winds following a cold front and nicely fits the standard "cigar shaped" concept often attributed to model plume calculations. In contrast, the pattern after release #3 from Dayton, shown in Fig. 2b, is associated with light winds

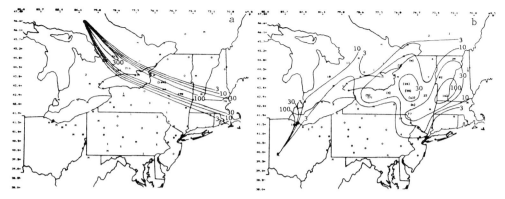

Fig. 2. Maximum tracer concentration (fL/L) measured at each sampling site after two releases. Values in parentheses may be low. (a) Release #5 from Sudbury. (b) Release #3 from Dayton.

and large vertical wind shear. Several features should be noted. The plume to the north-northeast that lies off the western edge of the sampling network is the result of winds at lower levels with strong southerly components. The absence of tracer at the surface to the northeast of the source and then the subsequent reappearance of the surface pattern farther east resulting from the upper-level winds with more westerly components indicates inhibited downward mixing for a period following the release.

Seven aircraft were used during CAPTEX to measure the vertical tracer distribution at various times after release. Each aircraft used an automatic sequential sampler modified to collect tracer over a short time interval (i.e., 6 to 10 minutes). In addition, one aircraft was equipped with a real-time dual-trap analyzer (ref. 12). Various flightpaths using two or more aircraft were flown primarily in the region outlined in Fig. 3a. The exact locations depended on many factors such as flightpath length to traverse the plume, total

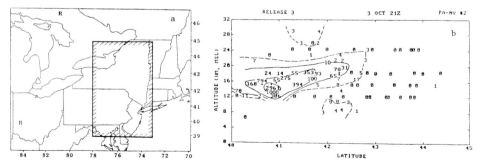

Fig. 3. (a) Location of majority of aircraft sampling flights, box, and release sites, R. (b) A "snapshot" at 2100 GMT of the tracer concentration, fL/L, cross section aloft looking back to the Dayton, OH release site, #3.

permissible sampling duration, number of sampling tubes, weather restrictions, and flight restrictions.

Figure 3b is a "snapshot" (2100 GMT) vertical cross-section showing the tracer measured by four constant-altitude and one spiraling aircraft for release #3. It was obtained by advecting sampled air parcels with the wind as measured on the spiraling aircraft or as estimated from nearby rawinsonde data.

Data files available for CAPTEX include the rawinsonde meteorological data for the regular network stations and additional EPRI stations, and ground-level and aircraft measured tracer concentration data.

2.3 Branching Atmospheric Trajectory (BAT) model

The BAT model is a three-layer Lagrangian model that reflects plume spread from vertical wind shear. The model incorporates the following features:

- Calculations made forward or backward in time (e.g., forward from a source for episodic or climatological calculations; backward from a receptor for source-receptor attribution studies).
- The plume is approximated by a series of puffs starting every 3 hours, each with a transport duration of 7 days.
- Three layers include a fixed-depth nighttime surface layer, a variable-depth boundary layer, and an upper layer of fixed height.
- Transport uses layer-averaged observed winds, inverse distance-squared wind weighting for advection, and modified Euler advection to approximate acceleration.
- Trajectories branch with changing boundary layer depths and day/night transitions, and each branch is followed independently.
- Dispersion results from vertical mixing and Gaussian horizontal diffusion.

The original definition of the three layers in BAT is shown pictorially in Fig. 4a. On the left, hypothetical observed rawinsonde data are plotted (potential temperature, θ, versus height, Z) and a critical inversion defined such that the lapse rate, $\Delta\theta/\Delta Z$, is greater than 0.005°K/m and the temperature change between the inversion top, θ_T, and bottom, θ_B, is greater than 2°K. The three resulting layers are shown on the right for daytime and nighttime. During the day, the boundary layer is defined from the surface to the critical inversion, and an upper layer from the critical inversion to a fixed height of 3000 m. During the night, a fixed depth surface layer of 300 m is established with the boundary layer and upper layer defined as during the day except that the nighttime base of the boundary layer is at 300 m.

BAT layer branching has recently been revised and is explained in Fig. 4b. The height above the surface is plotted on the vertical scale. A lowering inversion (left) branches the boundary layer into a boundary and an upper

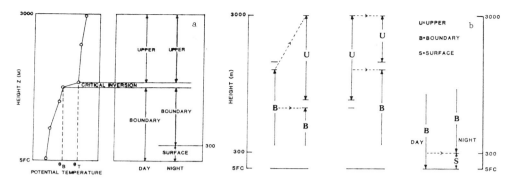

Fig. 4. Definition of the three BAT boundary layers. (a) BAT boundary layer determination on the left with the critical inversion criteria, $\Delta\theta/\Delta Z > 0.005°$ K/m and $\theta_T - \theta_B \geq 2°K$, and the original multiple layer definition on the right. (b) Revised direction of BAT layer branching.

layer, while a rising inversion (middle) branches the upper layer into an upper and a boundary layer. During a day-to-night transition (right), the boundary layer branches into a nighttime boundary and surface layer. It should be noted that the nighttime boundary layer does not contribute to calculated concentrations at the surface since it does not reach the surface.

An example of a BAT trajectory, obtained from one of several program output options, is shown in Fig. 5 for CAPTEX release #3 at Dayton, OH (asterisk). Daytime boundary and upper layers are coded B and U, respectively. A nighttime surface layer is coded plus (+) and a minus (-) indicates a nighttime boundary or upper layer. Thus, day transport (B and U) is easily distinguished from that at night (+ and -). It should be noted that only B and + layers contribute to surface concentrations in the BAT calculations. Numbers along the trajectories at day/night transitions indicate coded percent of start mass in 20% interval categories centered around the number (e.g., 5 = 40% to 60%).

2.4 BAT model evaluations

Model evaluation should include qualitative techniques such as comparing measured and calculated data patterns, and assessing over and undercalculation from scatter diagrams, as well as obtaining statistics for verifying calculations. A variety of statistical techniques have been utilized to determine confidence estimates for model calculations of pollution concentration and deposition when the calculations are applied to gain insight into environmental questions (ref. 13, 14). However, many of these techniques give inconclusive results when applied to the temporal and spatial scope of long-range model evaluation and model calculation verification. A statistical approach is outlined here for specific application to these topics.

   (i) Concentration patterns. Maps and diagrams for visual inspection of

444

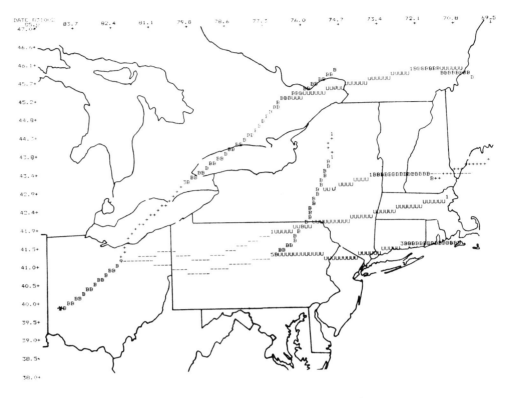

Fig. 5. BAT branching trajectories for CAPTEX release #3.

measured and calculated data are essential products to be used with the application of statistical comparison techniques. Often, seeing a measured and calculated concentration or deposition pattern gives insight into whether the modeled processes are reflecting reality in a reasonable manner. As an example, the BAT maximum calculated concentration pattern for CAPTEX release #3 is shown in Fig. 6, which can be visually compared to the measured pattern in Fig. 2b. The patterns are not dissimilar, both reflecting the large vertical wind shears. The initial low level transport to the north-northeast is evident in both patterns (even though it is further to the northeast in the calculated pattern), and the upper flow to the northeast and east-northeast with subsequent mixing to the surface at later times is also reasonably reflected in the calculations. In general, overcalculation is evident for this case.

(ii) Scatter diagrams. An essential tool in the process of model evaluation, Fig. 7 shows a typical scatter diagram, in this case for CAPTEX release #2. The axes give measured (horizontal) and calculated (vertical) concentration in fL/L plotted on a logarithmic scale. The slanted line across the

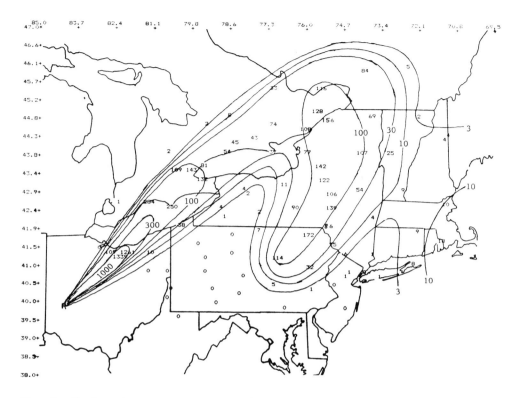

Fig. 6. Maximum tracer concentration (fL/L) calculated from BAT at each sampling site after CAPTEX release #3.

diagram is the one-to-one correspondence line. A paired calculated/measured concentration plotted above the line indicates overcalculation; below the line undercalculation. Plotted numbers indicate frequency of pairs at a point. Frequencies are given at the four corners of the diagram for the total number of pairs on the diagram (upper right), number of pairs along each axis of the diagram (upper left and lower right), and number of pairs when both measured and calculated values are less than a selected value (lower left)--3 fL/L in this case.

Of particular interest are the values along the axes either on or off the diagram, an important distinction necessary for statistical evaluation. For example, consider a pair with M (measured concentration) = 0 and C (calculated concentration) = 10. If a non-zero value is measured within 1° lat. (for CAPTEX statistics), this indicates the edge of the measured pattern to be close by. The zero concentration is then changed to the threshold axis value (1) and hence plotted along the axis on the diagram (one of the 4 pairs plotted 1/4 distance up the vertical axis). This is a reasonable procedure considering the

446

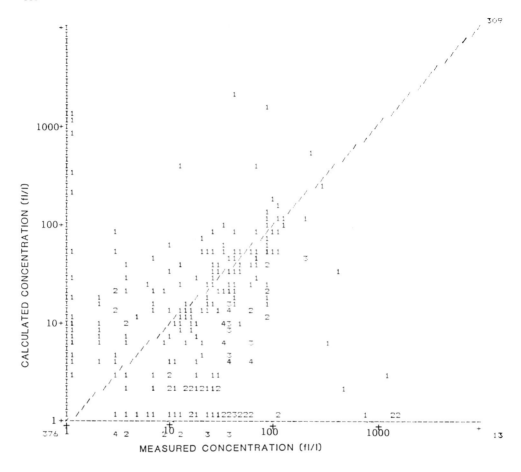

Fig. 7. Scatter diagram for CAPTEX release #2. Plotted numbers represent the frequency of pairs at that point.

uncertainties of pattern edges. If no non-zero values are measured within the scan radius, the concentration remains unchanged at 0 and is plotted on the axis off the diagram. When ratios C/M are considered in statistical evaluation, a reasonable distinction can then be made between C/M = 10/1 = 10 and C/M = 10/0 = ∞.

The overall features of the diagram show a wide scattering of points with some clustering along the one-to-one correspondence line and numerous points along the axis (these axis points seem to be the rule rather than the exception for model verification studies and hence the term "L-shaped" scatter diagram has been applied). The higher frequency of points below the one-to-one correspondence line illustrates a median undercalculation by the model for this case.

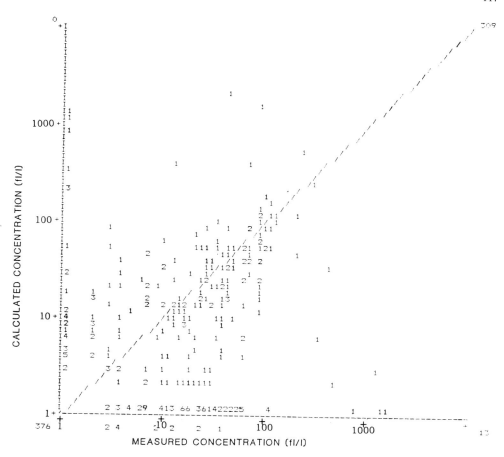

Fig. 8. Scatter diagram for CAPTEX release #2. Plotted numbers give the
spatial error (degrees of latitude) required to find a measured concentration
comparable to the model calculated concentration at that location.

Besides considering overcalculation and undercalculation, it is of interest
to determine a spatial error for a paired value. For example, given the pair
C = 10 and M = 1, how far in space would you have to go until you found a value
on the measured pattern where M = 10? If the distance is short, the model
calculation would be reasonable, even with a 10 to 1 ratio. The error might be
attributed to meteorological variability or small space and time differences on
a pattern with larger gradients. A big spatial error would indicate a model
problem. The scatter diagram of Fig. 8 shows the spatial error (in degrees
latitude) for each concentration pair for CAPTEX release #2 (the plotted points
are the same as in Fig. 7). Spatial errors of 1° or 2° lat. for CAPTEX
distances are not considered serious. Note that most of the axis pairs have

448

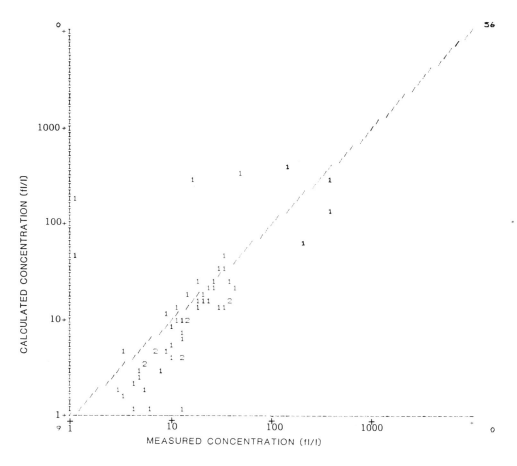

Fig. 9. Scatter diagram for CAPTEX release #2 (frequency of space- and time-averaged values).

this degree of error indicating that even though ratios of C/M might be large (or small), the model was performing in a reasonable manner (see, for example, the high calculation values near 1000 fL/L on the vertical axis where M < 1). Space and time averaging of the concentrations should decrease this error. A demonstration of this can readily be seen on the scatter diagram of Fig. 9 showing frequency of paired values where concentrations were averaged over the entire sampling period and within approximately 1° latitude intervals.

(iii) Boxplots. One of the most rewarding statistical techniques for verifying model calculations, evaluating model sensitivity runs, or comparing different models is the use of the boxplot (ref. 15). Fig. 10 shows an example of a boxplot, where the vertical axis on the diagram is the ratio of calculated (C) to measured (M) concentration plotted on a logarithmic scale. The solid

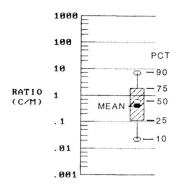

Fig. 10. Example of a boxplot
(Ratio is calculated/measured
concentrations).

dark line across the diagram corresponds to C/M = 1. The area above the line
indicates overcalculation; below the line undercalculation. The boxplot shows
the cumulative distribution of the ratios at percent values of 10, 25 (first
quartile), 50 (median), 75 (third quartile), and 90. In addition, the mean
ratio (solid dot) has been added to the boxplot. The boxplot depicted here is
for CAPTEX release #2 and reflects bias toward undercalculation by about a
factor of 2 (mean and median below the line at C/M = .5) with 90% confidence
limits within about a factor of 18 of the mean.

Fig. 11 gives two boxplots for each of the 7 CAPTEX releases using the BAT
model. For each release the boxplot on the right is for concentrations
averaged in time and space as discussed earlier; without averaging is shown on
the left. Included above each boxplot is the spatial error (°lat). It should
be emphasized that the ratio statistics in the boxplots presented here exclude
pairs where both measured and calculated concentrations are less than 3 fL/L
(lower left of the scatter diagrams). It has been determined that this concentration or larger is separable from analysis and contamination errors and
clearly indicates tracer pluming. Thus, only pairs with at least one value of
3 fL/L or more are included in these evaluation and vertification statistics.

Model calculation verification statistics differ significantly from release
to release. Release #2 is clearly the best case for both averaged and
unaveraged concentrations with relatively little bias (mean and median within
about a factor of 2) and confidence intervals encompassing the smallest ratio
limits. Release #4 is of interest in that model verification using unaveraged
concentrations is poor, but averaging greatly improves results (time averaging
was the greatest contributor to improvement in this case). Release #6 should
be viewed with caution since it has too few values to make reasonable verification conclusions.

450

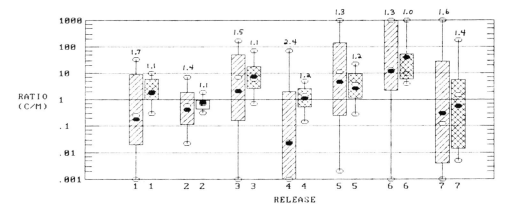

Fig. 11. Boxplots for the seven (7) CAPTEX releases of the BAT model-
calculated-to-measured concentration ratios (for each release, unaveraged on
the left and time- and space-averaged on the right). Spatial errors in degrees
latitude are given above each boxplot.

The boxplots in Fig. 12 are for all 7 CAPTEX releases combined (for each
pair, averaged in time and space on the right and unaveraged on the left). In
general, BAT calculated plume concentrations for CAPTEX (A--left pair of
boxplots in the diagram) are on the average within about a factor of 2 of the
measured values with 90% confidence intervals of about one order of magnitude
when time and space averaging is considered and two or more orders of magnitude
for unaveraged values. In addition, the average spatial error for calculated
concentrations is a little over 1 1/2° lat. and improves to a little over 1°

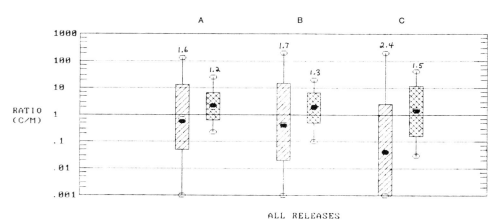

Fig. 12. Boxplots for all 7 CAPTEX releases combined (for each pair: unaveraged on the left and time- and space-averaged on the right). The left pair (A)
utilized all wind data, the center pair (B) did not use time-enhanced wind
data, and the right pair (C) used a single level wind with a constant mixing
height. Spatial errors in degrees latitude are given above each boxplot.

lat. with time- and space-averaging.

Of special interest are two sensitivity runs made with the BAT model and included in Fig. 12. The first (B--center pair of boxplots), uses only 00 GMT and 12 GMT winds (i.e., excluding the time enhanced rawinsonde data at 06 GMT and 18 GMT) with the exception of the release sites where the 06 GMT and 18 GMT rawinsonde data were included. It is interesting to note that these calculated concentrations do about as well as those including the time enhanced winds (A). This might be explained in part by the fact that tracer was released only when it was forecast to be transported over the sampling network, and the confidence for a forecast was greatest during relatively homogeneous, nonvariant wind conditions. The second sensitivity run (C--right pair of boxplots) was made by modifying BAT to reflect the often used constant mixing height, one layer model (in this case 1500 m above the surface), with transport winds at the 900 mb level only. The comparison with the standard BAT model is poor for unaveraged values but could be considered acceptable for averaged values.

3 COMPLEX TERRAIN TRACER MEASUREMENTS AND MODEL VALIDATION

The Department of Energy is currently sponsoring a program of Atmospheric Studies in Complex Terrain (ASCOT) to improve the technology needed to assess the air quality impacts of developing energy resources in areas of complex terrain. The program uses theoretical atmospheric physics research, mathematical models, and field experiments to help develop a modeling and measurements methodology that can be used to provide the air quality assessments in these areas.

With the program's initial focus being on the study of transport and dispersion of materials injected in or near nocturnal drainage flows, a series of exploratory field experiments of limited scope were conducted during July 1979 in the Anderson Creek valley of The Geysers geothermal area in northern California. The analyses of the results derived from those experiments provided initial insight into the structure of the drainage flows and permitted the design of a more comprehensive series of experiments that were conducted during September 1980 in the same valley. The experimental plan for the September 1980 studies consisted of five separate and identical experiments. Each experiment included multiple tracer releases that were coordinated with a series of meteorological measurements.

The experiments were conducted over an area of approximately 10×7 km^2 (enclosed area in Fig. 13a) in the California Known Geothermal Resources Area, approximately 130 km north of San Francisco. The area is surrounded by three mountains, Boggs Mountain (elevation approximately 1000 m) to the north, Cobb Mountain (1300 m) to the northwest, and Pine Mountain (1000 m) to the

452

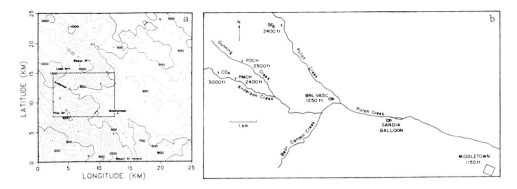

Fig. 13. Maps of the Geysers experimental area. (a) Topography of the region surrounding the experimental (enclosed) area. (b) Experimental area details showing locations of tracer release sites in the various creeks and the BNL and Sandia vertical sampling cables.

southwest, which form a valley opening toward the east-southeast. Surface vegetation is heterogeneous, varying from dense forest to almost bare soil.

Following a brief description of the meteorological and tracer experimental plans, measurements aloft and on the ground of the concentration of one of the tracers will be presented for the second of the five experiments, wind fields will be computed from a hydrodynamic model and compared to actual data, and surface tracer concentrations will be calculated from a random particle statistical model and tested for accuracy against the measured results.

3.1 Field meteorological and tracer system measurements and results

(i) Meteorological measurements. The systems used in these tests included four acoustic sounders, six tethersondes, five laser and optical anemometers, radiation sensors, soil temperature profile measurements, meteorological towers, an airborne multi-spectral scanner for surface temperature measurements, radiosondes, minisondes, and pressure sensors.

The acoustic sounder and tethersonde measurements were specifically designed to address the general program objectives of (1) defining the spatial and temporal characteristics of the drainage flows, (2) characterizing the pooling of drainage flows, and (3) an evaluation of the mechanisms responsible for draining the pool out into the Middletown area (Fig. 13b). Thus, most of the instrumentation was situated within the lower parts of the valley, while one tethersonde was dedicated to defining the larger scale flows over the ridge.

A network consisting of 27 surface meteorological stations, capable of telemetering the measurements to a centrally located base station, provided (1) real-time displays of the surface winds throughout the study area used for selecting the most desirable time to release the tracers, and (2) a data base of surface winds and temperatures to be used in the post analysis and modeling phase. The optical devices were designed to obtain spatially averaged wind speeds across the principal drainage areas, that is, the most likely tracer flow paths, and a long-term continously operating network of 10-m meteorological towers was augmented by a 60-m tower instrumented at several levels to provide definition of the wind and temperature structure within the shallow drainage flows. Most of these measurements were performed in the region to the west of the BNL VASC site (Fig. 13b).

In addition, a triangular upper air measurements network was established to tie the measurements within the valley to the regional scale flows and an airborne infrared sensing system was used for measuring surface temperature gradients along the sloped surfaces.

(ii) <u>Tracer measurements</u>. The specific objectives of these measurements were to evaluate the transport and dispersion of materials injected into each of the main drainage areas of the Anderson Creek Valley. As shown in Fig. 13b, the gaseous tracers released were two PFTs, PDCH (perfluorodimethylcyclohexane) in Gunning Creek and PMCH in Anderson Creek, two heavy methanes (CD_4) further up from the PMCH site, and sulfur hexafluoride (SF_6) in Putah Creek. Additional tracers used consisted of oil fog tracked by Lidar and tetroons tracked by radar.

The gaseous tracers and the oil fog were released simultaneously over a one hour period after the nocturnal drainage flows had been established. All of the releases were at ground level except for one of the heavy methanes which was released into the transition flows above the shallow drainage layer. Since this layout did not include sequential releases of tracers, the tetroon tracking system was used to evaluate the temporal variabilities in the transport characteristics throughout each experiment. Thus, tetroons were released from a variety of sites within the major drainage areas throughout the experimental periods.

An extensive network of surface samplers and two vertical profiling systems were deployed throughout the Anderson Creek Valley. Two sampling duration times were used: (1) samplers with 10-20 minute sample averaging times to provide plume passage information and (2) samplers collecting with averaging times of the order of two hours. In addition, real-time sulfur hexafluoride and perfluorocarbon samplers were also utilized. The two vertical profilers, balloon-borne sampling systems, were operated by Brookhaven (BNL VASC) and Sandia National Laboratory at the two sites shown in Fig. 13b. One, the BNL VASC (vertical atmospheric sampling cable; ref. 1), consisted of a sampling cable suspended from a tethered balloon to enable air samplers, located on the ground, to collect air from four 400-foot (122-m) integrated heights up to 1600 ft (488 m) above the surface. The other profiling system utilized on-board

454

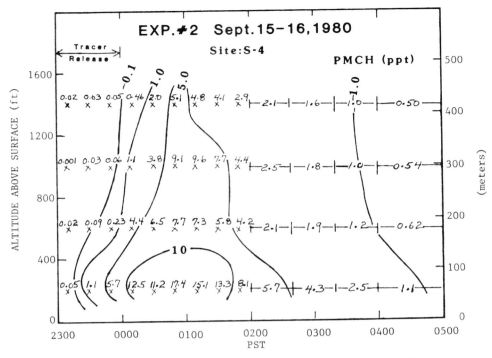

Fig. 14. PMCH concentrations aloft measured at the BNL VASC site (S-4).

samplers collecting samples at specific height intervals as the balloon was
hauled up and down. Several thousand tracer samples were collected from the
entire network during the five experiments, with about 1600 being analyzed for
the 2 PFTs; complete details of the meteorological and tracer measurement
results are available (ref. 16).

(iii) PMCH tracer results from experiment 2. The tracer release for this
second test took place from 2300 PST on September 15 to 0000 PST on September
16, 1980. A total of 493 grams of PMCH was released at the Anderson Creek site
(cf. Fig. 13b) in the one-hour period.

The PMCH concentrations of four elevations up to 1600 feet (488 m) at the
BNL VASC site (S-4) are shown as a function of time in Fig. 14. To obtain better time resolution, 20-min samples were collected for the first 3 h; 40-min
each for the next 2 h; and 60-min for the last hour (0400-0500). This experiment showed the most nearly uniform vertical PMCH concentration of the five
experiments. Although concentrations decreased with altitude, both PMCH and
PDCH exceeded 5 ppt (pL/L) in the 1200 to 1600 foot (366-488 m) layer, only a
factor of 3 to 4 times lower than the concentrations in the 0 to 400 ft (122 m)
layer. Thus, in experiment 2 (and also in experiment 5, but not shown), higher
additional layers should have been sampled. In the other three experiments,

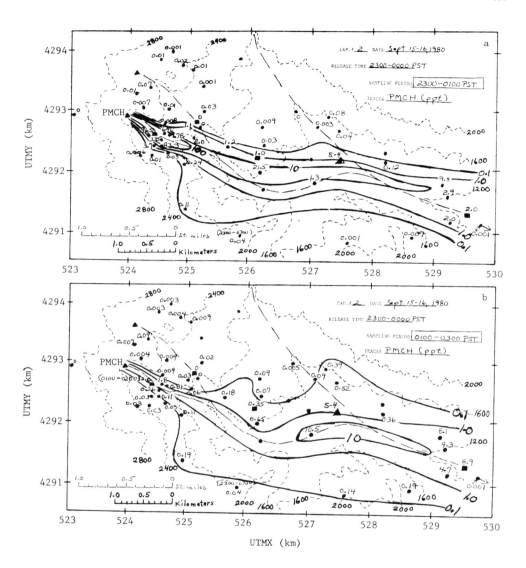

Fig. 15. Two-hour integrated PMCH concentrations on the ground surface of Anderson Creek Valley. Site S-4 (UTMX 527.6, Y 4292.1) vertical concentrations are shown in Fig. 14. Sampling period: (a) 2300-0100 PST, (b) 0100-0300 PST.

peak tracer concentrations in the highest layer sampled were about 0.2 ppt.

The PMCH ground level concentration patterns for the first two successive 2-h sampling periods commencing with the tracer release are shown in Fig. 15. Note that the tracer plume essentially follows the valley opening towards the east-southeast. In the section following the next, the measured tracer concentrations will be compared to model-generated results.

3.2 Comparison of measured and computed wind fields

A brief description of the meteorological modeling and calculational procedures will be given and then the results of the calculated wind fields will be compared to the measurements.

(i) Description of the meteorological model. The basic equations of HOTMAC (Higher Order Turbulence Model for Atmospheric Circulations) for mean wind, temperature, mixing ratio of water vapor, and turbulence are similar to those used by Yamada (ref. 17-18).

Surface boundary conditions are constructed from the empirical formulas of Dyer and Hicks (ref. 19) for the nondimensional wind and temperature profiles [see the Appendix of Yamada (ref. 17)]. Strictly speaking, the formulas are valid only for horizontally homogeneous surfaces. It is assumed, however, that the same relationships are fair approximations over nonhomogeneous terrain, provided that the formulas are applied sufficiently close to the surface. It should be noted that vegetation plays an active part in the apportionment of available heat energy between convective (sensible and latent) and conductive (into the soil) components. The technique discussed here is intended to address only the case of bare soil where the surface is conventionally characterized by roughness lengths. The complexity introduced by biological factors and drag forces due to tall trees (canopy flow) are beyond the scope of the present study.

Use of the similarity formulas requires a knowledge of the ground surface temperatures. The temperatures in the soil layer are obtained by solving the appropriate heat conduction equations; boundary conditions are the heat energy balance at the soil surface and specification of the soil temperature or soil heat flux at a certain depth. The lateral boundary values are obtained by integrating the corresponding governing equations except that variations in the horizontal directions are all neglected.

(ii) Computational procedure for experiment 2. An initial wind profile at the southwestern corner of the computational domain is first constructed by assuming a logarithmic variation (initially $u_* = 0.2$ m/s, and $z_0 = 0.1$ m) from the ground up to the level where the wind speed reaches the ambient prevailing value (3 m/s). Initial wind profiles at other grid locations are obtained by scaling the southwestern corner winds to satisfy the mass continuity equations. Wind directions in the upper layers observed by tethersondes were, in general, easterly. Thus, initial wind directions are assumed to be easterly everywhere.

The measurements by tethersondes indicate that vertical gradients of the potential temperature above the surface inversion layer are slightly positive ($0.5°K/1000$ m). Thus, the vertical profile of potential temperature is initially assumed to increase linearly with height, that is, $\theta = 295.2 + 0.0005 \ z$.

Initial potential temperatures are assumed to be uniform in the horizontal directions. Initial values for water vapor are constructed by using the initial potential temperature profiles, the pressure at the top of the computational domain (712 mb), and the observed relative humidity (50%). The turbulence kinetic energy and length scales are initialized by using the initial wind and temperature profiles and the relationships resulting from the level 2 model. These expressions are already given by Yamada (ref. 20) and are not repeated here.

The governing equations are integrated by using the Alternating Direction Implicit method and a time increment is chosen to satisfy the Courant-Friedrich-Lewy criteria. In order to increase the accuracy of the finite-difference approximation, mean and turbulence variables are defined at grids which are staggered both in the horizontal and vertical directions.

Mean winds, temperatures, and water vapor vary greatly with height near the surface. In order to resolve these variations, nonuniform grid spacings are used in the vertical direction. Grids of 21x26x16 (vertical) points are used to cover a computational domain of approximately $10 \times 8 \times 1.5$ km^3.

(iii) Comparison of calculated and measured wind fields. For experiment 2, initiation of the integration process began at 2200 PST and continued for eight hours. Since a staggered grid system was used, all variables were not computed for each of the grid points. However, for convenience in presenting the results, variables were interpolated to common grid points.

Initially, wind directions were assumed easterly everywhere. The temperature of the air close to the sloped surface becomes cooler during the nighttime (due to longwave radiation cooling) than the air at the same level but away from the surface. This temperature difference results in a horizontal pressure gradient which moves air down the slope, creating the so-called nocturnal drainage flows.

By 2330 PST, well-organized nocturnal drainage flows developed in the layer close to the surface (Fig. 16a). A grid point is located at the center of each wind vector, indicated by an arrow. The length of an arrow is proportional to the wind speed. An arrow of one grid length indicates 2 m/s. Directions of the surface winds would become parallel to the ground slopes if a large-scale pressure gradient was not imposed. The geostrophic wind (equivalent to a pressure gradient) at the top of computational domain ($z^* = 1460$ m) was easterly at 3 m/s.

Quantitative comparison of the computed results with observations encountered considerable difficulties for several reasons. For example, tethersonde data are instantaneous values, tower data are time averages, and laser anemometers are spatial averages over a distance between the transmitter and receiver. On the other hand, the computed values are an ensemble of means obtained over an entire grid volume. The number of observations was not

458

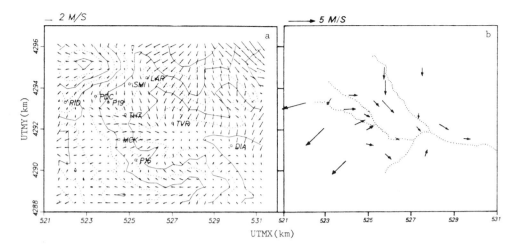

Fig. 16. Comparison of computed and observed wind fields for experiment 2.
(a) Computed wind vectors at 22 m above the terrain at 2330 PST. Symbols in
the figure indicate observational sites. Terrain is contoured by solid lines
with an increment of 200 m. The lowest contour is at 400 m above the mean sea
level. Dashed lines indicate contours halfway between the contours by solid
lines. (b) Observed wind vectors at 4 m above the ground. Wind speed and
direction are averaged for one hour between 2300 and 0000 PST.

sufficient to provide spatial averages comparable to the simulations. Nonethe-
less, some efforts were made to evaluate accuracy of the simulations.

Figure 16b represents horizontal wind vectors at 4 m above the ground
obtained by the network deployed by the National Center for Atmospheric
Research. Wind speed and direction were averaged for one hour between 2300 and
0000 PST. The computed (Fig. 16a) and observed (Fig. 16b) wind fields agree
qualitatively, although they are different in details. Separation between the
easterly ambient flows and westerly drainage flows occurred approximately at
the same locations for the observations and simulation. However, the computed
wind speeds are much smaller than the observed, particularly at the ridge locations. Observed wind speed and direction in the surface layer varied with time
although general flow pattern remained the same. On the other hand, the
computed wind fields in the surface layer had much smaller variations with time
than the observed.

Drainage flows develop first close to the ground, and increase in depth as
the air aloft cools by transferring heat energy toward the ground. Thus, it is
expected that the computed wind fields at 22 m would be of lower magnitude than
those measured at 4 m above the terrain (Fig. 16). In addition, computed wind
directions at 68 m above the surface (Fig. 17a) are still easterly at 2330 PST
except in a small area over the valley. As time progresses, drainage winds

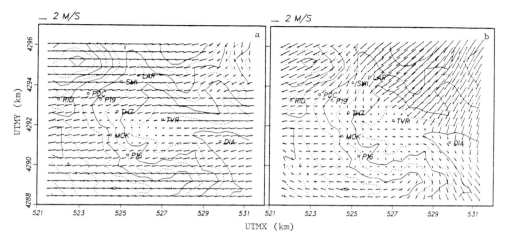

Fig. 17. Computed wind fields over the experiment 2 region at 68 m above the terrain. (a) At 2330 PST. (b) At 0200 PST.

develop at the higher level (Fig. 17b) and approach to a quasi-steady state.

Vertical profiles of the computed wind speed, wind direction, and potential temperature at 0200 PST at the ridge (RID) are shown in Fig. 18. Error bars on the wind speed profile indicate the computed variances of horizontal wind speeds. The computed wind directions are easterly except in a shallow layer (approximately 50 m) near the surface where the wind directions are north-easterly to northerly. The observed wind speeds at the ridge varied from 4 to 6 m/s between 0100 PST and 0320 PST.

3.3 Comparison of measured and computed tracer concentrations

Various tracer gases (perfluorocarbons, heavy methanes, SF_6, oil smoke) were used during the ASCOT experiment to simulate transport and diffusion of pollutants (e.g., H_2S) in the Geysers geothermal area. In this study, the distributions of the surface concentrations of PMCH shown earlier will be used to examine the performance of a random-particle statistical diffusion model. A brief description of this model will be given followed by computation of the model-predicted ground level PMCH concentrations and comparison with the actual measured ground level patterns.

(i) Model description. Locations of particles are computed from

$$x_i(t + \Delta t) = x_i(t) + U_{pi}\Delta t ,$$ (1)

where

$$U_{pi} = U_i + u_i ,$$ (2)

$$u_i(t + \Delta t) = au_i(t) + b\sigma_{ui}\zeta + \delta_{i3}(1 - a)T_{Lxi}\frac{\partial}{\partial x_i}(\sigma^2{}_{ui}) ,$$ (3)

460

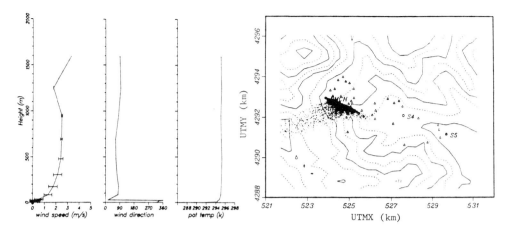

Fig. 18. Vertical profiles of the computed horizontal wind speed with error bars, wind direction, and potential temperature at the Ridge site.

Fig. 19. Trajectory of computed PMCH tracer particles projected on the surface at 0000 PST, 1 h into release. Ground samplers ($\triangle$) and vertical profilers (S4 and S5) are shown.

$$a = \exp\left(-\Delta t / T_{Lx_i}\right) \text{ , and} \tag{4}$$

$$b = (1 - a^2)^{1/2} \text{ .} \tag{5}$$

In the above expressions, U_{pi} is the particle velocity in x_i direction, U_i mean velocity, u_i turbulence velocity, ζ a random number from a Gaussian distribution with zero mean and unit variance, T_{Lx_i} the Lagrangian integral time for the velocity u_i, σ_{u_i} variance of velocity fluctuation u_i, and δ_{i3} is the Dirac delta. The last term on the right-hand side of Eq. (3) was introduced by Legg and Raupach (ref. 21) in order to correct accumulation of particles in inhomogeneous turbulent flows. The mean velocity U_i and velocity variances σ_{u_i} are obtained from the hydrodynamic model results discussed in Section 3.2(i).

(ii) Model-predicted ground level PMCH concentrations. Based on the steady release of PMCH from 2300 to 0000 PST and the mean wind fields from the hydrodynamic model, the random-particle model was used to compute the trajectory of particles on the surface at 0000 PST (Fig. 19). The model shows the main trajectory to be down the valley in an east-southeasterly direction, but with a significant amount of particles moving towards the west-southwest because of the easterly prevailing winds aloft. The particles displayed existed in a vertical column at each grid area up to the top of the computational domain (1500 m) and do not necessarily reflect potential ground level concentrations.

Comparing the total vertical column of particles (Fig. 19) with the actual

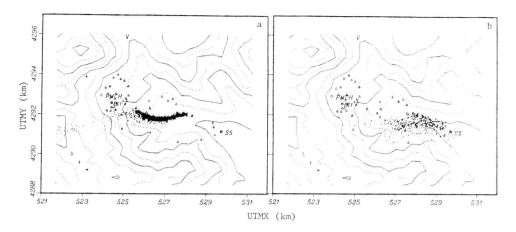

Fig. 20. Trajectory of PMCH tracer particles projected onto the surface during experiment 2 (similar to Fig. 19 at 0000 PST). (a) At 0100 PST. (b) At 0200 PST.

measured ground level PMCH concentrations measured from 2300 to 0100 PST (Fig. 15a), it is apparent that the tracer particles moving under the influence of the prevailing easterly winds were moving aloft to the west-southwest (Fig. 19) as there was no measurable concentration on the ground (Fig. 15a). By 0100 PST (Fig. 20a), the model still showed no tracer reaching site S5 at a UTX of about 529.7; but the ground samplers, sampling from 2300 to 0100 PST (Fig. 15a), showed a significant concentration (2.0 pL/L) at site S5.

The model shows that the plume bent slightly northward at site S4 by 0100 PST (Fig. 20a); as shown in the ground measured concentrations, a bulge in the tracer concentrations on the north slope of the valley was evident at site S4 in the 0100 to 0300 PST sampling period (Fig. 15b). The model-predicted maximum particle count at 0200 PST (Fig. 20b), located just before site S4 and stretching almost to site S5, did coincide with the maximum ground-level concentration isopleth of 10 ppt (Fig. 15b) at that time.

The actual comparison between the model-predicted and observed ground level concentrations of PMCH was obtained by considering an imaginary box of 10x10x4 (vertical) m^3 at each sampling location and counting the number of particles in the box for a two-hour period between 2300 PST and 0100 PST from which average concentrations were determined. A comparison between the computed and observed concentrations is shown in Fig. 21, where the solid line is the one-to-one correspondence line, the inner dashed lines show the range where computed and observed values are within a factor of 2 and the outer dashed lines for a factor of 10.

A relatively high (0.82) correlation coefficient was obtained between the

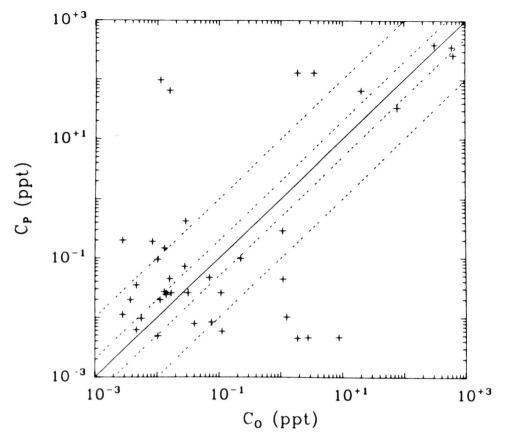

Fig. 21. Scatter diagram of the model-predicted (C_p) and observed (C_o) average PMCH tracer concentrations for the period between 2300 and 0100 PST at the sequential sampling sites.

observations and computations since high values (>10 ppt) of concentration were simulated well. On the other hand, a large disagreement between the observations and computations existed for low values (<10 ppt) of concentration which resulted in the discrepancy between the observed (46 ppt) and computed (61 ppt) average concentrations. Other statistics indicate that 20% of the computed concentrations were within a factor of 2 of the observation, 50% within a factor of 5, and 65% within a factor of 10.

For the model performance to be considered good, it must simulate well not only the high but also the low values of concentration. The low concentrations occurred, in general, away from the source or away from the plume center line. A small error in the computed wind direction could result in a large error in the computed plume location away from the source. The magnitude of plume spreading depends on the mixing due to atmospheric turbulence variables. A combination of errors in the computed wind and turbulence contributed to the largest portion of the scattering in Fig. 21. Considering the complexities (vegetation, topography, stability) involved in the present simulations, the model performance summarized in Fig. 21 might be encouraging. The spatial error and space- and time-integrated concentration scatter diagramming technique as well as the boxplot technique used in the CAPTEX '83 evaluations might prove useful in quantifying the performance of the complex terrain model, but this has not been done.

4 SUMMARY

PFT technology, including tracers and samplers, were deployed satisfactorily in both the long-range (CAPTEX) and complex terrain (ASCOT) terrain studies.

In CAPTEX '83, a single tracer, PMCH, was released seven times and ground level as well as aircraft samplers measured its concentration in the air from its background level of 4 fL/L up to maximum concentrations of typically less than 1 pL/L (ppt) over the range from 300 to 1100 km downwind of the two release sites -- Dayton, OH, or Sudbury, Ontario.

The BAT (branching atmospheric trajectory) long-range transport model was described and applied to each of the seven releases. The measured and calculated maximum ground level concentration patterns visually agreed for the example shown (CAPTEX release #3). Through the use of scatter diagrams of not only the measured and calculated concentrations but also the spatial error and frequency of space- and time-averaged concentrations, a clearer demonstration of the prediction capability of the model was presented. Boxplots were used to summarize the effectiveness of the model by presenting the model calculation verification statistics in a readily discernible form. Release #2, for example, showed a median bias toward undercalculation of about a factor of 2, with 75% confidence limits (50% of the points) within about a factor of 4 of the mean and within a factor of 18 of the mean for 90% confidence limits (80% of the points).

For ASCOT '80, a three-dimensional, time-dependent hydrodynamic primitive equation model was used to simulate the distribution of nocturnal drainage winds, temperature, and turbulence in the California Geysers area. The case studied here (experimental night 2) was under the condition that the ambient airflow was easterly and the direction of main drainage flow was approximately westerly. The computed depth of drainage flow was less than 50 m at the ridge, and increased to 200 m at the valley, which agreed well with the observations. The computed wind speed at the ridge was smaller than the observation, but it agreed well at the valley site. Vertical wind variances measured by a doppler

464

sodar indicated relatively large turbulence despite the fact that stable density stratification dominated throughout the night. The computed vertical wind
variances were also larger than those observed over flat terrain, but smaller
than the values recorded by the doppler sodar.

The computed wind vectors and turbulence were used to simulate the surface
concentration of a perfluorocarbon tracer gas. The model used was based on a
random-particle statistical method. One particle per second was released for
one hour at the source site, and the locations of particles were computed at
every 10 sec. An imaginary box of 10x10x4 (vertical) m^3 was considered at each
of the 39 surface sampling stations, and the number of particles passing
through each box for two hours was counted, from which the concentrations were
determined. A cross-correlation coefficient of 0.82 was obtained between the
computed and observed concentrations at the 39 sampling stations. Other statistics indicated that 20% of the computed concentrations were within a factor
of two of the observations, 50% within a factor of 5, and 62% within a factor
of 10.

ACKNOWLEDGMENT

The CAPTEX and ASCOT programs involved the contributions from many laboratories, universities, and groups, both experimentalists and modelers. Appreciation is expressed to the many colleagues who participated with resources,
time, and talent.

REFERENCES

1 R.N. Dietz, Perfluorocarbon tracer technology, Proc. Int. Course on
   Regional and Long-Range Transport of Air Pollution, Ispra, Italy, September
   15-19, 1986, Elsevier, Amsterdam, 1987.
2 G.J. Ferber, J.L. Heffter, R.R. Draxler, R.J. Lagomarsino, F.L. Thomas,
   R.N. Dietz and C.M. Benkovitz, Cross-Appalachian Tracer Experiment (CAPTEX
   '83) Final Report, January 1986, NOAA Technical Memorandum ERL ARL-142.
3 M.H. Dickerson, An overview, current status, and future plans for the DOE
   Atmospheric Studies in Complex Terrain (ASCOT) Program. Second conference
   on Mountain Meteorology, November 9-12, 1982, Steamboat Springs, Colorado,
   American Meteorological Society, 1981, pp. 10-13.
4 G.J. Ferber, K. Telegadas, J.L. Heffter and M.E. Smith, Air concentrations
   of krypton-85 in the midwestern United States during January-May 1974,
   Atmos. Environ., 12 (1982) 2763-2776.
5 R.R. Draxler, Measuring and modeling the transport and dispersion of
   krypton-85 1500 km from a point source, Atmos. Environ., 12 (1982)
   2763-2776.
6 G.J. Ferber, K. Telegadas, J.L. Heffter, C.R. Dickson, R.N. Dietz and P.W.
   Krey, Demonstration of a long-range atmospheric tracer system using
   perfluorocarbons, April 1981, NOAA Technical Memorandum ERL ARL-101.
7 J.L. Heffter, J.F. Schubert and G.A. Mead, Atlantic coast unique regional
   atmospheric tracer experiment (ACURATE), 1984, NOAA Technical Memorandum
   ERL ARL-130.
8 R.R. Draxler, J.L. Heffter and B.J.B. Stunder, Across North America Tracer
   Experiment (ANATEX) comprehensive plan, January 1987, NOAA Air Resources
   Labratory, Silver Spring, MD.

9 J.L. Heffter, A.D. Taylor and G.J. Ferber, A regional-continental scale transport, diffusion and deposition model, 1975, NOAA Technical Memorandum ERL ARL-50.

10 J.L. Heffter, Air Resources Laboratories atmospheric transport and dispersion model (ARL-ATAD), 1980, NOAA Technical Memorandum ERL ARL-81.

11 J.L. Heffter, Branching atmospheric trajectory (BAT) model, 1983, NOAA Technical Memorandum ERL ARL-121.

12 T.W. D'Ottavio, R.W. Goodrich and R.N. Dietz, Perfluorocarbon measurement using an automated dual-trap analyzer, Environ. Sci. Tech., 20 (1986) 100-104.

13 R.J. Londergan and N.E. Brown, Validation of plume model statistical methods and criteria, 1981, EPRI Summary Report EA-1673-SY, Project 1616-1.

14 D.G. Fox, Judging air quality performance: a summary of the AMS workshop on dispersion model performance, Bull. Amer. Meteorol. Soc., 62 (1981) 599-609.

15 R. McGill, J.W. Tukey and W.A. Larsen, Variations of box plots, Amer. Statistician, 32 (1978) 12-16.

16 P.H. Gudiksen, ASCOT data from the 1980 field experiment program in the Anderson Creek Valley, California, 1983, Lawrence Livermore National Laboratory, USID-88874-80, ASCOT-83-1, Vol. 1-3.

17 T. Yamada, A numerical simulation of nocturnal draininage flow, J. Met. Soc. Japan, 59 (1981) 108-122.

18 T. Yamada, Simulations of nocturnal drainage flows by a $q^2 \ell$ turbulence closure model, J. Atmos. Sci., 40 (1983) 91-106.

19 A.J. Dyer and B.B. Hicks, flux-gradient relationships in the constant flux layer, Quart. J. R. Met. Soc., 96 (1970) 715-721.

20 T. Yamado, The critical Richardson number and the ratio of the eddy transport coefficients obtained from a turbulence closure model, J. Atmos. Sci., 32 (1975) 926-933.

21 B.J. Legg and M.F. Raupach, Markov-chain simulation of particle dispersion in inhomogeneous flows: the mean drift velocity induced by a gradient in Eulerian velocity variance, Boundary Layer Met., 24 (1982) 3-13.

Regional and Long-range Transport of Air Pollution,
Lectures of a course held at the Joint Research Centre, Ispra, Italy,
15–19 September 1986, S. Sandroni (Ed.), pp. 467–478
© Elsevier Science Publishers B.V., Amsterdam — Printed in The Netherlands

EVALUATION PROCEDURE FOR LONG TERM AVERAGE MODELS

W. KLUG

1. INTRODUCTION

The last ten years or more have seen a great increase in the number of air pollution models. Although there are models developed for different applications, for different time- and space-scales there is also a number which were developed for the same purpose. If we have, as an example, 10 different models which claim that they can simulate SO_2 and SO_4 concentrations and depositions as a yearly average over Europe how can we determine quantitatively the quality of performance of these models? This seems to be a simple task but - as usual - the simple tasks often turn out to be very tricky if one tries to solve them. One way of attacking this problem - but only partly - is to compare the model results against each other. From this comparison we can already see - if the models are run on the same input data base - whether these models give more or less the same results or whether they differ to a large extent. This comparison can only give a preliminary answer because we are mainly interested in the answer how well does the model simulate reality or how well or bad compare the model results with observations, which we take - with some reservations - to represent reality. These reservations are justified because there are measurement errors, either instrumental or recording errors, but also - and this is very often the case - the observations are by their very nature not directly comparable with the model results. As an example we consider the results of a long term average box model, the boxes of which have horizontal dimensions of, say, 50 km and a vertical extension of a 1000 m. The results of this model are then concentrations <u>averaged</u> over a volume of 50 km x 50 km x 1000 m, whereas the depositions are averaged over an area of 50 km x 50 km. Observations are usually made at one certain geographical location and at the surface. It is obvious that these two values are not commensurable, although we are forced to take them because of lack of more suitable data.

2. SCATTER-DIAGRAM AND THE CORRELATION COEFFICIENT

The first and most simple step in evaluating the model results of a long term average model is to plot the observed and calculated concentrations /

depositions in one graph. At the same time it is helpful to calculate the regression line obtained after the method of least squares and the correlation coefficient. If c_0 stands for the observed value and c_c for the calculated, then the estimate of the mean values reads:

$$\overline{C}_0 = \frac{1}{N} \sum_{i=1}^{N} C_{oi} \quad , \quad \overline{C}_c = \frac{1}{N} \sum_{i=1}^{N} C_{ci} \quad . \tag{1}$$

It is obvious that c_0 is the measured value at the location of the observing station, but c_c must normally be obtained by interpolating from the computational grid values on to the location of the observing station. $\overline{c}_0$ and $\overline{c}_c$ are estimates of the mean values of the corresponding quantities and N the number of observations.

The estimate of the variances of c_0 and c_c are also needed. They are defined by:

$$S_0^2 = \frac{1}{N-1} \sum_{i=1}^{N} \left(C_{oi} - \overline{C}_0 \right)^2 \quad , \quad S_c^2 = \frac{1}{N-1} \sum_{i=1}^{N} \left(C_{ci} - \overline{C}_c \right)^2 . \tag{2}$$

The covariance of c_0 and c_c is given by:

$$S_{oc} = \frac{1}{N-1} \sum_{i=1}^{N} \left(C_{oi} - \overline{C}_0 \right)\left(C_{ci} - \overline{C}_c \right) . \tag{3}$$

With these definitions the estimate of the correlation coefficient r is obtained by:

$$r = \frac{S_{oc}}{S_0 S_c} . \tag{4}$$

The estimate of the regression line Y of C_0 on C_c is then given by:

$$Y = \overline{C}_0 + \frac{S_{oc}}{S_c^2} \left(C_c - \overline{C}_c \right) , \tag{5}$$

which can also be written as

$$Y = a_{oc} + b_{oc} C_c \tag{6}$$

with

$$a_{oc} = \bar{C}_o - b_{oc}\,\bar{C}_c \quad \text{and} \quad b_{oc} = \frac{S_{oc}}{S_c^2} \; . \tag{7}$$

The correlation coefficient r is a measure of the stochastic linear dependance
between the two variables considered. Its value is limited by $-1 \leq r \leq +1$.
If r is -1 or $+1$ there is a functional dependency between the two variables; if it is 0 the two variables are independent.

The first test is whether the computed correlation coefficient r is statistically significant different from 0 . The test quantity is given by

$$\frac{t_{2\alpha}}{(\nu + t_{2\alpha}^2)^{1/2}} \tag{8}$$

where $t_{2\alpha}$ is obtained from the student-t-distribution with error probability
2α and the degrees of freedom $\nu = N - 2$. Is $|r|$ smaller then (8), then it
can be assumed that the computed correlation coefficient does not differ
significantly from 0 .

3. DIFFERENCES IN THE CORRELATION COEFFICIENT

If we have two models and the corresponding estimated correlation coefficients are r_1 and r_2 , the question arises whether the difference between
these two correlation coefficients is statistically significant. The test
quantity to be considered is

$$\frac{|z_1 - z_2|}{\left(\frac{2}{N-3}\right)^{1/2}} \tag{9}$$

where

$$z_{1/2} = \frac{1}{2} \ln \frac{1 + r_{1/2}}{1 - r_{1/2}} \; . \tag{10}$$

These test limits C_α are given for the error probabilities

2α	0.05	0.01	0.001
C_α	1.9600	2.5758	3.2905

Is therefore the test quantity according to eq. (9) smaller than the C_α then it can be assumed that the difference between the two correlation coefficients is not statistically significant different from 0 .

4. THE MODEL'S BIAS

The first quantity which one has to consider when evaluating the performance of a model is the so called bias which is the difference in the mean values, i.e. $\bar{c}_0 - \bar{c}_c$. If this quantity is statistically significant different from 0 , which can be tested with the student's-t-test, this is already an important information on the performance of the model. However, if the difference is not statistically different from 0 , this does not mean that the model is performing well. After checking the model results in a scatter diagram mentioned before other statistical quantities have to be considered.

5. THE MODEL PERFORMANCE ERROR

The correlation coefficient was already mentioned, but again it is one more information more only and does not reveal all features of the model behaviour. Our interesting quantity which was proposed by Hanna (1984), is the normalized mean square difference between model and observation values. This number is called here the model performance error and denoted by M.

$$M = \frac{\frac{1}{N-1} \sum_{i=1}^{N} (c_{ci} - c_{oi})^2}{\left(\frac{1}{N} \sum_{i=1}^{N} c_{oi} \right)^2} \qquad (11)$$

It should be mentioned that in a more recent paper Hanna (1986) modified the definition slightly into a M^* :

$$M^* = \frac{\frac{1}{N-1} \sum_{i=1}^{N} (c_{ci} - c_{oi})^2}{\frac{1}{N} \sum_{i=1}^{N} c_{oi} \cdot \frac{1}{N} \sum_{i=1}^{N} c_{ci}} \qquad (12)$$

Obviously, the model performance error goes to zero for a perfect agreement between model results and observations. But it is also clear from the definition that the large differences determine the value of M , which is - by the way - also true for the correlation coefficient, where few large values with the remaining ones being small, determine the size of the correlation coefficient.

We are now confronted again with the problem that if we have two models one has a model performance error of M_1 and the other of M_2, how do we test that the difference $M_1 - M_2$ is statistically significant different from 0 . The significance test for the M-values cannot use the normal statistical procedure of the t-test since we have only one M_1-value and one M_2-value, but we need the standard deviations of M_1 and M_2 to perform the test. The standard deviation can of course only be obtained if there is an ensemble of M_1- and M_2-values. In order to overcome this difficulty a method has recently been developed by Efron (1982), which is called the 'boot strap method'. In order to obtain an ensemble of say, M_1-values, one takes the N pairs of observed and calculated values and calculates the first M_1-value. In the bootstrap procedure N new pairs are then selected randomly (with replacing the used ones back into the ensemble) from the original set and calculates a new M_1-value. This is done 100-1000 times and an ensemble of M_1-values is created. This data set has of course a $\bar{M}_1$ and a $\overline{M_1^2}$, an estimate of the mean value and of the standard deviation. We can now easily apply the t-test for statistically significant differences in the mean values.

A statistically even more powerful test is the test of $\overline{M_1 - M_2}$ against 0. If one applies this method one finds that the M-value is a very sensitive number and together with the bias and the correlation coefficient gives a good description of the model performance in a statistical sense.

6. PRACTICAL APPLICATION

In a recent report (Klug and Lüpkes (1985)) the results of 10 different long term average models were compared using the methods presented here. All models were run on an area which covers middle Europe excluding a part of the Mediterranean Sea and Southern France. The emission inventory taken was that of EMEP for 1978 with a grid length of 150 km. For the intercomparison project this was reduced by interpolation on a 127 km x 127 km grid. The wind rose (uniform for the whole computational area) was obtained from hourly measurements made at the top of the Belmont Tower in Lincolnshire, 389 m above ground during the period 1969-1974. To include the variations of probability of precipitation with mean wind direction a wind precipitation rose was used which Smith (1981) obtained from a limited analysis of statistics for Central England. The mixing layer height was taken to be constant and had a value of 1000 m. The chemical parameterization was linear. Results considered were SO_2- and SO_4-concentrations, SO_2- and SO_4- dry deposition, SO_2- plus SO_4- wet deposition and total S deposition.
In Fig. 1 - Fig. 3 we find the scatter diagrams for the Smith- and Klug-models for SO_2- and SO_4-concentration and SO_2- plus SO_4- wet deposition. In each graph

472

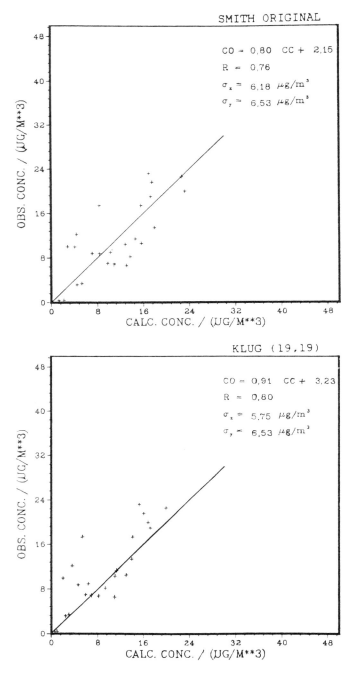

Fig. 1: Observed and calculated SO_2-concentrations in units of SO_2

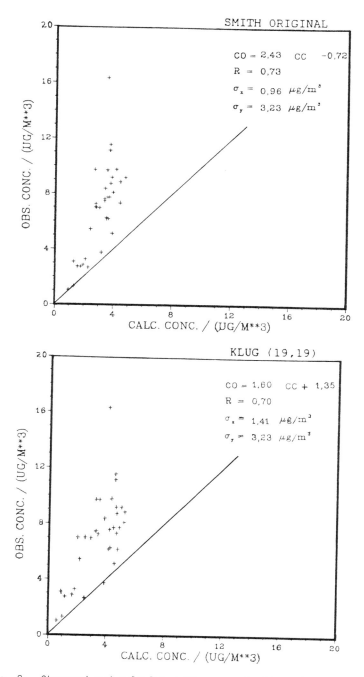

Fig. 2: Observed and calculated SO_4-concentrations in units of SO_4

474

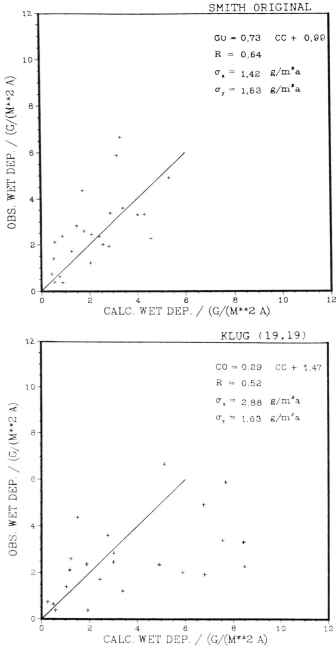

Fig. 3: Observed and calculated wet $SO_2 + SO_4$ deposition

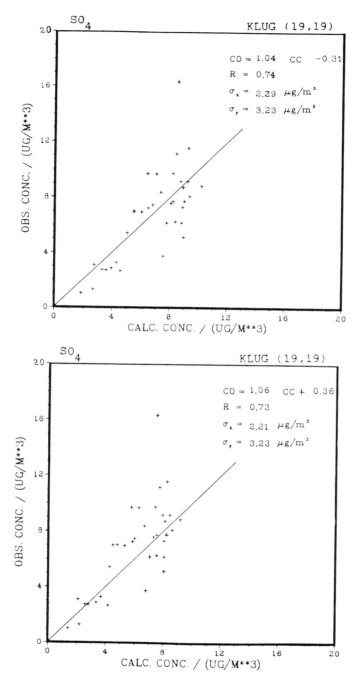

Fig. 4: SO$_4$-concentrations for KLUG model with mean wind velocity 7.9m sec^{-1} and removal rate 10%/hr above 15%/hr below

	SO$_2$ CONCENTRATION				SO$_4$ CONCENTRATION				SO$_2$ + SO$_4$ WET DEPOSITION					
	M.V.	S	R	M	M.V.	S	R	M	M.V.	S	R	M	ΣM	R̄
KLUG	8.40	5.36	0.78	0.19	3.31	1.37	0.69	0.39	3.65	2.53	0.48	0.86	1.44	0.65
RIVM	9.39	5.98	0.78	0.16	3.51	1.35	0.70	0.36	4.73	3.20	0.47	1.75	2.27	0.65
SMITH OR	11.16	6.18	0.76	0.15	3.11	0.96	0.73	0.44	2.24	1.42	0.64	0.26	0.85	0.71
SMITH SI	8.32	5.31	0.77	0.20	3.03	1.18	0.70	0.45	4.17	2.94	0.61	1.11	1.76	0.69
VENK CO ★	9.78	5.51	0.80	0.14	4.21	1.00	0.74	0.29	1.78	1.23	0.51	0.40	0.83	0.68
VENK SI ★	11.24	6.78	0.79	0.15	4.95	1.52	0.73	0.19	2.07	1.20	0.47	0.36	0.70	0.66
FISHER	14.52	8.88	0.80	0.30	7.89	2.20	0.76	0.12	2.07	1.24	0.51	0.34	0.76	0.69
TNO	6.73	3.98	0.78	0.29	2.41	0.99	0.65	0.57	4.07	2.14	0.48	0.84	1.70	0.64
MPA	9.09	6.02	0.77	0.18	3.36	1.47	0.69	0.38	4.62	3.30	0.45	1.80	2.39	0.63
KLUG ★	8.67	5.75	0.80	0.17	3.42	1.41	0.70	0.38	3.95	2.88	0.52	1.10	1.65	0.67
EMEP	11.17	8.41	0.83	0.17	6.30	2.95	0.77	0.10	3.23	1.34	0.83	0.17	0.44	0.81
OBSERVED	11.11	6.53			6.84	3.23			2.62	1.63				

M. V. = MEAN VALUE   S = STANDARD DEVIATION   R = CORRELATION COEFFICIENT

M = MODEL PERFORMANCE ERROR

Fig. 5: Comparison with measurements, units are for concentrations in $\mu g\, SO_{2/4}/m^3$, for depositions in $g\, SO_4/(m^2 a)$.

INVENTORY	WIND /m/s	WR /hr	VD4 m/s:%	SO2 CONCENTRATION M.V.	S	R	M	SO4 CONCENTRATION M.V.	S	R	M	SO2 + SO4 WET DEPOSITION M.V.	S	R	M	ΣM
(21,22)	9.	36	0.2	8.40	5.36	0.78	0.19	3.31	1.73	0.69	0.39	3.65	2.53	0.48	0.86	1.44
(21,22)	9.8	10	0.2	11.20	6.44	0.77	0.15	5.25	1.62	0.73	0.17	1.51	0.86	0.47	0.47	0.79
(21,22)	9.8	10	0.0	11.20			0.15	5.90			0.13	1.55			0.46	0.74
(21,22)	9.8	5	0.02	11.20			0.15	5.66			0.14	1.36			0.52	0.81
(19,19)	9.8	36	0.02	8.67	5.75	0.80	0.17	3.42	1.41	0.70	0.38	3.95	2.88	0.52	1.10	1.65
(19,19)	9.8	10	0.02	11.60	6.93	0.80	0.15	5.42	1.68	0.75	0.15	1.62	0.97	0.50	0.43	0.73
(22,26)	9.8	10	0.02	11.90	7.01	0.80	0.15	5.94	1.72	0.76	0.12	1.72	1.00	0.51	0.40	0.67
(19,19)	7.9	10	0.02	12.50	7.89	0.80	0.19	6.85	2.29	0.74	0.10	1.80	1.14	0.50	0.39	0.69
(19,19)	7.9	15	0.02	11.66	7.55	0.80	0.16	6.09	2.21	0.73	0.11	2.45	1.63	0.51	0.37	0.64
EMEP				11.17	8.41	0.83	0.17	6.30	2.95	0.77	0.10	3.23	1.34	0.83	0.17	0.44
OBSERVED				11.11	6.53			6.84	3.23			2.62	1.63			

M. V. = MEAN VALUE   S = STANDARD DEVIATION   R = CORRELATION COEFFICIENT
M = MODEL PERFORMANCE ERROR   VD4 = SO_4 DRY DEPOSITION VEL.   WR = WET REMOVAL RATE

CONCENTRATION IN μG $SO_{2/4}$ / M^3 DEPOSITION IN G SO_4 / (M^2 A)

Fig.6: Sensitivity analysis for KLUG - model

the regression line is given by the linear equation $c_0 = b_{oc}c_c + a_{oc}$, the correlation coefficient r and the standard deviations s_c and s_0 . In Fig. 2 it stands out clearly, that both models were deviating from the observed SO_4-concentrations in a systematic way. This deviation could be made much smaller in the Klug-model if the mean wind velocity was decreased to 7.9 m sec^{-1} and the wet removal rate to 10%/h respectively 15%/h. The tables represented in Fig. 5 and Fig. 6 obtain the statistical quantities for the different models and for a sensitivity analysis with the Klug-model.

7. CONCLUSIONS

A framework for evaluating the performance of long term average models has been suggested above. It has been shown that scatter diagrams, mean bias, correlation coefficient and the model performance error are suitable statistical performance measures. With the help of a sensitivity analysis the parameters could be optimized in order to get better performance of one of the investigated models.

8. REFERENCES

- Efron, B. (1982), The Jackknife, the Bootstrap, and other Resampling Plans, CBMMS-NSF-38, Soc. Ind. and Appl. Math., Philadelphia, 92pp.

- Hanna, S. (1984), A Simple Method of evaluating Air Quality Models, Environmental Research Technology Inc., Massachusetts.

- Hanna, S. (1986), A review of Air Quality Model Evaluation Procedures, paper presented at WMO International Conference on Air Pollution Modelling and its Application, Leningrad, USSR, May 1986.

- Klug, W. and C. Lüpkes (1985), A Comparison between Long Term Interregional Air Pollution Models, Final Report for the Umweltbundesamt, Berlin.

- Smith, F.B. (1981), The Significance of Wet and Dry Synoptic Regions on Long-Range Transport of Pollution and its Depositions, Atmosph. Environment 15, p. 67-98.

Regional and Long-range Transport of Air Pollution,
Lectures of a course held at the Joint Research Centre, Ispra, Italy,
15–19 September 1986, S. Sandroni (Ed.), pp. 479–501
© Elsevier Science Publishers B.V., Amsterdam — Printed in The Netherlands

THE RESPONSE OF LONG-TERM DEPOSITIONS TO NON-LINEAR PROCESSES INHERENT IN THE WET REMOVAL OF AIRBORNE ACIDIFYING POLLUTANTS.

F. B. SMITH

ABSTRACT

This paper explores the importance of non-linear processes that are inherent in the in-cloud oxidation and removal by rain of acidifying species on the dry, wet and total depositions of these species averaged over a long period of time, such as a year or longer.

Because of the immense difficulties in providing the correct input data, no attempt is made to represent the highly complex nature of these non-linear processes in any exact sense. Instead a simple parametrisation is used which conveys the correct spirit of these non-linearities, and is amply sufficient to determine the qualitative response of long-term depositions to them.

Equations are formulated and solved which model the fate of a primary pollutant, like sulphur dioxide, emitted from a source. The equations, being stochastic in nature, can simulate the overall behaviour appropriate to a large ensemble of situations. They include the effects of transport, plume-growth, dry deposition, oxidation, exchange between "dry" and "wet" regions, intermediary additional sources, and the parametrised wet deposition referred to above. The results strongly indicate that the related total (wet plus dry) depositions downwind from a particular source at any receptor beyond a few hundred kilometres are always approximately proportional to the magnitude of the source-emission. This conclusion has marked relevance to the discussion on how best to achieve a reduction in environmental damage.

Closer in to the source, however, marked non-proportionality is sometimes evident, and any reductions in the emission will result in a significantly less than proportional reduction in the total deposition. Precise details then become related to the nature of the non-linear processes involved and cannot be predicted by the simple parametrised form used in this paper.

1.   INTRODUCTION

Everyone is aware that Europe has an "acid rain" problem. The media now and then turn our attention to the forests of Germany, Austria, Czechoslovakia and elsewhere which show varying degrees of pollution damage. They remind us of the fishless lakes of southern Norway and Sweden, of the build-up of heavy metals in some artesian waters, and of other worrying problems.

It would be wrong to suggest that these are anything other than serious problems but the question arises as to how best to set about solving them. It is easy to say, "Cut emissions", but of what species, and by how much, and what will the inevitable side-effects be?

Looking more deeply into the problem, it soon becomes clear that there isn't just one problem but several rather loosely connected problems. Forest damage in Czechoslovakia, for example, seems to result from the direct effect of sulphur dioxide, when in rather stable atmospheric conditions in winter, plumes are carried from power stations in the Polish-Czech border region into the wooded Tara Mountains. In Scandinavia the acidification of lakes and streams seems to be linked primarily to sulphuric and nitric acids obtained through oxidation of sulphur dioxide and nitrogen oxides emitted perhaps thousands of miles away and washed out in rain or snow. In Germany, on the other hand, the major factors seem to be ozone (produced by the action of strong sunlight on polluted atmospheres containing nitrogen oxides and hydrocarbons), climatic stress on the trees (very cold winters or long dry periods in summer), with perhaps wind-blown acidic fogs thrown in for good measure. To make matters even more complicated some of the effects take place quite quickly whilst others take a long time.

Cut emissions? Well yes, if we had no emissions at all that would in time solve the problem. But that is an impossible dream. A more modest reduction sounds more reasonable. But will it work? Air chemists tell us that there are circumstances which are not all all unrealistic, in which a reduction of nitrogen oxide emissions could actually result in an increase in ozone concentrations and potentially cause more damage to trees .... and it may only require one or two bad ozone episodes to do irreparable harm to a tree given the right circumstances. Clearly then, we have to proceed with deep understanding and care even though the problem is desperately urgent in some areas. Sadly the base-rock store of good field data on which to build our understanding is often missing - the problem has come upon us too quickly to acquire the data we urgently need.

It is natural to turn for help in these circumstances to modellers who can try to simulate what they believe is going on in the real world, with certain simplifications and assumptions, and come up with numerical predictions of what should be happening. Of course the results are only as good as the physics, the assumptions, and the input data that have gone into the model; and I say that as a modeller myself! In practice, real field data of the highest quality and models, properly construed for the job, need to go hand-in-hand. Each needs the other.

Cut emissions? There is of course a big cost to pay. Initially the cost may seem to be solely financial; but often this will have repercussions in other areas: in employment, in social well-being, in other environmental areas and so on. Fitting bulky desulphurisation equipment to a modern power station would almost always entail a major rebuilding programme at enormous cost. Britain is building very few power stations in the 1980s although come the middle of the next decade many power stations built in the 1950s will be ending their expected lifetimes and will be up for replacement. Logically this would be the ideal time to include efficient desulphurisation units at a fraction of the cost of fitting to existing stations. But can we wait that long? Those suffering damage are inclined to say "No". Most countries in western Europe have already dramatically cut their emissions over the last decade. Even the UK which is sometimes accused of being slow in these matters has reduced its emissions from well over 6 million tonnes of SO_2 in 1970 to around 3.6 million tonnes last year. These countries have achieved this by a change to less polluting fuels, such as nuclear or natural gas. However the general decline in traditional heavy industry in response to world pressures has also been a major factor.

All this should not make us complacent. We should all be dedicated to trying to minimise environmental damage whereever it occurs, especially since all of us in the West share some responsibility in affecting ecologically sensitive areas.

Let us consider in a very broad-brush way where emissions have their greatest impact. It is probably true to say that every source, wherever it is, must make some contribution to every ecological problem, no matter how small or large that contribution may be. But in broad terms I believe it can be argued that the "German" ozone problem is largely a regional-scale problem associated with stagnant summer anticyclones with light winds, the air "sloshing" around and becoming increasingly polluted in time. Since anticyclones are generally associated with low-level outflow for good dynamical reasons, the pollution eventually works its way outwards from the centre, rather than the reverse, and so emissions from countries beyond the anticyclone are unlikely to be deeply involved: they are pushed away from the anticyclone, not drawn into it. Similarly emissions from western countries must yield concentrations of SO_2 that are generally much too small by the time they reach Czechoslovakia to contribute in any measurable way to tree damage there.

Scandinavia is rather a different problem. I believe all the evidence points to a significant contribution from several western European countries. The wet deposition of sulphate over the Norwegian mountains is often associated with mobile depressions moving northeastwards drawing in air from the UK, and from much of western Europe in polluted "fingers" of air in the pre-cold-frontal conveyor belts. Models and daily measurements point to a UK contribution, for example, which is about 15-20% of the total for the southern most badly-afflicted parts of Norway, and about 6-12% for Norway as a whole.

At this point, it is well to remember that even in southern Norway not all lakes are fishless by any means. Many other factors of considerable importance are relevant: the local composition of the soils and rocks, the incidence of fish diseases, the effect of afforestation has on surface-water acidity, the availability of heavy metals in the soil, and the preservation or destruction of spawning areas in the streams, to mention but a few.

Moreover much of the sulphur deposition in Norway comes from the contribution of so-called "background" sulphate at low concentrations; that is from airborne sulphate which cannot be immediately attributed to known sources within the model, which often comes in off the Atlantic, the origins of which may include dimethyl sulphide from plankton in the ocean, and some may have travelled all the way from America; whilst some may be of European origin which has travelled outside the model area. Clearly changes in European emissions are only going to have a modest effect where the background contribution dominates over the direct European contribution. However we do have an obligation to make further cuts.

To decide whether is can be done reasonably or not, the question arises, "If we reduce our emissions by x%, will the depositions be reduced by something of the same order (say within 0.7X to 1.3X%) and will the effects respond roughly pro rata?" The last bit of the question is very difficult to answer at present, and so I will ignore it! Nobody would be happy with the situation where a reduction of emissions by X% led to a reduction in deposition of something greatly less than X%; the money could be better spent.

The first bit of the question sets off a second question: "Am I concerned with short-term episodic depositions or only with long-term depositions?". The question is an important one because we know that in any one situation the uptake of sulphur into rain is so complex and depends on so many "non-linear" physical and chemical relationships that no simple relationship can be expected between how much sulphur is in the air and how much comes out in the rain at a specified site. As an example, when the concentration of sulphur in the air is high, raindrops may become saturated and may not carry out significantly more than when the airborne concentration is much lower. This means that single eposodes will not simply reflect changing emissions. If this is so, is the story any different for depositions averaged over much longer times? I believe the answer is "Yes", as I will show later. And this is important because long term depositions seem to many to be what matters overall in the Scandinavian problem, even though short term episodes may sometimes have important consequences. Over a long period of time there appears to be an almost exact balance between the amount of sulphur deposited and the amount getting into lakes and rivers.

(At this point it is well to remember that sulphuric acid is not the only problem: nitric acid, weaker acids, released aluminium and other heavy metals are also important. But I choose to consider only sulphur species since most is known of these, they are dominant and to some extent they are symptomatic of the other species).

2. THE PHYSICAL PICTURE

We are going to construct a model that tries to reflect the major processes which go on in the atmosphere that help to determine the link between emissions and long-term depositions. Figure 1 shows these processes in a schematic way. Almost all the sulphur will be emitted as sulphur dioxide. Emission takes place from a wide variety of sources, some near the ground, others at 100 metres or more height. The wind carries the plume away downwind, and the probability distribution of wind-directions will be one of the most important bits of input data. The airflow will almost always be turbulent,

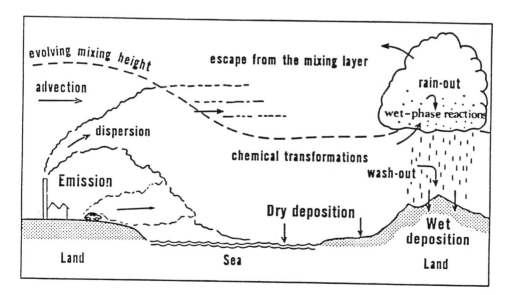

Processes involved in the deposition of atmospheric pollutants

Figure 1. A schematic representation of the various
           processes affecting plumes in the atmosphere.

and this is very important. It means the plumes will normally diffuse to fill
the whole of the so-called mixing layer - a layer of variable depth which is
anything up to about 2 kilometres deep. This mixing not only spreads some of
the sulphur (we will use the term "sulphur" loosely to mean all related
sulphur species) to ground level where it can be taken up by the surface -
a process called dry deposition - but it also enables the sulphur dioxide to
mix and be converted by oxidants of various kinds originally present in the
surrounding ambient atmosphere to sulphate aerosol.

At somw stage the plume may be drawn into the influence of rain clouds.
Some of the "sulphur" may be washed out below cloud as the rain falls through
the plume, although this process is not particularly efficient. More
effective removal occurs when the sulphur is drawn into cloud and gets into
growing droplets. The model will be very cavalier about all this - it cannot
hope to treat the highly complex micro-physical cloud processes properly.
Instead it will assume some rather simple, but broadly representative, empirical
relationship between what is in the air and how much comes out in the rain.
The justification for this is that "sulphur" appears to be very rapidly removed
in rain at normal concentrations. Empirical evidence suggests the removal
time is usually less than the typical duration of a rainy period, as evidenced
by the "clean" look of the atmosphere after rain. It should be said that air-
chemists find this rapid cleansing somewhat difficult to explain at present
when they try to simulate the detailed behaviour of what is going on in cloud,
which may suggest that some efficient uptake route has still to be discovered.

Overall, dry deposition exceeds wet deposition over most of Europe
averaged over a long period of time. This is because even though rain is a
very efficient remover of "sulphur" it occurs only about 7% of the time. In
more remote areas, like northern Scandinavia, the reverse is true and the
reason for this is that since the sulphur dioxide is gradually oxidised to
sulphate the loss to the ground by dry deposition decreases (sulphate is only
very slowly taken up by the ground) whereas, relatively, wet deposition
increases (sulphate is more rapidly taken up by cloud droplets than SO_2).

The model we will develop will be a statistical model. It will describe
the probability of there being a specified mix of SO_2 and sulphate left in the
plume after a given distance (or time) of travel downwind of the source, and
the implied wet and dry depositions associated with the mix. To help in this,
we need to include information on how rain is distributed in time and space.
Obviously the probability of rain is not the same everywhere all the time:
rain is more likely near an active warm front than in the middle of a summer
anticyclone. We therefore conceive the existence of dry regions (where the
probability of rain is essentially zero) and of wet regions (where the
probability is high). These regions move both relative to the ground and

relative to plumes in the mixing layer. They have typical time-scales (we choose 40 hours for dry regions and 8 hours for wet regions based on studies following low-level trajectories) which determine the probability that an element of the plume will advect from a dry region into a wet region (or vice versa) in any short time interval.

Summing up over all possibilities (all conceivable mixes duly weighted by their probability of occurrence) yields the deposition field downwind of the source.

3. NON-LINEARITIES

It is generally agreed by scientists that dry deposition is proportional to the concentration of "sulphur" near the ground, when due allowance is made for the mix. Wet deposition may not depend so simply on the concentration, as we argued earlier. Normally one would expect that high airborne concentrations might not yield as much wet deposition in pro rata proportion as lower concentrations.

When a polluted plume enters a rain system it leaves a "footprint" of deposition. If the air concentration is rather low the footprint would be sharply peaked and subsequently fall off to near zero within a travel time of a few hours. If however the concentration is much higher, so that saturation effects occur, the early rain cannot bring down more than a certain burden and the footprint is extended out over a longer travel time or distance. Ultimately the integrated deposition must be identical to the airborne input: what goes up must come down! But where it comes down must change.

The basic qualitative argument is that if a receptor is sufficiently far away from the source, the variability in where rain occurs along the plume track means that on many different occasions the receptor sees all parts of the footprint and the time integrated deposition equates to the space-integrated area under the footprint which in turn equals the effective emission. The exact shape of the footprint (and thus the details of the non-linearities) then becomes unimportant. The relationship thus reduces to one of expected proportionality between emissions and depositions.

3.1 The Model

The basic equations are reasonably familiar, being four in number and representing the conservation of sulphur dioxide and of sulphate in wet and dry regions (see for example Smith, F. B., Atmos. Environ., 15, p 863-874 1981). The only essential difference is that for an isolated plume, account has to be made to the changes in concentration arising from plume growth with downwind distance. A more subtle difference, but one which is the very essence of this paper, is the nature of the wet removal terms by which the rainout of sulphur dioxide and sulphate are represented in the equations. As already indicated in

486

the Abstract these terms, unlike in Smith (1981), are non-linear in
character, and it is the implications of this character that are explored in
the analysis.

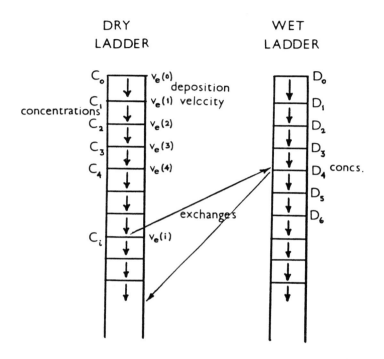

Figure 2. The two ladders by which the conservation
equations for total sulphur in dry and wet regions
are solved. The model incorporates the following
processes: emission into wet and dry periods, dry
deposition, wet deposition in wet periods only,
chemical transformation of sulphur dioxide to
sulphate, different deposition rates for the two
species, transfer of matter between wet and dry
periods.

Because of the non-linear nature of the wet removal terms it is impossible
to find exact analytical solutions to the equations. Consequently resort is
made to an approximate method which is summarised in Figure 1. A single
source and its subsequent depositions are studied. It is considered that the
emissions at source take place, over a long period of time, into both dry and
wet periods. Simplifying the meteorology to the extent of assuming a constant

mixing depth, the initial air concentrations are the same in both dry and wet periods (in Figure 2 this means $C_o = D_o$). Considering the emissions to be purely in terms of sulphur dioxide, the deposition velocity is taken to be that of SO_2 and is given a constant value of 1 cm s^{-1}. In the first time step, the concentration in the air as it advects downwind falls due to dry deposition, to plume width growth (where this is relevant), and may increase due to the addition of new sulphur dioxide due to the air flowing over new sources. As a result the concentration changes from C_o to C_1. Some conversion of sulphur dioxide to sulphate has also taken place, and this results in a lowering of the effective deposition velocity in the second time-step. Exactly the same changes are taking place on the dry ladder, except that a bigger fall in concentration can be anticipated as a result of rainout. Consequently D_1 is significantly smaller than C_1. The magnitudes of the concentrations C_i and D_i at the ith time-step can be evaluated very simply, and is the first step in the computer programme that is used to study the question of proportionality between emissions and depositions.

The amount of sulphur on the dry ladder and on the wet ladder is initially defined by the emission and by the fraction of time the source is within dry conditions, compared to wet conditions. However in later time-steps allowance has to be made for the advection of boundary layer air, carrying the pollution, from dry regions into wet, and vice versa. These exchanges are known only statistically, but that does not matter because the model itself is a statistical representation of the fate of sulphur emissions from the source. The statistical exchange rates are defined by two timescales, taken from Rodhe and Grandell (1972). The timescales are 8 hours (the average time spent in wet regions), and 40 hours for dry regions. At each time step then a multiple exchange between "rungs" on the two ladders takes place so that a defined fraction on one ladder at one rung is transferred to a rung on the other ladder at a concentration that is as near as possible the same. This exchange means that as time proceeds the emission is no longer held on a single rung on each ladder, but rather on an increasingly large number of rungs. In order not to have an excessive number of rungs to cope with in the program, a finite number is chosen in which the concentration associated with the lowest rung is so small that material arriving on this rung is not passed any further down, thereby invoking a small, hopefully negligible error. Another small error arises because on exchanging material between one ladder and the other there is no way of ensuring proper account is made of the different mix of SO_2 to SO_4 in the incoming air and what is assumed at that rung.

The chart summarises some of the parameter values used in the model. Not shown in the next Section is the result of a rather thorough sensitivity study which looked in detail at the response of the deposition values to changing the

SOME DETAILS :

WIND SPEED : CONSTANT $\approx 10 \cdot 4$ m s^{-1}

MIXING DEPTH : CONSTANT $= 800$ m

RATE OF RAIN : $= 0 \cdot 34$ mm h^{-1}
in wet periods only

TIMESTEP : 15 mins

WET and DRY PERIODS : experienced with statistical timescales of 8 and 40 hours respectively

DRY DEPOSITION VELOCITIES :
1 cm s^{-1} for SO_2 & $0 \cdot 2$ cm s^{-1} for SO_4

EMISSIONS : uniform and steady over
150 x 150 km EMEP gridsquare

NON-LINEAR RAIN-OUT :
removal rate in rain $\propto \dfrac{C}{1 + \epsilon C}$
where C = concentration
ϵ = non-linearity parameter

parameter values. For good physical reasons it showed that whilst the
magnitudes of the depositions varied as some of the parameters were changed,
as one would expect, the implications regarding proportionality were hardly
affected at all. The model turns out to be extremely robust. The reason for
this can be seen in Figure 3. Consider the middle footprint. The shape of the
footprint depends on the detail of the non-linear processes involved in wet
removal. Therefore on any one occasion when a receptor R is being affected,
what it receives in wet deposition depends on where it is relative to the
footprint; but provided R is sufficiently far away from the source, over a
large number of rain events it will find itself in a wide variety of positions
in the footprint, and effectively integrates over the whole footprint which
in turn must be equated to the constant emission. The long-term depositions
then become independent of the details of the non-linear processes, and it is
only much nearer to the source where the whole footprint cannot be sampled that
these processes become very evident. At the end of each timestep the total
dry deposition DD(t) is given by:

$$DD(t) = v_e(t) \times \sum_i (A_i C_i + B_i D_i) \, \Delta t$$

where A_i is the amount of sulphur at concentration C_i on the dry ladder and B_i
is the amount of sulphur at concentration D_i on the wet ladder.

Δt = the timestep length

The wet deposition WD(t) is given by:

$$WD(t) = \lambda \, \Delta t \sum_i ((B_i D_i)/(1 + \epsilon D_i))$$

All these simple calculations are carried out with great speed and efficiency
on a small computer.

3.2   Results

Figure 4 shows the three types of plumes considered. The first is a very
wide diffuse plume coming from a large number of small sources. Typical
concentrations in such a plume are about 30 $\mu g \, m^{-3}$. The second plume is from
an isolated large source, such as a modern power station. The initial
concentration may be as high as 1000 $\mu g \, m^{-3}$, but because of plume growth
(which is assumed linear in time in the model) the concentration drops rather
quickly. The third plume comes from a group of large sources such as can be
found in the north Midlands of the UK as shown in the Figure. Here the
average concentration could be 600 $\mu g \, m^{-3}$ and will fall much more slowly since
plume growth will not be very important over the distances of concern.

The model is used to explore the response of the long-term downwind
deposition fields to halving emissions at source. Depositions are said to be
"approximately proportional" to emissions if a 50% reduction in source-strength

Three possible footprints of wet deposition during a rain event

PROPORTIONAL MODEL in which rain persists for long enough to remove most airborne material

Rate of wet deposition

Start of rain   End of rain   → Time

NON-PROPORTIONAL MODEL in which rain persists for long enough to remove most airborne material

Rate of wet deposition

Start of rain   End of rain   → Time

NON-PROPORTIONAL MODEL in which rain does NOT persist for long enough to remove most airborne material

Rate of wet deposition

Start of rain   End of rain   → Time

Figure 3.

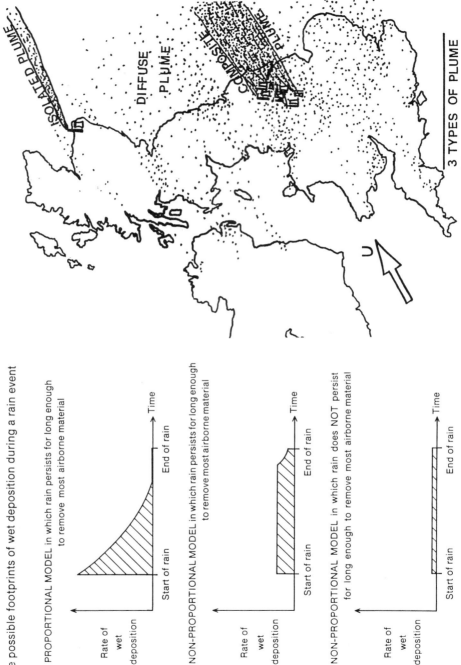

ISOLATED PLUME

DIFFUSE PLUME

COMPOSITE PLUME

3 TYPES OF PLUME

Figure 4.

produces a reduction in deposition at a receptor in the range 40 to 60%.

Figure 5 shows the response in depositions for the first diffuse plume to halving emissions. As can be seen, for dry, wet and total deposition the relationship is approximately proportional beyond 400 km, and in fact very soon becomes such that enhanced benefits are achieved by cutting emissions.

Figure 6 shows very similar results for the expanding power station plume. In slight contrast, Figure 7 shows the response for the third type of plume: the composite plume. Here the dry and the total depositions are approximately proportional at all distances, but because of the high concentrations in the plume the non-linear effects persist for a very large distance, and the wet deposition is suppressed and makes a relatively small contribution to the total deposition. Thus even where non-linear processes are important in the wet deposition, the overall influence remains negligible.

Table 1 summarises the results.

In conclusion, beyond a few hundred kilometres from a source area the effects of non-linearity are rather small on the response of depositions to cutting emissions, at least in the long-term depositions. The effect on the magnitudes of episodes has to be still explored, and since episodes can have very significant effects on the ecosystems in sensitive areas, we cannot claim that modest reductions in emissions will necessarily eliminate damage in such areas.

TABLE 1

Model results showing the distance d beyond which deposition is "approximately proportional" to emission. Although the model is remarkably robust to changes in the values of the various input parameters, these values of d, should be taken to indicate only in a broad sense values that should apply in reality.

SOURCE TYPE	INTERMEDIATE SOURCES	Values of d for:	
		WET DEPOSITION	TOTAL DEPOSITION
Broad dispersed sources	No	300 km	250 km
	(a)	150 km	50 km
	Yes (b)	2000 km	50 km
Single isolated large point source	No	500 km	200 km
	(a)	values not currently available: probably similar to those for the broad dispersed plume.	
Composite plume from cluster of large sources	No	3000 km	0 km
	(a)	1400 km	0 km
	Yes (b)	2000 km	0 km

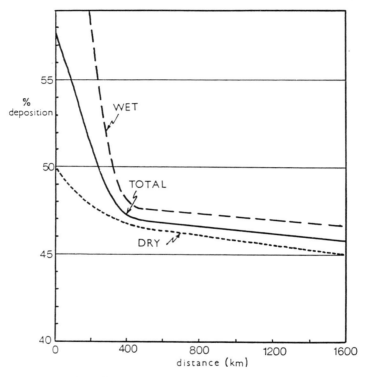

Figure 5.

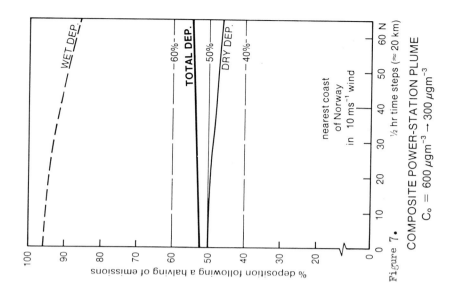

Figure 7. COMPOSITE POWER-STATION PLUME
$C_o = 600\ \mu gm^{-3} \rightarrow 300\ \mu gm^{-3}$

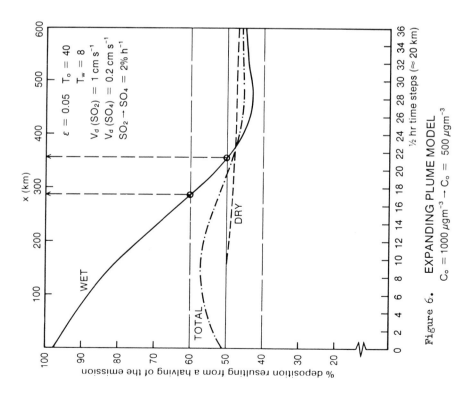

Figure 6. EXPANDING PLUME MODEL
$C_o = 1000\ \mu gm^{-3} \rightarrow C_o = 500\ \mu gm^{-3}$

(a) intermediary source-strengths unaltered when initial source-strength reduced.

(b) initial & intermediary source-strengths reduced by same proportion. Distance from East Midlands source area to sensitive lakes of S Norway is about 800 km.

Figures 8 and 9 show the results of an investigation into the effect of a range of mountains, such as the Norwegian mountains, on the issue of proportionality. The mountains were represented in the model by a trebling of the rainfall rate and an increase in the probability of rain by extending the time-scale of wet-periods there by 50%. Figure 8 shows a large response in the amount of wet deposition, increasing it by about 4 on the upstream edge, but a rapid decline to near normal values within 200 km. Downwind of the range, although no reduction of rain from normal was assumed, a dip in both the wet and the dry deposition is predicted due to the large losses upwind over the mountains. Figure 9 shows that in spite of these large effects, proportionality is still virtually achieved within the mountainous area.

Figure 10 shows the results of an investigation into the effect of varying the conversion rate of SO_2 to sulphate on the relative importance of dry and wet deposition. The faster the conversion rate the more rapid the SO_2 diminishes and dry deposition therefore becomes relatively less important than wet deposition.

4. EPISODES

Whilst, as indicated above, reductions in emissions will reduce long-term total depositions in approximate proportion, the same is unlikely to be true in episodes of concentration of deposition in rain or snow. Although the effect is unquantified at present, it is conceivable that equally-serious episodes will still occur following a reduction in emissions, although perhaps rather less frequently. Thus if any aspect of the environment is prone to significant damage from short-lived episodes, this damage may still occur unless the reduction in emissions is very substantial indeed. These conclusions are however very tentative at present, and further work in this area is urgently required.

APPENDIX

An agreed document between working scientists at the Meteorological Office and the Central Electricity Research Laboratories in the UK on the "Present State of knowledge regarding the Proportionality of Emissions and the Long-term Average Deposition of Sulphur Oxides".

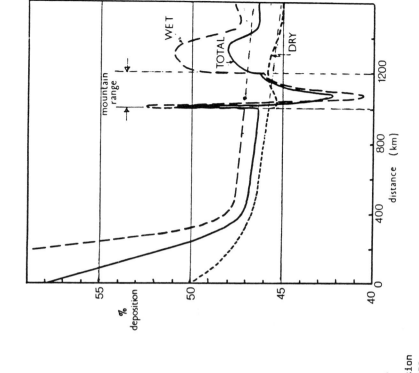

Figure 9. The same plot for the same parameter values as in Figure 5 but now with the influence of the mountain range described in Figure 8. Although marked perturbations are to be seen, proportionality is still virtually achieved.

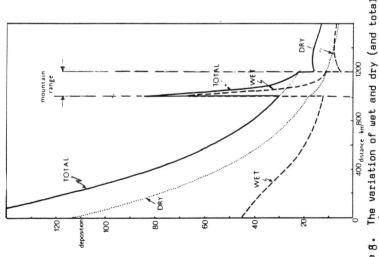

Figure 8. The variation of wet and dry (and total = wet + dry) with distance downwind from the source. A mountain range between 1000 km and 1200 km with markedly more rain causes an interesting perturbation to the wet deposition profile. Parameter values are as follows:

outside the mountains:

C_0 = 50 ug m^{-3}, = 0.05, $v(SO_2)$ = 1 cm s^{-1}, $v(SO_4)$ = 0.2 cm s^{-1}, T_d = 40 hours, T_w = 8 hours, conversion rate $SO_2 - SO_4$ = 2% h^{-1},

inside the mountains:

T_d = 36 h, T_w = 12 h, rainfall rate trebled.

496

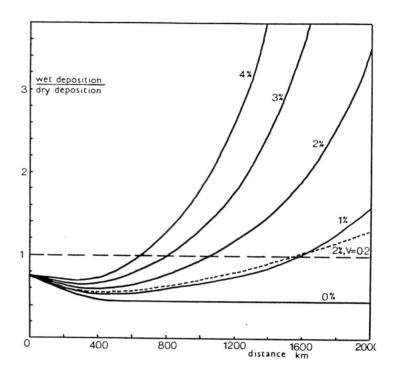

Figure 10. The ratio of wet deposition to dry deposition
for various conversion rates of SO_2 to sulphate. Wet
deposition begins to exceed dry at distances greater
than a few hundred kilometres (the exact value depending on the conversion rate). The full curves are all
for zero deposition velocity of sulphate; the pecked
line is for a deposition velocity of 0.2 cm s^{-1}.

FRAMEWORK FOR DISCUSSION

1.   The issue of non-proportionality concerns the long-term average relationship between wet, dry and total sulphur deposition and the emissions of SO2 in Europe. The relationship is said to be proportional if a reduction in a given source would produce a porportional reduction in the deposition from that source over a specified receptor region.

2.   Modelling studies of wet deposition over Europe have shown the need to introduce a so-called background concentration of sulphate in rain, the origin of which is not known so that it cannot be attributed to any source region or type. This component of wet deposition which can be a sizeable fraction of the total deposition in remote high rainfall areas where problems with aquatic ecosystems have been reported, is not addressed in the conclusions following.

3.   The present discussion is confined to the proportionality between emissions and the deposition attributed to those emissions.

4.   Recent studies of long-range transport models have included terms which allow the relationship between wet deposition and source strength to be non-proportional. These have been restricted to situations in which there is a single large source region and a well-defined receptor area downwind, with and without additional sources in between.

5.   The studies have been concerned with different types of sources (a few large point sources compared with a large number of small distributed sources). The behaviour of sources of equal strength but of different type may not be the same.

6.   The criterion for describing a relationship as "approximately proportional" is taken to be that a 50% reduction in source strength would produce a reduction in deposition in the range 40 to 60%. The following conclusions are restricted to defining the range over which the relationship is approximately proportional.

7.   It is not possible to determine from deposition measurements at a particular receptor whether or not a change in emission strength of a particular source area has produced a proportional change in that area's contribution to the deposition. This is because (i) there is no straight-forward way of distinguishing the contribution from a specific set of sources from all the other sources contributing to deposition at that receptor and (ii) deposition will vary on seasonal, annual and even decadal timescales due to meteorological and climatic factors which change the relative contribution of individual sources to the deposition. It is only by using transport models, which include these effects that the proportionality issue can be properly

addressed.

8.   In the statistical modelling studies contributing to our conclusions a simple, general form is used to represent the relationship between the rate of wet deposition and airborne SO2 concentration. This parametrisation allows the shape of the response in wet deposition to a change in emissions to be investigated. More confidence may be shown in results which are insensitive to changes in parameter values. The range of parameter values which are considered reasonable have been taken from previous modelling studies which for long-term averages show a fair (to within a factor of two) agreement with observation.

9.   It is therefore believed that the following conclusions may be applied generally, but where specific numbers are given these are based on recent modelling work.

EXISTING KNOWLEDGE

1.   Long-term average dry deposition is approximately proportional to source strength.

2.   Dry deposition at long distances is reduced by the oxidation of sulphur dioxide to sulphate-containing aerosol, which dry deposits more slowly than sulphur dioxide. The oxidation process during dry conditions proceeds at a rate which is determined by the ambient level of pollutants and meteorological conditions. However, the rate is generally low compared with the average rate of removal of SO2 by dry deposition.

3.   The nature of wet deposition is determined by the efficiency with which material is removed by rain events over the receptor. The average rate of wet removal of aerosol sulphate is believed to be quite fast, resulting in efficient wet removal. The contribution to wet deposition from sulphate formed during dry conditions is thought to be roughly proportional to the source strength.

4.   The rate at which sulphur dioxide is removed by rain is believed to be a process which could lead to strong non-proportionality. Direct washout of sulphur dioxide by rain is thought to be slow, because the raindrops saturate with dissolved sulphur dioxide. More rapid wet removal requires the oxidation of sulphur dioxide to sulphate in rainwater and cloud water in precipitating clouds. This is thought to proceed at a rate which is non-proportional in sulphur dioxide concentration, in the sense that as the sulphur dioxide concentration increases a smaller proportion of the sulphur dioxide is oxidised during a given time.

CONCLUSIONS FROM RECENT MODELLING STUDIES

1.   Let us consider an air parcel which is incorporated into a rain system for a period of time. The parcel's exposure to rainfall, which we shall simply call the duration of the rain period, may not persist for long enough for all the sulphur to be removed. In addition the rate of sulphur removal during the period is not a simple function of sulphur dioxide concentration and time of exposure. These factors must be considered in determining whether, on average, the wet deposition at any receptor will be proportional to the emissions from a given source.

2.   The sulphur from a given source deposited at a receptor will have been exposed to rain for a duration in the range from zero up to the travel time from the source to the receptor, depending on where between the source and receptor it first contacted rain. Receptors located at travel times from the source greater than the average duration of rain will, over many occasions, tend to experience deposition at all stages of the rain period assuming no preferential locations for rainfall inception.

3.   Whether all or only part of the sulphur is removed during the period of rain depends on the airborne concentration of SO2 at the start of the period. This will have been determined by source strength, dispersion, oxidation to sulphate, the rate of dry deposition prior to the rain and previous rain encounters. If the concentration is low enough for most of the sulphur to be removed and the receptor is far enough away from the source to have experienced rain of all possible durations, then the wet deposition will be proportional to emission. In the next paragraphs these statements are made more quantitative.

4.   Travel times in excess of about 7 hours (that is greater than the mean duration of wet periods) are sufficient to ensure that the average wet deposition does not depend on details of the wet removal process provided that most of the sulphur is removed during the rain period. Preliminary results based on known chemical mechanisms indicate that for SO2 concentrations above about 5 ppb most of the sulphur may not be removed. If the concentration of sulphur dioxide drawn into precipitating cloud is above this level even receptors beyond seven hours travel time will be expected to experience some non-proportionality. In the range where the wet deposition behaves non-proportionally, the degree of non-proportionality depends sensitively upon details of the removal process which are not accurately defined at present.

5.   As mentioned in paragraph 3 airborne sulphur dioxide concentrations decrease with distance from the source, and hence, the degree of non-proportionality of wet deposition is expected to decrease on average with distance from the source.

6.   With removal rates based on the results mentioned in paragraph 4 and typical meteorological conditions, the following conclusions are obtained from long-range transport models. The long-term average wet deposition of emissions from widely dispersed sources, such as urban areas, is likely to be approximately proportional to emission strength once a travel distance of 200 km is reached. For a single, large point source, such as a power station, this distance is about 500 km. For groups of power stations, such as those in the Yorkshire and Midlands areas, the combined plume being less subject to dilution because of intermixing, wet deposition may only be roughly proportional to the emission strength beyond a distance of 1000 km.

7.   These distances depend on a number of assumptions and only apply in cases where there are no additional sources between the source and the receptor. It is possible to include in the model intermediate distributed sources between a large source and the receptor. The presence of substantial intermediate sources influences the proportionality of wet deposition beyond these distances. Halving the strength of the large source and that of intermediate emissions produces a less than proportional decrease in the wet deposition out to at least 2000 km.

8.   Total (wet plus dry) deposition from a given source behaves more proportionally than wet deposition. This occurs because of the near proportionality of dry deposition and the reduced contribution of wet removal to the total deposition associated with conditions leading to highly non-proportional wet deposition. It is not likely for wet deposition from a given source to be simultaneously a dominant proportion of total deposition and also highly non-proportional to the emission strength.

9.   These results are not inconsistent with observations in high wet deposition, high rainfall areas, since the observed wet deposition may be derived from several different sources, some of which contribute proportionally, and others which contribute non-proportionally. Some high rainfall regions receive a high background wet deposition, the reponse of which to emission reductions is not known.

10. Applying current linear models to explain annual wet deposition measurements over Europe as a whole indicates that, on average, half or more of the airborne sulphur is removed during a rain event. Lower efficiencies probably correspond to events in areas close to large sources where the current linear models overestimate wet deposition and modifications are required. Higher efficiencies correspondingly refer to events in less polluted areas where linear models are internally self consistent and accurate enough for many purposes. The detailed study of rain events is required to evaluate these

differences in greater detail.

11. The changes that would be brought about in annual average source-receptor relationships for total sulphur deposition simply by the inclusion of non-linear effects close to major source areas are likely to be within the overall accuracy of the models, given the uncertainties that arise from the meteorological assumptions, source inventories, variation in dry deposition, and measurement accuracies. Larger non-linear effects would be expected in the wet component of deposition, with significant changes occurring in the estimated source-receptor relationship, particularly close to large sources. These effects might not be observable in the high rainfall areas where background deposition can dominate.

SUBJECT INDEX